Green Mobile Devices and Networks

Energy Optimization and Scavenging Techniques

Green Mobile Devices and Networks

Energy Optimization and Scavenging Techniques

Hrishikesh Venkataraman and Gabriel-Miro Muntean, Editors

CRC Press
Taylor & Francis Group
Boca Raton London New York

CRC Press is an imprint of the
Taylor & Francis Group, an **Informa** business
AN AUERBACH BOOK

CRC Press
Taylor & Francis Group
6000 Broken Sound Parkway NW, Suite 300
Boca Raton, FL 33487-2742

Version Date: 20111020

International Standard Book Number: 978-1-4398-5989-6 (Hardback)

Library of Congress Cataloging-in-Publication Data

Green Mobile Devices and Networks: Energy Optimization and Scavenging Techniques / editors, Hrishikesh Venkataraman and Gabriel-Miro Muntean.
p. cm.
Includes bibliographical references and index.
ISBN 978-1-4398-5989-6 (hardback)
1. Mobile communication systems--Energy conservation. 2. Green electronics. I. Venkataraman, Hrishikesh. II. Muntean, Gabriel-Miro, 1972-

TK5103.2.E545 2012
621.3845'6--dc23 2011037685

Visit the Taylor & Francis Web site at
http://www.taylorandfrancis.com

and the CRC Press Web site at
http://www.crcpress.com

Contents

SECTION II SCAVENGING TECHNIQUES

Preface

Wireless communications are evolving rapidly toward "beyond 3rd generation (B3G) and 4G systems." At the same time, multimedia transmissions, video-on-demand, gaming, etc. are becoming increasingly popular among the growing number of users. Additionally, over the past couple of years, the demand for multimedia communications and, particularly, video streaming to handheld mobile devices has grown by leaps and bounds. In particular, it is expected that by 2013, mobile phones and other browser-enabled mobile devices will overtake PCs as the most common access device worldwide. As the technology progresses, wireless devices, such as smartphones, iPhones, PDAs, etc., are offering a large number of sought-after features to customers and support for increasingly complex applications. With each passing year, the functionality and computing power of mobile devices is increasing exponentially, with more and more applications and communication technologies being added consistently to handheld wireless devices. The data rate required for supporting these services is also increasing significantly. This implies a high power requirement at the transmitting and, especially, the receiving wireless devices. However, there is an annual power improvement of only 6 percent over past years and this has not grown in tune with processing and communication technologies. This has a serious impact on the practical use of the mobile devices, especially when accessing rich media-based services. For example, the battery of an iPhone 4GS lasts a mere five hours during Internet connectivity on a 3G network.

Given the stringent requirements and the current limitations of the battery powering of mobile devices, serious efforts are required not only to improve the battery quality, but also improve the battery life. In order to achieve greater success from mobile technology over the next decades, the concept of battery recharging every one to two days has to be completely revamped. This is, of course, easier said than done. A very important question that needs to be investigated and would challenge the researchers/handset manufacturers/network operators is what kind of improvement in the battery can be achieved without significantly altering the overall performance? This is a very interesting yet a very difficult proposition. Recently, there have been several efforts to optimize the energy consumption in both devices and networks. At the level of a device receiving video content via wireless networks, the content's bit rate, frame rate, and color depth could be altered seamlessly depending on the current battery power status. Such a periodic yet a dynamic adaptive mechanism would significantly optimize the battery consumption. An important thing to ponder is that energy optimization schemes can only reduce the consumption and thereby, increase the battery life by a certain but limited amount. There should be alternate mechanisms that need to be used or proposed in order to improve the self reliance of the devices or at least significantly extend the power in the devices by generating or harvesting energy from the environment. An interesting, but challenging aspect is to look at different energy-harvesting techniques and their adaptability to be used by wireless/mobile devices and networks. In order to achieve this, significant changes have to be made in both the hardware mechanisms and software policies to adapt energy use to user requirements for the tasks at hand and to enable automatic recharge from the environment.

Significantly, at the heart of all the technology platforms and handsets introduced are networking and radio communications, thereby enabling base stations/routers/devices to support rich media services, regardless of where the users are physically located. With the latest extensive demand for high-speed Internet browsing and multimedia transmissions over the wireless networks, the focus of mobile networking has been mainly on increasing the data rate and, importantly, the system processing capacity. However, recently it has become quite evident that data rate increase and throughput maximization are not the only objectives in the next generation of wireless systems. *Tomorrow's networks should be optimized for performance and for energy efficiency as well.* A network optimized for both performance and energy implies a very different design and architecture and this is what is needed for high data rate communication to be sustainable in the future. To dramatically reduce the energy consumption of today's wireless networks, a radical new approach needs to be initiated. Hence, the next wave of energy efficient networks will not come simply from more traditional research on single aspects, such as physical layer research, but will require *holistic, system-wide, breakthrough thinking that challenges basic assumptions.*

Harvesting energy from the environment is an important aspect that can create a significant impact in the working pattern of current wireless networks. Energy harvesting can be done at the transmitters, receivers, routers, etc. However, energy harvesting in networks/base stations, etc. is still in a very nascent stage, as compared to energy harvesting in devices. This is primarily because of two reasons. Firstly, the amount of energy required by the wireless networks is very high and it is not possible to harvest such a large amount of energy at the moment. Secondly, the networks/base stations are located at one place and operated by mobile network operators, which are run by big companies. Hence, it becomes easier to power the base station through the existing electricity grid rather than harvesting energy from the environment. However, at the same time, given the increasing computational complexity and the power requirement of the base stations, extracting energy from the environment to power the operations of the base stations is an extremely relevant issue in the decades to come. In fact, in the sensor network domain, given the critical power requirement, energy harvesting for wireless sensor networks is already being carried out. It is an interesting research challenge to extrapolate the energy-harvesting mechanisms from sensor networks to wireless cellular networks.

Energy harvesting in devices is a relatively easy challenge. This is primarily because of the low power requirement of wireless devices. Further, a wireless device is exposed to different sources of energy in the environment, such as heat, light, mechanical keys, electromagnetic waves, audio, etc. Hence, a holistic approach would be to optimize energy harvesting through each individual mechanism and then integrate these different aspects.

This book is a first of its kind focusing solely on energy management in mobile devices and networks. It provides a detailed insight into the different energy optimization techniques and energy harvesting mechanisms in both wireless devices and networks. A unique aspect of the book is the detailed and integrated coverage of different optimization and energy scavenging techniques by different experts. This has not been dealt with before and offers a unique platform for the readers. The book is divided into two parts. The first part describes various energy optimization techniques, whereas the second part presents the energy-harvesting mechanisms.

The first part has seven chapters that focus on energy optimization techniques. Of these, the first three chapters focus on "energy optimizations in devices," while the next four chapters deal with "energy optimization in wireless networks." Chapter 1 talks about energy management and energy optimization techniques for location-based services in mobile devices. Chapter 2 explains the mechanism for energy efficient supply for mobile devices. Chapter 3 models the energy costs of different applications in wireless devices/handsets and is an extension of their previous proposed work in the same domain. In case of wireless networks, the energy consumption for the components across different wireless networks remain the same. However, the pattern of the energy consumption varies across different types of networks. Given the importance of voice communication in cellular networks, Chapter 4 talks about exploiting on–off characteristics of human speech for energy conservation in WiMAX-based systems. Further, given the amount of voice over Internet protocol (VoIP) IP services, Chapter 5 provides an insight into the quality of experience-based energy conservation techniques for VoIP services in Wireless LAN. Notably, a distributed ad hoc network represents a highly complex network in terms of both implementation and deployment. Hence, Chapter 6 explains the importance of considering multiple criteria (minimum energy, multiple relay, etc.) in a mobile ad hoc network and extends their previous work in this field. Above all, given the amount of energy optimization techniques already developed for wireless sensor networks, Chapter 7 provides a comprehensive overview of energy optimization in wireless sensor networks and how it could be potentially extrapolated for a generic wireless network.

The second part of the book includes six chapters that focus on energy harvesting techniques. Given the importance and the amount of research work being carried out for energy harvesting in wireless devices, four out of the six chapters in this section are dedicated to factors and mechanisms for different energy harvesting solutions for wireless devices. The last two chapters talk about common energy harvesting techniques in wireless networks. Chapter 8 evaluates CMOS RF-DC rectifiers for electromagnetic energy harvesting in mobile devices. Further, Chapter 9 explains in detail energy scavenging techniques using a magneto inductive method, while Chapter 10 discusses the mixed signal low power techniques in energy harvesting systems. In Chapter 11, we look at designing wireless sensors with intelligent energy-aware middleware and how could this be extrapolated into futuristic wireless devices. Similarly, the last two chapters of the book, Chapter 12 and Chapter 13, provide an energy consumption profile for energy harvested wireless sensor networks and radio frequency energy harvesting/management for wireless sensor networks, respectively.

Green Mobile Devices and Networks: Energy Optimization and Scavenging Techniques can serve as a benchmark for postgraduates, future engineers, and designers in developing energy-optimal solutions and at the same time provide a deeper insight for the next generation of researchers to harvest energy from the environment for developing the next generation telecommunication systems.

The editors would like to wish the audience a happy reading time and would be happy to receive any queries from the readers.

Hrishikesh Venkataraman
Gabriel-Miro Muntean

The Editors

Hrishikesh Venkataraman, PhD, is a senior researcher and Enterprise Ireland (EI) principal investigator with Performance Engineering Laboratory at the Irish national research center—The RINCE Institute, at Dublin City University (DCU), Ireland. He obtained his PhD from Jacobs University Bremen, Germany, in 2007, for his research on wireless cellular networks. He obtained his master's degree from Indian Institute of Technology (IIT) in Kanpur, India, in 2004, and did his master's thesis from Vodafone Chair for Mobile Communications, Technical University Dresden, Germany, in 2003–2004 under the Indo-German DAAD (*Deutscher Akademischer Austausch Dienst*) Fellowship. His research interests include mobile multimedia, wireless communications, and energy in wireless. Dr. Venkataraman has published more than 30 papers in journals, international conferences, and book chapters, and has won a Best Paper Award at an international conference at the University of Berkeley, California, in October 2009. Currently, Dr. Venkataraman is an executive editor of *European Transactions on Telecommunications* (*ETT*) and is a founding member of the UKRI (United Kingdom/Republic of Ireland) chapter of the IEEE (Institute of Electrical and Electronics Engineers) Vehicular Technology Society.

Gabriel-Miro Muntean, PhD, has established a strong track record in the areas of quality-oriented and performance-aware adaptive multimedia streaming and data communications in heterogeneous wireless environments. Dr. Muntean has been the co-director of a 10-person research laboratory since 2003, which is a state-of-the-art facility at the Dublin City University (DCU) Engineering building and well equipped for multimedia delivery research. He has successfully supervised three PhD and three masters for research students, and is currently supervising seven postgraduate researchers and one postdoctoral researcher. Dr. Muntean has received more than 1 million Euro of funding, having been principal investigator on two EI (Enterprise Ireland), one SFI (Science Foundation Ireland), and five IRCSET (Irish Research Council for Science, Engineering, and Technology) grants and collaborator on two other major Irish grants. In addition, he has been leading Samsung- and Microsoft-funded research projects. Dr. Muntean has authored one book, edited two, and has published five book chapters as well as 25 journal articles and more than 60 conference papers. He has been awarded four Best Paper Awards and is an associate editor for *IEEE Transactions on Broadcasting*.

The Contributors

Dr. Mehran Abolhasan is a senior lecturer at the School of Computing and Communications within the faculty of Engineering and Information Technology (FEIT) at the University of Technology Sydney. He has authored over 50 international publications and has won over one million dollars in research funding over the past 5 years. His current research interests are in Wireless Mesh, 4th Generation Cooperative Networks and Body Area and Sensor networks.

Prof. Johnson Ihyeh Agbinya (PhD La Trobe University) is an associate professor of Remote Sensing Systems Engineering at La Trobe University in Melbourne Australia. He also is Professor (Extraordinaire) of Computer Science at the University of the Western Cape, Cape Town and Professor (Extraordinaire) of telecommunication at Tshwane University of Technology (French South African Technical Institute) Pretoria, South Africa. He was Principal Engineer at Vodafone Australia from 2000 to 2003 managing Vodafone Australia research in mobile communications. Prior to that, he was a Senior Research Scientist at CSIRO Division of Telecommunications and Industrial Physics (CSIRO ICT) from 1993 to 2000. At CSIRO he focused on biometric R&D, specifically, face, voice and palm print recognition and compression systems including VoIP. He is widely published with more than 200 peer-reviewed authored and co-authored journal and conference papers and six technical books in telecommunications and sensing. He is the Editor of the African Journal of Communication and Information Technology (AJICT) and its founder. He is also founder of various conferences including Auswireless, BroadCom and IB2Com and international committee member of the International Conference on Mechatronics, SETIT and AfroCom. His current research interests include remote and short range communications and sensing, nano-networks and applications of metamaterials in electronic communications and radar, personal area networks, inductive embedded medical devices and wireless power transfer.

Prof. Li-Minn Ang is currently with the Centre for Communications Engineering Research at Edith Cowan University. He received his PhD and Bachelor degrees from Edith Cowan University, Australia in 2001 and 1996 respectively. He was a lecturer at Monash University (Malaysia Campus) and Associate Professor at Nottingham University (Malaysia Campus). His research interests are in the fields of visual information processing, embedded systems and wireless sensor networks.

Dr. Yassine Hadjadj Aoul is an associate professor at the University of Rennes 1, France, where he is also a member of the IRISA Laboratory. He received a B.Sc. In computer engineering with high honors from Mohamed Boudiaf University, Oran, Algeria, in 1999. He received his Master's and PhD degrees in computer science from the University of Versailles, France, in 2002 and 2007, respectively. He was an assistant professor at the University of Versailles from 2005 to 2007, where he was involved in several national and European projects such as NMS, IST-ATHENA, and IST-IMOSAN. He was also a post-doctoral fellow at the University of Lille 1 and a research fellow, under the EUFP6 EIF Marie Curie Action, at the National University of Dublin, where he was involved in the DOM'COM and IST-CARMEN projects, which aim at developing mixed Wi-Fi/WiMAX wireless mesh networks to support carrier grade services. His main research interests concern the fields of wireless networking, multimedia streaming, congestion control and QoS provisioning, and satellite communications. His work on multimedia and wireless communications has led to more than 25 technical papers in journals and international conference proceedings.

Prof. Labros Bisdounis was born in Agrinio, Greece, in 1970. He received the Diploma and PhD degrees in Electrical Engineering both from the Department of Electrical and Computer Engineering, University of Patras, Greece, in 1992 and 1999, respectively. From 2000 until mid-2008 he was with the Research & Development Division of INTRACOM S.A. (INTRACOM TELECOM S.A. since January 2006), Athens, Greece, working as a project manager of European and national research projects regarding the design and development of VLSI circuits and embedded systems for telecom applications. Currently, he is an associate professor and the head in Electrical Engineering Department of Technological Educational Institute of Patras, Greece and the head of the Electronics and Measurements Technology laboratory of the department. In addition, starting from September 2007, he is with the School of Science and Technology of the Hellenic Open University as an external tutor. His main research interest is on various aspects of electronic circuits and systems such as: low-power and high-speed digital circuits and embedded systems design, system-on-chip design, CMOS circuits timing analysis and power dissipation modeling, sensors. Prof. Bisdounis is an author of more than 25 papers in international journals and conferences, as well as of book chapters, teaching notes and technical reports on the above-mentioned areas, and has received more than 350 citations. He is a member of IEEE (Institute of Electrical & Electronic Engineers) and Technical Chamber of Greece.

Sonali Chouhan received her PhD degree in Electrical Engineering from the Indian Institute of Technology Delhi, India, in 2009. Since March 2010 she has been with the department of Electronics and Electrical Engineering at Indian Institute of Technology Guwahati, India, where currently she is an Assistant Professor. Her research interests include wireless sensor networks, error control codes, energy optimization, embedded systems, and genetic algorithms. Recently, Dr. Chouhan received the Microsoft Outstanding Young Faculty award in 2010.

Prof. Gianluca Cornetta obtained his MSc Degree from Politecnico di Torino (Italy) in 1995 and his PhD from Universidad Politécnica de Cataluña (Spain) in 2001 both in Electronic Engineering. In 2003 he joined Universidad CEU-San Pablo in Madrid (Spain) where he is presently an associate professor. Prior to joining Universidad CEU-San Pablo, he has been a lecturer in the Departement of Electronic Engineering of Universidad Politécnica de Cataluña (Spain), a digital designer at Infineon Technologies Gmbh (Germany), and an ICT consultant at Tecsidel SA (Spain) in the field of real-time embedded systems. In 2004 he founded the Department of Electronic System Engineering and Telecommunications that he chaired until February 2008. He is also a research fellow at the Vrije Universiteit Brussel and an invited professor at the Institut Superieur d'Electronique de Paris (ISEP) where he teaches wireless system design in the Advances in Communication Environment (ACE) Master. His current research interests include RF circuit design for wireless sensor networks with special emphasis on IEEE 802.15.4 (ZigBee), digital communication circuits, software radio, and distributed real-time embedded systems.

Prof. Swades De received his BTech in Radiophysics and Electronics from the University of Calcutta, India, in 1993, MTech in Optoelectronics and Optical Communication from the Indian Institute of Technology (IIT) Delhi, in 1998, and PhD in Electrical Engineering from the State University of New York at Buffalo, in 2004. Before moving to IIT Delhi in 2007, where he is currently an Associate Professor of Electrical Engineering, he was an Assistant Professor of Electrical and Computer Engineering at NJIT (2004–2007). He also worked as a post-doctoral researcher at ISTI-CNR, Pisa, Italy (2004), and has nearly five years industry experience in India in telecommunication hardware and software developments (1993–1997, 1999). His research interests include performance study, resource efficiency in multihop wireless and high-speed networks, broadband wireless access, and communication and systems issues in optical networks.

Mr. Komlan Egoh received his diplome d'ingenieur in Electrical Engineering from the Ecole Nationale Supérieure d'Ingénieurs in 2001 at the University of Lome in Togo. In 2005 he received his MS degree in Internet Engineering from the department of Electrical and Computer Engineering at the New Jersey Institute of Technology. He is currently a PhD candidate in the same department. His research is in the general areas of communication networks, wireless mesh and ad hoc sensor networks. From 2001 to 2004, and 2007-2008, he has work software engineer in various technology companies in Europe, Africa and North America.

Dr. Nikolaos Fragoulis received his BSc degree in Physics in 1995, his MSc in Electronics & Computer Science in 1998 and the PhD in Microelectronics in 2005, all from the Electronics Laboratory (ELLAB), Dept. of Physics, University of Patras (UoP), Greece. He has worked as postdoctoral researcher or project manager in several national and European-funded R&D projects, in the fields ofmicroelectronics, analogue and digital signal processing. He has worked in the private sector as a software engineer and as a silicon-based systems engineer. Dr Fragoulis has authored or co-authored more than 30 journal and conference papers, and 3 book chapters. He is currently Vice President of technology in IRIDA Labs, Ltd., Greece.

Mr. Philipp M. Glatz received his BS and MS in Telematics specializing in System-on-Chip-Design and Computational Intelligence from Graz University of Technology, Austria, in 2005 and 2007, respectively. As a university assistant with the Institute for Technical Informatics at Graz University of Technology, Austria, he is assigned research and teaching duties. Currently, he is a PhD candidate in electrical and computer engineering. His research interests include wireless sensor network middleware and power awareness with a focus on network coding, energy harvesting with a focus on energy efficiency and measurement systems as well as tool chain and development environment integration. He is a member of the Institute of Electrical and Electronics Engineers and is author and co-author of more than 20 publications.

Mr. Leander B. Hörmann is a PhD candidate in electrical and computer engineering at the Institute for Technical Informatics, Graz University of Technology, Austria. As a university assistant at this institute, he teaches undergraduates and conducts research. He received his BSc and Dipl-Ing in Telematics specializing in Technical Informatics and Autonomous Robots from Graz University of Technology, Austria, in 2008 and 2010, respectively. His research interests include system architecture of energy harvesting wireless sensor networks, low power techniques and the simulation of software and hardware. He is a member of the Institute of Electrical and Electronics Engineers.

Prof. H. S. Jamadagni received his bachelors degree in Electrical Engineering from Bangalore University, India in 1970. In 1972 he received ME degree and PhD in 1986 degrees from Indian Institute of Science, Bangalore, India. From 1972 to 1974 he was a Deputy Engineer at Indian Telephone Industries (ITI), Bangalore. From 1974 onwards he is associated with the Indian Institute of Science and has been a professor since 2001. He was appointed as the Chairman, Centre for Electronics Design and Technology (CEDT) IISc in 1996. He served as its chairman until June 2009. He is Principal Investigator (PI) in several major National and International collaboration and sponsored research projects including a few European Union (EU) Framework projects. His research interests are Telecom, VLSI, embedded systems, wireless sensor networks, pedagogy, energy harvesting, cognitive radio, E-Learning, and technologies for wildlife conservation. Currently, he is a member of Telecom Regulatory Authority of India.

Dr. Mikkel Baun Kjærgaard is a postdoctoral researcher at the Department of Computer Science in Aarhus University, Denmark. His current research interests are within the area of pervasive positioning: positioning anywhere, anytime, of anything. His interests spans from innovative applications within this area to technical challenges such as energy efficiency. He holds a PhD in Computer Science from Aarhus University based on research on indoor positioning using radio location fingerprinting.

Dr. Adlen Ksentini received the MS degree in telecommunications and multimedia networking from the University of Versailles and the PhD degree in computer science in 2005 from the University of Cergy, Pontoise. His PhD dissertation focused on QoS provisioning in IEEE 802.11-based networks. Since 2006, he is an associate professor at the University of Rennes 1, France, and member of the IRISA Laboratory. His research interests include QoS and QoE support for multimedia content, congestion control in LTE and Green Networks. He is a coauthor of more than 20 technical journal papers and international conference proceedings. He is a member of the IEEE.

Prof. Xiao-Hui Lin received his BS and MS degrees in Electronics and Information Science from the Lanzhou University, in 1997 and 2000, respectively. He got his PhD degree in Electrical and Electronic Engineering from the University of Hong Kong in 2003. He is now an associate professor in the Faculty of Information Engineering, Shenzhen University in Guangdong, China. His research interests include mobile computing, wireless networking, and multimedia communication. In these fields, he has published more than 40 papers in international leading journals and refereed conferences.

Ms. Ling Liu received her BS in Electronic Engineering from Shenzhen University in 2008. After that, she began her MS study in the same university. She is expected to get her MS degree in telecommunication systems in 2011. She is now a system engineer in UTStarcom, Shenzhen. Her research interests include wireless networks, signal processing and information theory.

Ms. Mehrnoush Masihpour completed her BS degree in Computer Sciences from University of Najafabad, Esfahan in Iran in 2007 and the Graduate Certificate in Telecommunication Networking from University of Technology, Sydney in

Australia in 2009. She started her PhD at the University of Technology, Sydney in 2009. Her research interests are wireless communication networks, magnetic induction communication, personal area network and cooperative communications in mobile networks. She has authored 14 international publications. She also has been involved in the industry for more than three years.

Prof. Kshirasagar Naik is an associate professor in the Department of Electrical and Computer Engineering at the University of Waterloo. His current research interests include energy cost modeling and analysis of smartphones, energy cost modeling and analysis of cloud computing architectures, wireless communication systems, intelligent transportation systems, vehicular networks, sensor networks, communication protocols, and application software. He has published numerous research articles in high quality international conferences and journals. He is a co-author of a textbook entitled "Software Testing and Quality Assurance: Theory and Practice" published by John Wiley in 2008. His second book entitled "Software Evolution and Maintenance" will be published by John Wiley in April 2012. He is on the program committees of several IEEE international conferences.

Dr. Ignas G. M. M. Niemegeers got a degree in Electrical Engineering from the University of Gent, Belgium, in 1970. In 1972 he received a MScE degree in Computer Engineering and in 1978 a PhD degree from Purdue University in West Lafayette, Indiana. From 1978 to 1981 he was a designer of packet switching networks at Bell Telephone Mfg. Cy, Antwerp, Belgium. From 1981 to 2002 he was a professor at the Computer Science and the Electrical Engineering Faculties of the University of Twente, Enschede, The Netherlands. From 1995 to 2001 he was Scientific Director of the Centre for Telematics and Information Technology (CTIT) of the University of Twente, a multi-disciplinary research institute on ICT and applications. Since May 2002 he holds the chair Wireless and Mobile Communications at Delft University of Technology, where he is heading the Centre for Wireless and Personal Communication (CWPC) and the Telecommunications Department. He was involved in many European research projects, e.g., the EU projects MAGNET and MAGNET Beyond on personal networks, EUROPCOM on UWB emergency networks and, eSENSE and CRUISE on sensor networks. He is a member of the Expert group of the European technology platform eMobility and IFIP TC-6 on Networking. He is also chairman of the HERMES Partnership, an organization of leading European research institutes and universities in telecommunications. His present research interests are 4G wireless infrastructures, future home networks, ad-hoc networks, personal networks, cognitive networks.

Mr. Rajesh Palit received his BS degree in Computer Science and Engineering from Bangladesh University of Engineering and Technology (BUET), and MS degree in Computer Engineering from University of Manitoba in 2000 and 2004, respectively. He is now a PhD candidate in the Department of Electrical and

Computer Engineering at University of Waterloo, Canada. He also served 2 years in industry and 2 years in academia as a faculty member in North South University, Dhaka, Bangladesh. His current research interests include energy efficient wireless networks, pervasive and green computing.

Prof. S. R. S. Prabaharan is currently a full professor of Electronics in the University of Nottingham in its Malaysia Campus. Prabaharan graduated from The American College, an autonomous affiliate of Madurai Kamaraj University, India and all of his degrees are from Madurai Kamaraj University. He earned his PhD in Solid State Devices in 1992. He has worked at CSIR research labs (National Aerospace Laboratory) in Bangalore, India on Ionic/electronic devices using Raman spectroscopy, CECRI, another CSIR research lab in Karaikudi, India. Later, he became a Research Scientist under DST (Department Science and Technology) Young Scientist Research Award program. He also taught at Universiti Malaya, Malaysia and later joined the Applied Sciences Faculty at Universiti Teknologi Petronas. In 1999, he joined the Faculty of Engineering, Multimedia University as a Senior Lecturer before moving to The University of Nottingham Malaysia Campus as an Associate Professor. Dr Prabaharan has been invited as visiting Senior Fellow/Scientist in different academic institution to foster research and initiate joint research activities with academic/research institutions which include: UPMC, Paris (1998) and Tokyo Institute of Technology; Japan (2003), Southern University, USA (2005) and University of Sheffield, UK (February 2006). He is also a visiting research consultant to Southern University, Baton Rouge, USA where he was invited under US Army Research Project. He is a peer reviewer for numerous international journals published by Elsevier, Springer and Hindawi, and has been the guest editor for a Solid State journal published by Springer (in press, 2007). His research interest are in the field of supercapacitors; hybrid power sources; lithium-ion batteries; nanotechnology of clean energy; solid state devices; semiconductor gas sensors; modeling and simulation of power electronic circuits (UPS, Mini power grids and Solar PV MPP devices); supercapacitors for power assist applications.

Dr. T. V. Prabhakar received his Bachelors degree in Science from Bangalore University, India in 1983. In 1987 he received MSc (Physics) degree, and obtained MSc (Engg) degree in 2004 from Indian Institute of Science (IISc), Bangalore, India. He joined IISc in 1985 and served in various capacities. He is currently a Senior Scientific Officer in Centre for Electronics Design and Technology (CEDT), IISc, Bangalore. He is the founder member of the Zero Energy Networks (ZEN) laboratory in CEDT. He is serving as a technical lead in several National and International collaboration sponsored research projects. Some of the areas include Telecom, Embedded systems, Wireless sensor networks and Education projects. He is associated with many Industrial Research/Developments projects in the areas related to Wireless sensor networks, Embedded systems, and Energy harvesting.

His present research interests are Communication networks, Wireless networks, Energy harvesting Wireless sensor networks, Embedded systems and application of technology for development application of technology for development application of technology for development.

Dr. R Venkatesha Prasad received his bachelors degree in Electronics and Communication Engineering and MTech degree in Industrial Electronics from University of Mysore, India in 1991 and 1994. He received a PhD degree in 2003 from Indian Institute of Science, Bangalore, India. During 1996 he was working as a consultant and project associate for ERNET Lab of ECE at Indian Institute of Science. While pursuing the PhD degree, from 1999 to 2003 he was also working as a consultant for CEDT, IISc, Bangalore for VoIP application developments as part of Nortel Networks sponsored project. In 2003 he was heading a team of engineers at the Esqube Communication Solutions Pvt. Ltd. Bangalore for the development of various real-time networking applications. Currently, he is a part-time consultant to Esqube. From 2005 till date he is a senior researcher at Wireless and Mobile Communications group, Delft University of Technology working on the EU funded projects MAGNET/MAGNET Beyond and PNP-2008 and guiding graduate students. He is an active member of TCCN, IEEE SCC41, and reviewer of many Transactions and Journals. He is on the TPC of many conferences including ICC, GlobeCom, ACM MM, ACM SIGCHI, etc. He is the TPC co-chair of CogNet workshop in 2007, 2008 and 2009 and TPC chair for E2Nets at IEEE ICC-2010. He is also running PerNets workshop from 2006 with IEEE CCNC. He is the Tutorial Co-Chair of CCNC 2009 & 2011 and Demo Chair of IEEE CCNC 2010. He is an invited member of IEEE ComSoc Standards Board.

Prof. Roberto Rojas-Cessa received the M Comp Eng degree and the PhD degree in Electrical Engineering from Polytechnic Institute of New York University, Brooklyn, NY. He also received an MSc degree in Electrical Engineering from the Research and Advanced Studies Center (CIVESTAV), Mexico. He received his BS in Electronic Instrumentation from Universidad Veracruzana, Mexico. Currently, he is an associate professor in the Department of Electrical and Computer Engineering, New Jersey Institute of Technology, Newark, NJ. He was an adjunct professor and a research associate in the Department of Electrical and Computer Engineering of Polytechnic Institute of New York University. He has been involved in design and implementation of application-specific integrated-circuits (ASIC) for biomedical applications and high-speed computer communications, and in the development of high-performance and scalable packet switches and reliable switches. He was part of the team designing a 40 Tb/s core router in Coree, Inc., in Tinton Falls, NJ. His research interests include high-speed switching and routing, fault tolerance, quality-of-service networks, network measurements, and distributed systems. He was a visiting professor in Thammasat University, Rangsit Campus, Thailand, in 2010. His research has been funded by U.S. National Science Foundation and Industry.

He was the recipient of the Advance in Research Excellence of the ECE Dept. in 2004. He has served in several technical committees for IEEE conferences and as a reviewer for several IEEE journals. He has been a reviewer and panelist for U.S. National Science Foundation and the U.S. Department of Energy. He has more than 10 years of experience in teaching Internet protocols and computer communications. Currently, he is the Director of the Networking Research Laboratory at the ECE Department and the Coordinator of the Networking Research Focus Area Group of the same department.

Prof. David J. Santos obtained his MSc and PhD Degrees both from Universidad de Vigo, Spain (in 1991 and 1995 respectively). From 1995 to 2005 he has been a professor at Universidad de Vigo and a visiting scholar to University of Rochester (USA) and University of Essex (UK). Since 2005 he is an associate professor at Universidad CEU-San Pablo in Madrid (Spain) where he also chairs the Division of Engineering of the Escuela Politécnica Superior. His research interests include: quantum information processing, quantum optics, optical communications, communication circuits, and applied mathematics problems related with process modelling and optimisation, and data mining.

Prof. Kah Phooi Seng received her PhD and Bachelor degree (first class honors) from University of Tasmania, Australia in 2001 and 1997 respectively. She is currently an associate professor in the School of Electrical & Electronic Engineering at The University of Nottingham Malaysia Campus. Her research interests are in the fields of intelligent visual processing, biometrics and multi-biometrics, artificial intelligence and signal processing.

Prof. Ajit Singh received the BSc degree in electronics and communication engineering from the Bihar Institute of Technology (BIT), Sindri, India, in 1979 and the MSc and PhD degrees from the University of Alberta, Edmonton, AB, Canada, in 1986 and 1991, respectively, both in computing science. From 1980 to 1983, he worked at the R & D Department of Operations Research Group (the representative company for Sperry Univac Computers in India). From 1990 to 1992, he was involved with the design of telecommunication systems at Bell-Northern Research, Ottawa, Canada. He is currently an associate professor at Department of Electrical and Computer Engineering, University of Waterloo, Waterloo, ON, Canada. His research interests include network computing, software engineering, database systems, and artificial intelligence.

Prof. Dr. Christian Steger received 1990 the Dipl-Ing degree (equivalent to the American Master of Science) and 1995 the Dr. Techn. degree (equivalent to the American PhD degree) in Electrical Engineering, Graz University of Technology, Austria. Graduated from Export, International Management and Marketing course in June 1993 at Karl-Franzens-University of Graz. From 1989 to 1991 Software Trainer and Consultant at SPC Computer Training Ges.m.b.H., Vienna. From 1990 to 1991 research engineer at the Institute for Technical Informatics, Graz University of Technology. Since 1992 he is an assistant professor at the Institute for Technical Informatics, Graz University of Technology. In summer 2002 he was a visiting researcher at the Department of Computer Science at the University College Dublin (Ireland). He heads the HW/SW co-design group (8 PhD students) at the Institute for Technical Informatics. His research interests include embedded systems, HW/SW co-design, HW/SW co-verification, SOC, power awareness, smart cards, UHF RFID systems, multi-DSPs. He is currently working with industrial partners on heterogeneous system design tools for system verification and power estimation/optimization for RFID systems, smart cards and wireless sensor networks. Christian Steger has supervised and co-supervised over 73 master's thesis and co-supervised 8 PhD students, and published more than 70 scientific papers as author and co-author. He is member of the IEEE and member of the ÖVE (Austrian Electro-technical Association). He was member of the organizing committee of the Telecommunications and Mobile Computing Conference 2001, 2003, and 2005.

Dr. Christos Theoharatos was born in Athens in 1973. He received a BSc degree in Physics in 1998, an MSc degree in Electronics & Computer Science in 2001, and a PhD degree in Image Processing and Multimedia Retrieval in 2006, all from the Electronics Laboratory (ELLAB), Dept. of Physics, University of Patras (UoP), Greece. He is currently and R&D manager for IRIDA Labs. He is also involved as a Post-Doc researcher to the Digital Information Processing group, at ELLAB – UoP. During the last five years he has been acting as a Technical Manager in a number of European and National R&D projects in the fields of signal and image processing, multimedia services and information technology. He has published more than 30 journal and conference papers in the fields of his expertise. His main research interests include pattern recognition, multimedia databases, image processing and computer vision, data mining, and graph theory.

Prof. Abdellah Touhafi obtained his MSc Degree in Electronic Engineering from Vrije Universiteit Brussel (Belgium) in 1995 and his PhD from the Faculty of Engineering Sciences from Vrije Universiteit Brussel (Belgium) in 2001. In 2001 he became post-doctoral researcher at Erasmushogeschool Brussel where he researched on environmental monitoring systems. In 2003 he became professor and founded his research group on reconfigurable and embedded systems. Since 2009 he is the program coordinator in the Industrial Sciences Department. His current research interests include embedded real-time systems, high performance and reconfigurable computing, Sensor Webs for localization and environmental monitoring, security, Software Defined Radio and digital communication circuits.

Dr. José Manuel Vázquez obtained his MSc and PhD Degrees both from Universidad Politécnica of Madrid. He has over thirty years experience in the IT sector, designing and developing a variety of innovative projects for market-leading companies. During his career he has played different roles and positions of responsibility in various areas of business for which he worked such as production, sales, marketing, communication and R&D. He is currently a lecturer at University CEU-San Pablo in Madrid and managing partner of a consultancy company focused on the implementation of change management and BPR for new companies in the digital economy. It has also been evaluating research projects of the European Union and has served on various national and international committees related to marketing and regulation in the field of IT.

Prof. Dr. Reinhold Weiss is professor of Electrical Engineering (Technical Informatics) and head of the Institute for Technical Informatics at Graz University of Technology, Austria. He received the Dipl-Ing degree, the Dr-Ing degree (both in Electrical Engineering) and the Dr-Ing habil degree (in Realtime Systems) from the Technical University of Munich in 1968, 1972 and 1979, respectively. In 1981 he was as a visiting scientist with IBM Research Laboratories in San Jose, California. From 1982 to 1986 he was professor of Computer Engineering at the University of Paderborn (Germany). He is author and co-author of about 170 scientific and technical publications in Computer Engineering. For E&I (Elektrotechnik & Informationstechnik, Springer-Verlag) he served several times as a guest editor for special issues on Technical Informatics and Mobile Computing, respectively. In 2001 and 2003 he organized two Workshops on Wearable Computing. His research interests focus on Embedded Distributed Real-Time Architectures (parallel systems, distributed fault-tolerant systems, wearable and pervasive computing). He is a member of the International Editorial Board of the US- journal "Computers and Applications" (ISCA). Further, he is a member of IEEE, ACM, GI (Gesellschaft für Informatik, Germany), and ÖVE (Österreichischer Verein für Elektrotechnik, Austria).

Mr. Adamu Murtala Zungeru received his BEng degree in Electrical and Computer Engineering from the Federal University of Technology (FUT) Minna, Nigeria in 2004, and MSc degree in Electronics and Telecommunication Engineering from the Ahmadu Bello University (ABU) Zaria, Nigeria in 2009. He is a Lecturer Two (LII) at the Federal University of Technology Minna, Nigeria in 2005-to date. He is a registered engineer with the Council for the Regulation of Engineering in Nigeria (COREN), and member of the Institute of Electrical and Electronics Engineers (IEEE). He is currently a PhD candidate in Electrical and Electronics Engineering at the University of Nottingham Malaysia Campus. His research interests include Energy Efficient routing, Energy harvesting, Storage and Management in Wireless and Visual Sensor Networks.

List of Reviewers

1. Mr. Leander Hörmann, *Technical University (TU) Graz, Austria*
2. Mr. Philipp Glatz, *Technical University (TU) Graz, Austria*
3. Dr. Rajesh Palit, *University of Waterloo, Canada*
4. Dr. Mikkel Baun Kjærgaard, *Aarhus University, Denmark*
5. Dr. Yassine Hadjadj-Aoul, *INRIA, University of Rennes 1, France*
6. Dr. Nikos Fragoulis, *IRIDA LABS, Greece*
7. Dr. Sonali Chouhan, *Indian Institute of Technology (IIT) Guwahati, India*
8. Dr. Hrishikesh Venkataraman, *Dublin City University (DCU), Ireland*
9. Dr. Gabriel-Miro Muntean, *Dublin City University (DCU), Ireland*
10. Dr. Sean Marlow, *Dublin City University (DCU), Ireland*
11. Ms. Ramona Trestian, *Dublin City University (DCU), Ireland*
12. Mr. Adamu Murtala, *University of Nottingham, Malaysian campus, Malaysia*
13. Mr. Mehrnoush Mashipour, *University of Technology, Sydney, Australia*
14. Dr. Otto Andersen, *Western Norway Research Institute (WNRI), Norway*
15. Dr. Ranga Venkatesha Prasad, *Technical University (TU), Delft, Netherlands*

OPTIMIZATION TECHNIQUES

Chapter 1

Energy Management for Location-Based Services on Mobile Devices

Mikkel Baun Kjærgaard

Contents

1.1 Introduction

Location-based services (LBS) that utilize the position of mobile devices to provide user functionality, such as services for navigation, location-based search, social networking, games, and health and sports trackers, are becoming more and more important. Research has investigated such services for more than a decade now, and, recently, they also have become commercially important as they claim a large share of the mobile applications deployed on mobile phones (Skyhook Wireless, 2010).

A successful LBS must not excessively drain the battery of mobile devices. Battery capacity is a scarce resource in mobile devices because it is not increasing at the same pace as the new power-demanding features that are added to mobile devices. If users experience that a specific LBS drains the battery, they might stop using the service. It is, however, not a simple task to build low-power-consuming LBSs because such services make heavy use of many power-consuming features of mobile devices, such as the radio to receive and send data, the screen to display maps, or positioning sensors. Today's mobile devices contain several positioning sensors, e.g., a built-in GPS receiver or a WiFi radio that can be used for positioning. For a general introduction to positioning technologies, we refer the reader to LaMarca and Lara (2008). Therefore, an LBS has to take great care in how it uses device features to minimize the power consumption, especially if the service is to run continuously.

In this chapter, we characterize the power consumption of location-based services and consider profiling and modeling the power consumption of mobile device features, which is a prerequisite for most methods for minimizing the power consumption and for their evaluation. Then, we present methods for minimizing power consumption where we divide the methods into sensor management strategies and position update protocols. For example, we will present our software system EnTracked that implements several novel sensor management strategies and position update protocols that can lower the power consumption of many types of LBSs by 64 percent for a continuous moving device and by up to 93 percent for an occasionally moving device.

1.2 Power Consumption and Location-Based Services

How crucial it is for an LBS to save power depends on the usage pattern, battery recharge options, and how the service uses the phone's features. With regard to the usage pattern, an important parameter is how long a service is expected to run on a phone. The most important LBSs to minimize power consumption are those that are long running for hours or days; however, such services also provide many opportunities for applying power-saving methods. The importance of minimizing the power consumption also depends on users' recharge options because a service can be allowed to consume a lot of power if a user is able to recharge the phone when finished using the service (Banerjee et al., 2007). Due to such considerations, it might be a situation that is dependent on how important it is that a service consumes minimal power. In regards to the phone feature usage, the consumption

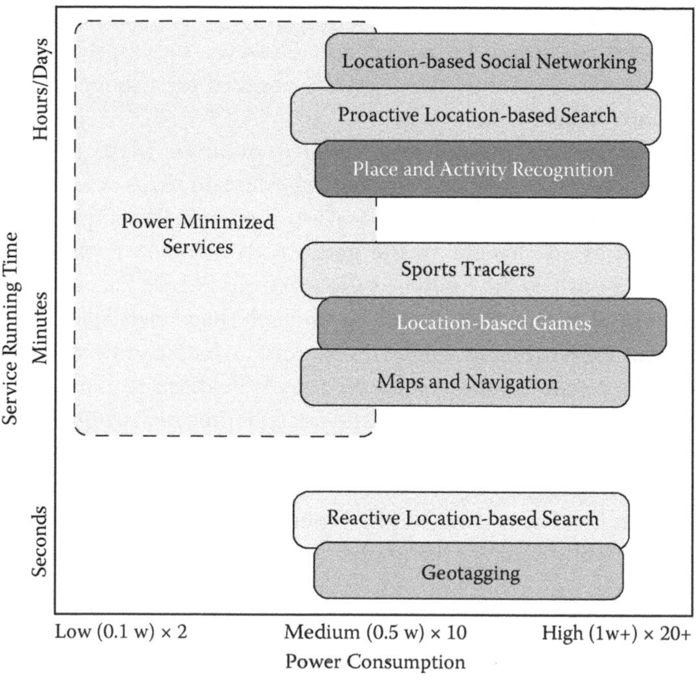

Figure 1.1 Service types grouped by service running time and power consumption with multiplicity factors for power consumption compared to a 0.05 watt standby consumption. (From Kjærgaard, M. B. 2011. *IEEE Pervasive Computing.* Forthcoming. With permission.)

impact depends on the power consumption of the individual features. Later sections describe how to profile the power consumption of individual phone features and give some values for a typical mobile phone.

A classification of the power consumption for different types of LBSs is shown in Figure 1.1, originally presented in Kjærgaard (2011). The classification types are inspired by the service types introduced by Bellavista, Küpper, and Helal (2008). The figure classifies service types with respect to their running time and power consumption. Running times are classified into second-long, minute-long, and hours/days-long, and power consumption into low-, medium-, and high-consuming services; a factor is given indicating the impact on the battery lifetime compared to a standby battery consumption of 0.05 watt.

The figure shows two service types that only run for seconds. Geotagging subsumes services that attach location information to other digital material, e.g., pictures, and reactive location-based searches are services that, when requested, search for information related to the user's location, for instance, about the nearest subway stations. The consumption of such services is medium to high due to the fact that the screen, communication, and positioning features are all used. Furthermore, the power consumption of such services is difficult to minimize by software means because a short, well-defined task has to be carried out. However, the impact on the battery lifetime is not significant because these services are used for a short amount of time and not frequently rerun.

Three service types are given that run for minutes. Maps and navigation involves services that can show people where they are on maps or satellite imagery and provide navigation directions to a location. Location-based games are games that use location as an element in the game play, e.g., the finding of physical caches using GPS positioning known as Geocaching or Live Pac Man, where real persons run around as monsters intending to catch the player. Sports trackers are services that can log where and when you exercise, which can be shared and analysed. Again, the consumption of such services is medium to high, but, because they run for minutes, their impact on the battery lifetime is higher. When services run for minutes, it is an advantage that they have a low consumption, but maybe not necessary, e.g., if a user is able to recharge the phone when finished. However, one problem that users might experience is that if they forget to turn a service off, it might discharge the phone without them noticing before it is too late. To avoid these issues, methods for minimizing the power consumption can be used to lower the power consumption to prolong the battery life.

In Figure 1.1, three services are shown that run for hours or days. Place and activity recognition are services that can register the whereabouts and activities of a user to, e.g., construct a daily diary or calculate a CO_2 footprint from the user's behavior. Proactive location-based searches are services that can push information to the user in the form of query results, e.g., if a user registers a search for free city bikes, the user will be notified when in the proximity of them. Location-based social networking is a service that enables the user to link location to social networking, e.g., to be notified when near friends or events. Again, the consumption of such services is medium to high, but, because the services run for hours or days, it is very important that they consume a minimal amount of power because they would otherwise have a major impact on how fast the battery will discharge, e.g., 20 times faster with a high consumption compared to standby consumption. Therefore, for long-running services, it is crucial to apply methods for minimizing the power consumption.

1.3 Profiling and Modeling the Power Consumption of Mobile Devices

As a first step in understanding the power consumption of mobile devices, one could consult their specifications. However, these will often not give the full picture because values are missing (e.g., power consumption values for central processing unit (CPU) usage) and dynamic aspects are not considered. The dynamic aspects are caused because features do not instantly power on or off, e.g., a 3G radio needs several seconds to power on before it has established a connection and the same is true

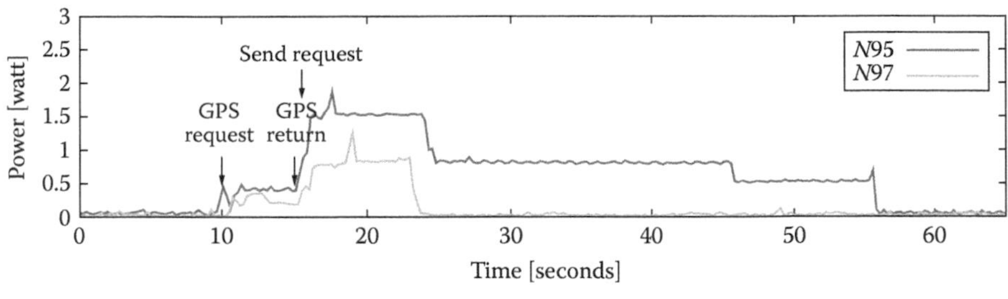

Figure 1.2 Power consumption on Nokia N95 and N97 phones for requesting a GPS position and sending the position to a remote server.

when it powers off. Therefore, the power consumed when sending data is not simply modeled as a single value for the consumption. A solution for capturing dynamic aspects is to power profile a device. To use this information to proactively minimize the power consumption of LBSs or to evaluate different design options requires that one has adequate models of the power consumption. This section discusses the profiling and modeling of the power consumption of mobile devices.

To truthfully model the power consumption of a phone, one has to consider dynamic aspects in addition to the consumption of individual features. To illustrate these aspects, Figure 1.2 shows two power profiles of a Nokia[â] N95 and a Nokia N97 phone running a Python script that every 60 seconds invokes the GPS to produce a single position fix, opens a transmission control protocol (TCP) connection to a server over the 3G radio, sends the position fix, and then closes the connection. It can be seen from the figure that the single steps are not executed instantly, and that it takes some seconds for the GPS to produce a position fix and to send the position fix. Furthermore, after sending, both the 3G radio and the GPS keep consuming power for a while. Finally, the different phones also have different delays and power levels associated with their features.

The ability to accurately model the power consumption and delays is important for three reasons. Firstly, without a model, we cannot make informed decisions about what actions to take to minimize the power consumption. Secondly, if we do not have a model of the delays, we do not know how much time to reserve for delays at runtime to update positions within required accuracy limits. Thirdly, it might be too laborious to evaluate different options for power saving for each step in the design process by deploying the software on a phone, mimic real behavior (e.g., a walking tour outdoors), and measure the power consumption. As an alternative, a model for power consumption allows simulation of the power consumption without deploying the software, which enables a faster development process. A drawback of such models is that they depend on the estimation of device-dependent parameters as already illustrated in Figure 1.2. Therefore, there is a tradeoff between the model's accuracy with regards to the number of parameters that the model takes into account and the practicability of using the model, in terms of the effort to profile the parameters for a new device.

1.4 Device Model

In the following, we present a device model originally proposed in our previous work (Kjærgaard et al., 2009) consisting of two parts: (1) a power model that describes the power usage of a phone, and (2) a delay model that describes the delays, for instance, when requesting a phone feature, e.g., the time it takes for a GPS to return a position. The model considers a subset of the phone features relevant for position tracking using GPS and inertial sensors. If needed, the model could be easily extended with additional variables to also consider WiFi, Bluetooth, and GSM positioning. Furthermore, the basic model assumes that no CPU heavy tasks have to be considered, but they could be factored in given a mapping between, e.g., the size of the input of the task and the resulting power consumption. For interactive user applications, one also would need to take into account the power usage of features such as the computations for the application logic, keystrokes, camera use, and screen use.

In the models, we consider the following phone features:

- Accelerometer (*a*)
- Compass (*c*)
- GPS (*g*)
- Radio idle (*r*)
- Radio sending (*s*)
- Background (I_p)

For each feature, the variable used to reference the feature later in the text is given in brackets. Background is not strictly a feature, but is included in the power model to cover the background consumption of the phone.

The power model consists of two functions defined in the equations below: the power function *power* and the consumption function $c_{d,p}$ where d is a feature's power-off delay and p its power consumption.

$$power(a_t, c_t, g_t, s_t, c_t) = I_p + c_{gd,gd}(g_r) + c_{rd,sd}(r_t) + c_{rd,sd}(s_t)$$

$$c_{d,p}(x) = \begin{cases} p & if\ x <= d \\ 0 & if\ x > d \end{cases}$$

The equation uses the variables a_t, c_t, g_t, r_t, s_t for the different features listed in the above list to denote their last usage. Each variable denotes at time step t, the number of seconds since the feature was last powered off (a variable is zero if the feature is in use in the current time step t). Since the idle power consumption is constant, no variable i_t is introduced. Furthermore, the parameters a_p, c_p, g_p, r_p, s_p, I_p denote the power consumption of a feature, e.g., 0.324 watt for a Nokia N95 internal GPS. The parameters a_d, c_d, g_d, s_d, r_d denote the number of seconds until a feature is powered off after last usage, e.g., 30 seconds for a Nokia N95 internal GPS. More example values for the different features will be provided in a later example for a Nokia N95 phone, but also can be found in Kjærgaard (2010).

The delay model contains functions that capture the delay for any feature that has a significant associated delay. Features that have none or negligible delays are modeled as they instantly perform their task. In a mobile phone, it is mainly the GPS and the radios that have request delays when powering on associated with them, which can be modeled as two functions—$req_g(g_t)$, $req_s(s_t)$—that describe the request delays for the GPS and for activating the radio for sending.

1.4.1 Example: Modeling the Nokia N95 Phone

In the following, we take a Nokia N95 phone as an example and explain how we parameterize the model for this phone and present results for how well the model fits with actual device measurements.

The Nokia N95 8GB is a 3G phone with an internal GPS module and a triaxial accelerometer, both of unspecified brand, and a 1,200 mAh battery. The phone runs the Symbian 60 operating system version F1. To measure the power consumption of the phone, we used the Nokia-developed tool Nokia Energy Profiler version 1.1 (Nokia, 2011). The Nokia Energy Profiler tool has been built by Nokia to enable developers to analyze the power consumption of mobile applications and it supports a power sampling rate of up to 4 Hz. To measure the delays and power consumption of different features, several Python scripts have been developed that enable and disable features and measure various delays. The Python scripts run on the N95 with the aid of the Python Interpreter for S60, version 1.4.4 (Pys60 Community, 2011) and the included libraries that provide access to phone features, such as the internal GPS and the triaxial accelerometer. The internal GPS supports a sampling rate of 1 Hz and the triaxial accelerometer, a sampling rate of around 35 Hz. To make measurements involving sending data using the phone's 3G radio, a TCP/IP server was implemented in Java and deployed on a server connected to the Internet with a public IP address to which the phone was able to connect.

To determine the power parameters a_p, g_p, r_p, s_p, I_p, we have collected a number of power consumption traces with a N95 phone with different features enabled and disabled. Before each trace collection and before all of our other experiments, the phone was fully charged to counter the influence of the nonlinear voltage decrease of batteries (Brown et al., 2006). First, the Nokia Energy Profiler application was started, then the Python interpreter was started with a Python script that enabled or disabled certain features for a specific amount of time. The total script running time was five minutes for these measurements. Then the Python interpreter was closed and the Nokia energy profiler was stopped. The power consumption trace

Table 1.1 Power Consumption for Features of the Nokia N95

Feature	*Average Power [milliwatt]*
Background (l_p)	62
Accelerometer (a_p)	50
GPS (g_p)	324
Radio idle (r_p)	466
Radio sending (s_p)	645

collected with the Nokia energy profiler was exported to a file. These traces were trimmed to remove the consumption logged while the Python script was not running and when the screen was powered on. The average feature consumptions were calculated from the trimmed traces and are listed in Table 1.1. In the model, we use the average values for the parameters.

The request delays modeled by the two functions $req_g(g_t)$ and $req_s(s_t)$ have been measured using the same experimental setup. Firstly, the GPS request delay for assisted GPS was measured as the time between requesting a GPS measurement and the moment when a position was returned. The radio request delay was measured as the difference between the GPS timestamp and the reception timestamp on a remote server. A more detailed discussion of the measurements can be found in (Kjærgaard et al., 2009).

$$req_g(d)\begin{cases} 1 & \text{if } x <= 30 \\ 6 & \text{if } x > 30 \end{cases}$$

$$req_g(x)\begin{cases} 0.3 & \text{if } x <= 6 \\ 1.1 & \text{if } x < 6 \end{cases}$$

Following a similar experimental approach, the power-off delay, which is the time a feature takes to power off after the last usage, also has been measured and is listed in Table 1.2. The results indicate that the power-off delay for the GPS and

Table 1.2 N95 Power-off Delays for Features

Feature	*Average Time [seconds]*
GPS	30.0
Radio idle	31.3
Radio sending	5.45

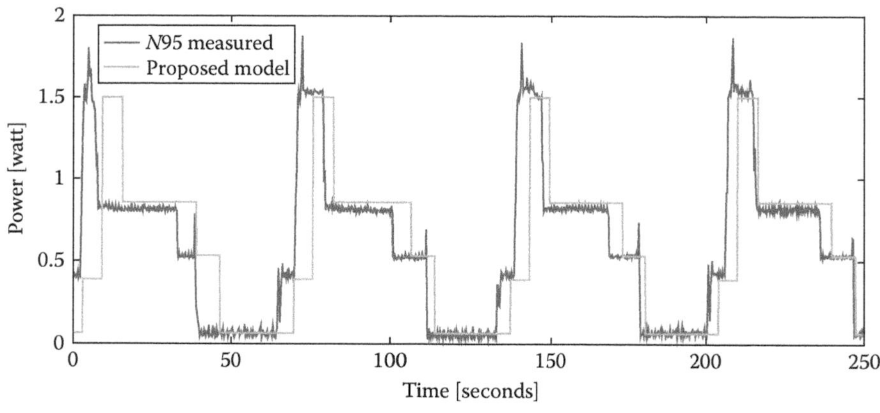

Figure 1.3 Comparison of consumption measured on a Nokia N95 with the modeled consumption. To improve readability, the curve for the proposed model has been shifted in time to not directly overlap with the measured curve.

for radio idle is around 30 seconds and a little below 6 seconds for radio sending. The power-off delay for radio idle is relative to when radio sending has powered off to idle mode.

To validate the proposed device model, we now compare the power consumption for periodic tracking calculated with the device model to the power consumption of traces collected on an N95 phone. Figure 1.3 plots data from the collected trace for 60 seconds periodic tracking, overlaid with the predicted power consumption of the device model. We can see how the proposed model closely matches the real power consumption. Therefore, this model can be used to inform the design of our tracking techniques toward minimizing the power consumption.

1.5 Methods for Minimizing the Power Consumption

This section reviews methods for minimizing the power consumption of LBSs. When considering methods for minimizing the power consumption, we have to consider how the services are distributed. Figure 1.4 outlines a conceptual model, which differentiates between local services running on mobile devices and remote services running in the cloud (Hayes, 2008). The local services will request positions from an API (application program interface) on the device, which means that the used power for positioning can primarily be linked to on-device sensors and processing. Exceptions to this are positioning methods that depend on server assistance, e.g., A-GPS and WiFi positioning. The remote services, on the other hand, will request positions from an API residing in the cloud, which means that the used power for positioning in addition to the on-device consumption results

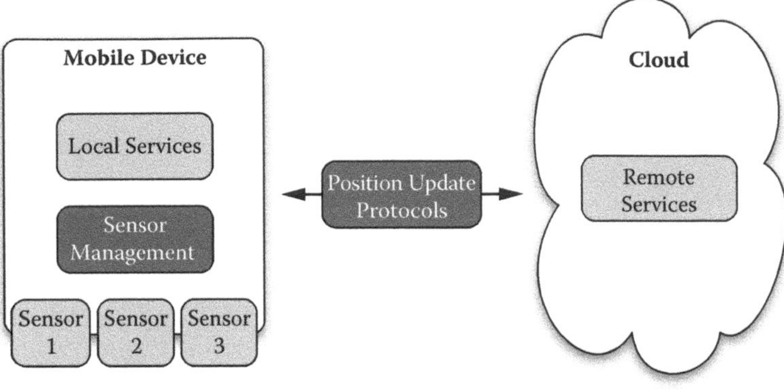

Figure 1.4 Overview of sensor management strategies and position update protocols.

from the radio's consumption for wireless connectivity. Therefore, it makes sense to differentiate between remote and local services, which also affect properties, such as position latency and privacy. If many remote services are interested in monitoring the position of a device, a dedicated tracking service might be deployed that is responsible for monitoring the device and for forwarding position updates to other remote services.

We, furthermore, divide the responsibility of handling service requests into on-device sensor management strategies and position update protocols (residing both on the device and in the cloud). Sensor management strategies decide how to use available position sensors to estimate the current position. Position update protocols control the interaction between the device and remote services. Such a division enables a flexible combination of different sensor management strategies and position update protocols and better overall performance by optimization of either subproblem.

After presenting relevant sensor management strategies and position update protocols, as a case, we will present our software system EnTracked that implements several novel sensor management strategies and position update protocols that can lower the power consumption of many types of LBSs by 64 percent for a continuous moving device and by up to 93 percent for an occasionally moving device.

1.5.1 Sensor Management Strategies

Sensor management strategies decide how to use available position sensors to estimate the current position. Sensor strategies could be implemented considering relevant properties, such as power consumption, positioning accuracy, the positioning availability in different environments (e.g., outdoor versus indoor), security (e.g.,

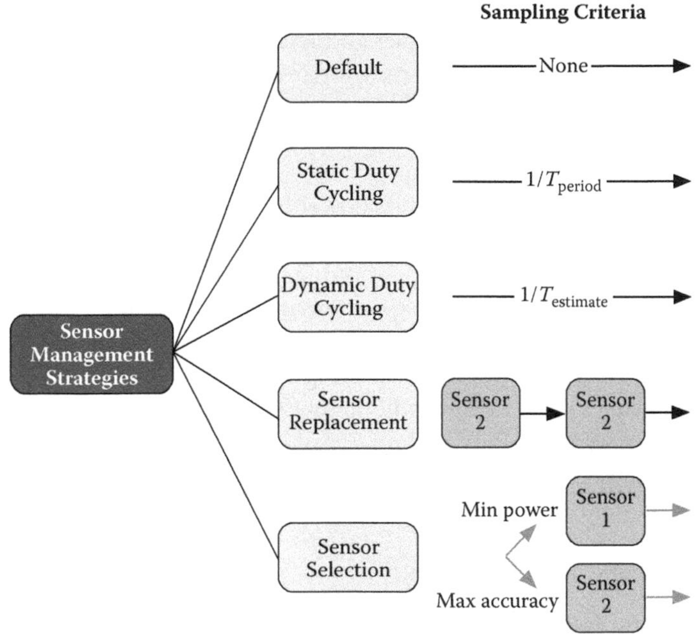

Figure 1.5 Overview of sensor management strategies.

spoofing attacks (Tippenhauer et al., 2009)), and privacy (e.g., WiFi positioning reveals a target's existence (Kjærgaard, 2007)).

In this section, we consider five types of sensor management strategies illustrated in Figure 1.5. The basic sensor management strategy is the default strategy that delivers positions as provided by a sensor. This strategy is relevant if a sensor's internal management already does a good job with regards to minimizing power consumption. The strategy of static duty cycling saves power by interleaving sampling with sleeping periods, which saves power because the sensor can be powered off during the sleeping periods. A static threshold T_{period} in seconds defines the length of the sleeping periods and, therefore, the resulting sampling frequency. This strategy is relevant for services where a lower frequency than the one supplied by a given sensor is enough to meet the requirements of the respective LBSs. The strategy of dynamic duty cycling also interleaves sampling requests with sleeping periods, but dynamically increases and decreases the sleeping periods to save power while ensuring that service requirements for the positioning accuracy are satisfied. The strategy continuously estimates a threshold $T_{estimate}$ in seconds for the sleeping period from properties, such as the speed and heading of a target. This strategy is relevant in cases where an adequate static duty-cycling threshold cannot be selected, e.g., for tracking targets with changing motion patterns.

The strategy of sensor replacement supervises the usage of high consuming sensors by events generated by a simpler and less consuming sensor. The positioning using a high consuming sensor can, for instance, be requested only when a simple motion sensor senses motion. This strategy is relevant in cases when a target has changing motion patterns that can be sensed by simpler and less consuming sensors.

The strategy of sensor selection saves power by switching between sensors with the goal to use the sensors, which use the least power to provide positions that satisfy the service requirements for positioning accuracy. This strategy is relevant in cases when services have changing requirements to positioning accuracy and several sensors are available, e.g., WiFi, GSM, or GPS with different properties with respect to power consumption and positioning accuracy.

In the following subsections, we will present concrete methods for applying the three strategies of dynamic duty cycling, sensor replacement, and sensor selection for location-based services.

1.5.1.1 Dynamic Duty Cycling

To apply dynamic duty cycling a model is needed for how to relate service requirements for positioning accuracy to sampling frequency. In the following, we present a model that relates the requested positioning accuracy to time and estimated accuracy and speed. The model consists of two steps: (1) to calculate the current accuracy and (2) to use the calculated current accuracy to calculate an estimate for the sleeping threshold $T_{estimate}$.

The first step takes into account the sensor-estimated accuracy a_{pos} as well as the time t_{pos} of the most recent position sample and the sensor estimated speed v_{pos}. The model then calculates the current accuracy $a_{current}$ with respect to the most recently delivered position as defined in the equation below:

$$a_{current} = a_{pos} + (t_{current} - t_{pos}) \times v_{pos}$$

The second step is to calculate the estimate for the sleeping period threshold $T_{estimate}$ from the service required positioning accuracy $a_{service}$, the current accuracy $a_{current}$, and the sensor estimated speed v_{pos}. The threshold $T_{estimate}$ is estimated using the equation below to calculate the time it will take a target to move beyond the service required limit, considering the current accuracy with respect to the last delivered position.

$$T_{estimate} = \begin{cases} \dfrac{a_{service} - a_{current}}{v_{pos}} & \text{if } a_{service} > a_{current} \\ 0 & \text{if } a_{service} =< a_{current} \end{cases}$$

Systems can then use this model to continuously estimate a new threshold to dynamically decrease or increase the sleeping period. Other models exist and also extensions that make the models able to handle delays, e.g., the time to first fix for GPS receivers (Kjærgaard et al., 2009).

1.5.1.2 Sensor Replacement

The different sensors in current mobile devices enable the usage of simpler sensors to supervise the usage of more consuming ones. The primary example, which we will discuss here, is to use an accelerometer as a simple sensor to sense motion. Most modern devices include a triaxial accelerometer, which provides acceleration measurements in three dimensions; the Nokia N95's accelerometer consumes only 0.05 watt compared to 0.32 watt for its GPS. Therefore, we can save power by using the accelerometer to sense motion and only use the GPS when the target is actually moving. Thus, we have to detect the two motion states, i.e., standing still and moving, relying on accelerometer readings. As the detection should not hurt the robustness of the positioning, we are interested in a detection scheme that has a low tolerance for movement, which will ensure that we detect movement very well. To implement such motion detection, the following simple scheme can be applied. First, an acceleration measurement is collected for each of the three axes, then, for each axis, the variance of the last 30 measurements is calculated and the three variance values are summed. Finally, the summed value is compared to a threshold that determines if motion is sensed or not. To optimize robustness or power consumption, the threshold can be chosen to favor either detecting motion or stillness. A drawback of this scheme is that a person walking with the device in the hand, and keeping the device steady, might lower the acceleration enough for the variance not to reach the threshold for movement detection. This poses a problem and can only be solved by using more clever movement detection schemes, such as the ones proposed by Reddy et al. (2010) or motion sensing from radio signals proposed by King and Kjærgaard (2008). Another sensor replacement strategy is to use the compass for sensing direction changes (Kjærgaard et al., 2011).

1.5.1.3 Sensor Selection

The common types of positioning both have different levels of power consumption, coverage, and positioning accuracy. Therefore, depending on the usage situation, power can be saved by selecting the optimal sensor at runtime. Recent measurements comparing GPS, WiFi, and GSM positioning for the N95 reported an average positioning accuracy of 10 m, 40 m, and 400 m, respectively, and a depletion time for a fully charged battery of 9, 40, and 60 hours, respectively (Constandache et al., 2009). Therefore, it is evident that power can be saved by switching to less accurate positioning methods when possible. The selection of which method to use can be based on parameters such as the service-requested positioning accuracy, e.g., using one of the computational frameworks proposed in Constandache et al. (2009) or Kjærgaard et al. (2009).

1.5.2 Position Update Protocols

Position update protocols control the interaction between the device and remote services, which have to consider relevant properties, such as server-side requested position accuracy, power consumption, data carrier availability, and privacy.

In terms of position update protocols, we restrict ourselves to the four device-controlled reporting protocols illustrated in Figure 1.6. Please consult Leonhardi and Rothermel (2001) for a description of other types of protocols and for an analytical analysis of the protocols in terms of their accuracy guaranties and communication efficiency. All the protocols assume that some sensor management strategy will manage the position sensors to continuously provide adequate positions to the protocols as the strategies would to any local service.

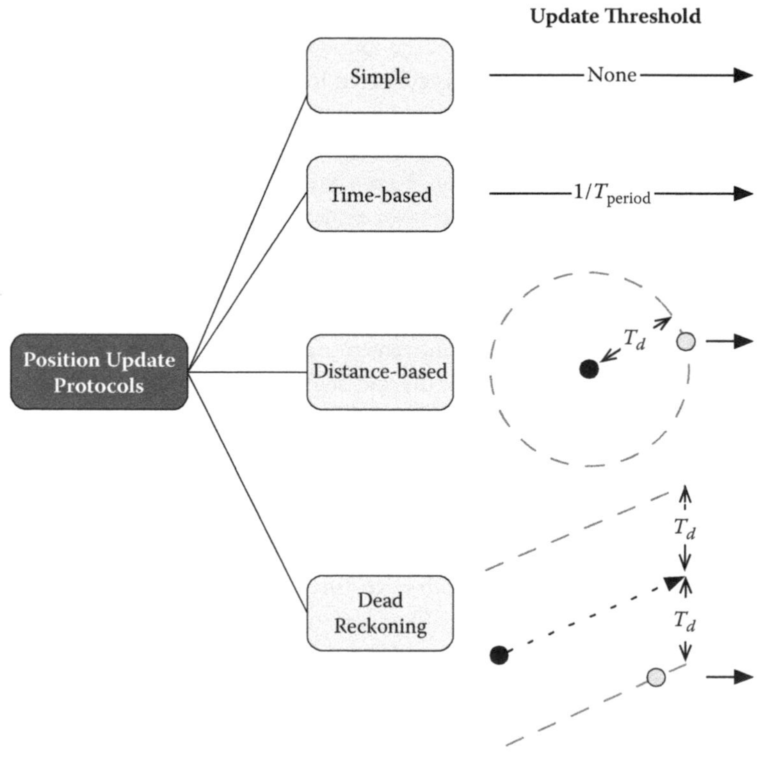

Figure 1.6 Overview of position update protocols.

The simple reporting protocol sends an update to the remote service each time a position sensor provides a new position. The advantage of this protocol is that it is simple to implement, but it results in many unnecessary position updates. Time-based reporting sends an update each time a certain time interval of T_{period} seconds has elapsed. Compared to the simple reporting protocol depending on T_{period}, this protocol can decrease the number of updates. However, because the protocol depends on a static time threshold, the protocol would produce the same number of updates regardless of whether the target is moving or not.

Distance-based reporting sends an update when the distance between the current position and the most recently reported position becomes greater than a given threshold $T_{distance}$ in meters. The advantage of this protocol is that it takes motion into account and, therefore, does not produce any updates if the device is not moving. However, during continuous movement, the protocol would still produce many updates. Dead-reckoning reporting is the most complex protocol of the four and optimizes reporting for continuous movement by not only sending the current position to the remote service, but also the current speed and heading. If the remote service at any time after the update needs the current position of the target, it should extrapolate it from the most recently sent position using the provided heading and speed. To keep the remote service's information up to date, the protocol will send an update from the device when the distance between the current position and the one extrapolated by the remote service becomes greater than a given threshold $T_{distance}$ in meters. The advantage of this protocol is that it can minimize the number of updates during both periods of continuous and no motion, but it has the disadvantage that it is more complex to implement than the other protocols. Further extensions to the dead-reckoning protocol exist, which, for instance, in the case of tracking of vehicles, make use of the road network to further reduce the number of updates (Civilis, Jensen, and Pakalnis, 2005).

1.6 Case: EnTracked

As a case, we will in this section consider the system EnTracked (Kjærgaard et al., 2009) built with the goal to dynamically track mobile devices in a both energy-efficient and robust manner. Thus, robust position updates have to be delivered to applications within service-specified accuracy limits, where accuracy refers to the distance between the known position of the application and the real position of the device. The realized system focuses on tracking pedestrian targets equipped with GPS-enabled devices. The system implements several of the presented sensor management strategies and provides all of the presented position update protocols. The system has more recently been extended to trajectory tracking and other modes of transportation (Kjærgaard et al., 2011).

1.6.1 System Description

To use EnTracked, location-based services have to provide service requirements for positioning accuracy for target tracking. In practice, location-based services do not always require the highest possible positioning accuracy as relevant occupancy limits can be calculated for many services. For example, a map service that shows the positions of a number of mobile devices can use the zoom level to determine relevant accuracy limits (such as 25 meters for street-level view, 100 meters for a suburb, and 200 meters for a city-wide view). Another example is the many types of social networking services that focus on relationships between the positions of devices, for instance, to detect when people come into proximity or when they separate. Methods have been proposed to efficiently track devices to reveal relationships, such as the ones proposed by Küpper and Treu (2006). The methods work by dynamically assigning tracking jobs with changing accuracy limits that they calculate based on the distance between the targets. Such methods produce tracking accuracy limits ranging from 10 meters to several kilometers, depending on the distance between the devices.

When a remote location-based service requests to use EnTracked, the steps illustrated in Figure 1.7 are carried out. Firstly, a service issues a request for the tracking of a device with an accuracy limit (1). Secondly, the server side of EnTracked propagates the request to the client side part of EnTracked (2). Thirdly,

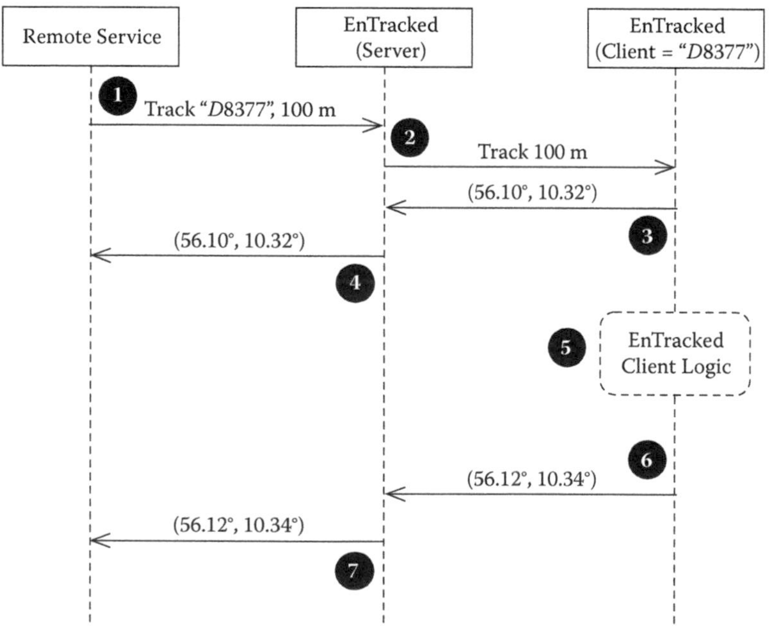

Figure 1.7 The steps of EnTracked when used by a location-based service.

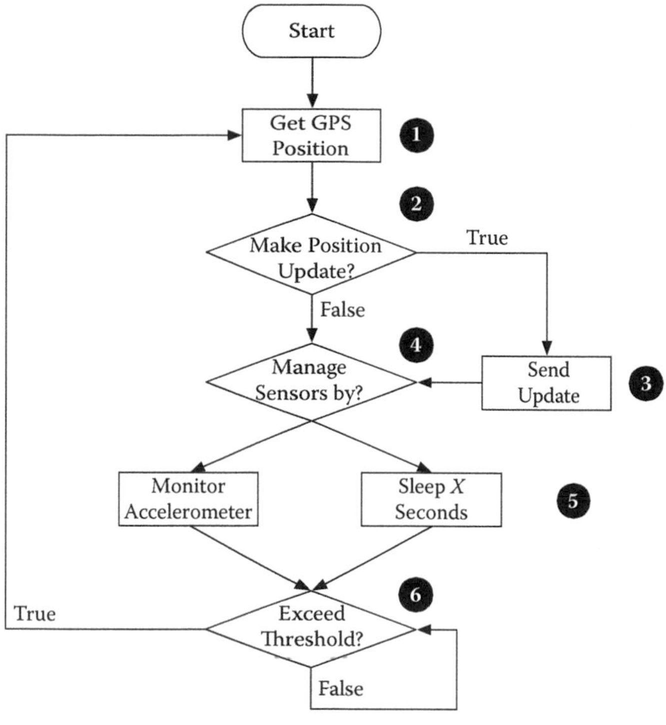

Figure 1.8 Flow chart of the EnTracked client logic.

the client finds an initial position and returns it through the server to the service (3)+(4). Fourthly, the EnTracked client logic schedules sensor management strategies and position update protocols to deliver the next position within the accuracy limit (5). Fifthly, at some point later, EnTracked determines that a new position has to be delivered to the client through the server (6)+(7). If several remote services request tracking for the same device, EnTracked configures the device for tracking with the highest requested accuracy to fulfill all of the services' limits. When a local service uses EnTracked, requests are passed directly to the client side logic.

Whenever the EnTracked client, as described above, has received a request, the client handles the request following the steps illustrated in Figure 1.8. To get an initial position, a GPS position is requested (1) that is in the remote case then provided to a position update protocol to evaluate whether a position update should be sent; in the local case, it is sent directly to the local service (2). If a position update is scheduled, the update is sent to the server (3). Then, the system applies the sensor management strategies of dynamic duty cycling, sensor selection, and sensor replacement to schedule the least power-consuming sensor task based on the current requirements (4). The scheduled sensor tasks to pick from could involve, e.g., monitoring the accelerometer or to sleep for a certain period (5). The process is restarted, once a task determines that a new GPS position is needed (6).

1.6.2 Results

This section presents evaluation results that characterize the magnitude of power savings that can be obtained by using EnTracked with different position update protocols to update a remote service. Furthermore, we also consider robustness with regards to the required positioning accuracy.

The previous presentation of EnTracked might indicate that out of the box any sensor strategy can be combined with any protocol. However, one has to take care of a number of implementation pitfalls. An example is the dead-reckoning protocol, which assumes that the server can extrapolate the position as long as it does not receive new updates from the mobile device. In the classic protocol, the threshold is tested continuously because a default sensor management strategy is implicitly assumed. The problem lies in what to do when an accelerometer-based sensor management strategy avoids providing new updates because the device is detected not to move. In this case, the server will continue to extrapolate the position, which might violate the threshold. To address this issue, we have extended the dead-reckoning protocol to test periodically if the server-predicted position is about to violate the threshold, and, in this case, send an extra position update with the last reported position and zero speed to stop erroneous extrapolation.

Another problem with the distance-based and dead-reckoning protocols is the limited robustness they provide because they might not be able to keep the maximum error below the required positioning accuracy due to delays and positioning errors. To improve protocol robustness, we use the GPS receiver's estimates of its current accuracy a_{pos} in meters and take this into account when evaluating if the protocol threshold has been passed, e.g., for distance-based reporting, the threshold equation then becomes: $d_{traveled} + a_{pos} < T_{distance}$ where $d_{traveled}$ is the distance between the last reported position and the current position.

To provide results for EnTracked with different position update protocols we will consider the following dataset collected for pedestrian movement patterns in an urban area with no stops, presented previously in Kjærgaard (2010). The dataset has been collected with Nokia N95 phones for three pedestrian targets walking a 4.85 km tour in an urban environment. The dataset consists of ground truth positions and 1 Hz GPS and 35 Hz acceleration measurements collected from the built-in sensors. The ground truth was collected at 4 Hz with a high accuracy u-blox LEA-5H receiver with a dedicated antenna placed on the top of a backpack carried by the collector. The ground truth measurements were manually inspected to make sure they followed the correct route of the target. Using an urban setting resulted in a rather high magnitude of average GPS errors of 29.1 meters on the Nokia N95 phones. That the dataset does not include any stops makes it more difficult to save

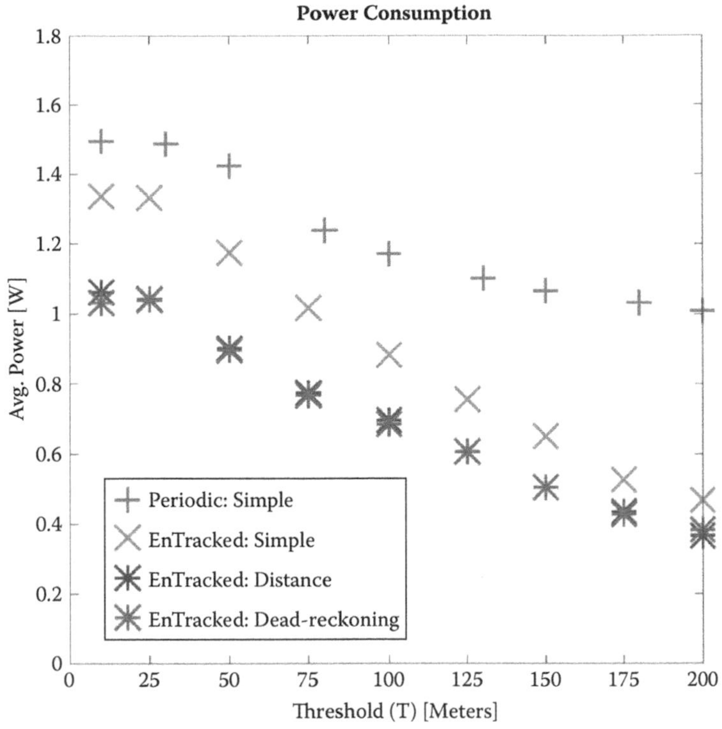

Figure 1.9 Comparison of average power consumption for Periodic: Simple (T = 10 m/s *T_{period}), EnTracked: Simple (T = $a_{service}$), EnTracked: Distance (T = $a_{service}$ = $T_{distance}$), and EnTracked: Dead-reckoning (T = $a_{service}$ = $T_{distance}$).

power since it implies that EnTracked cannot use the sensor replacement with the accelerometer, but only the dynamic duty cycling to save power.

The power consumption results from running different combinations of sensor management strategies and protocols are shown in Figure 1.9. To denote what position update protocol is used, we use the following notation EnTracked:{Protocol}. One can from the results notice how the increase of T_{period} for Periodic: Simple only provides small savings compared to the three EnTracked combinations. Of the three EnTracked combinations, the combination with the simple protocol provides the smallest savings ranging from 159 mW to 542 mW compared to Periodic: Simple depending on the threshold. The combination of EnTracked with the distance-based protocol provides a decrease in power consumption between 433 mW to 645 mW or in percentage of savings between 29 to 64 percent compared to Periodic:Simple and depending on the threshold. Comparing distance-based and dead-reckoning, there is only a small difference where the dead-reckoning version is a few mW better for the threshold of 10 meters and a few mW worse for the

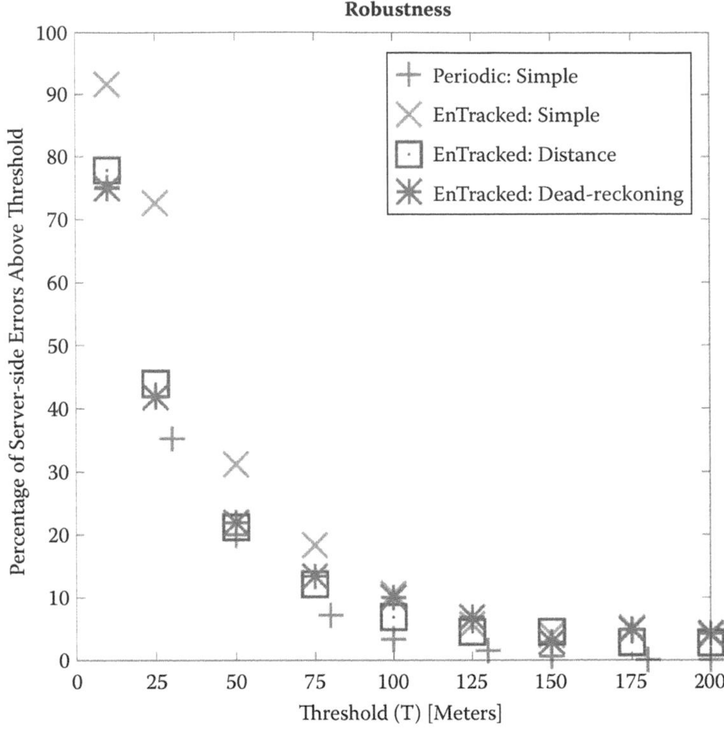

Figure 1.10 Comparison of robustness for Periodic: Simple (T = 10 m/s $*T_{period}$), EnTracked: Simple (T = $a_{service}$), EnTracked: Distance (T = $a_{service}$ = $T_{distance}$), and EnTracked: Dead-reckoning (T = $a_{service}$ = $T_{distance}$).

200 meters threshold. One reason for the negligible improvement of dead-reckoning over the distance-based protocol is that, if one compares with ground truth the average accuracy for the speed and the heading, estimates are low given the urban area and, therefore, the server predictions will often be extrapolated in an erroneous direction. Furthermore, it also can be linked to the movement style of an urban pedestrian, which is expected to include many and sharp turns. If the usage had included periods of still time, the savings could have dropped down to 93 percent with the help of accelerometer-based sensor replacement.

Figure 1.10 shows a robustness plot to analyze the robustness of such systems, e.g., to evaluate if the magnitude of GPS errors makes small thresholds irrelevant. The robustness is here defined as the percentage of time that the distance between the real position and the server known position is greater than the threshold. In all cases, the Periodic: Simple combination has the lowest values, in half of the cases below five percent. For the smaller thresholds, the percentage is higher because the GPS errors alone often are enough to violate the smaller thresholds, as the average

GPS error for the dataset is 29.1 meters. Comparing to the three EnTracked combinations, they all have a higher percentage of errors, but for most thresholds the difference is only a few percent points. The only major outliner is the EnTracked: Simple combination that has trouble at lower thresholds. Therefore, we can conclude that the system can save power without having a severe impact on robustness.

1.7 Summary

Location-based services have to pay careful attention to their power consumption in order not to drain the batteries of mobile devices. In this chapter, we characterized the power consumption of location-based services. Furthermore, we considered profiling and modeling the power consumption of mobile device features, which is a prerequisite for most methods for minimizing the power consumption and the evaluation thereof. Afterwards, we presented methods for minimizing power consumption where we separated the methods into sensor management strategies and position update protocols. As a case, we presented a software system named EnTracked that implements several novel sensor management strategies and position update protocols that can lower the power consumption of many types of LBSs with 64 percent for a continuous moving device and up to 93 percent for a periodically moving device.

ACKNOWLEDGEMENT

The author acknowledges the financial support granted by the Danish National Advanced Technology Foundation for the project Galileo: A Platform for Pervasive Positioning under J.nr. 009-2007-2.

REFERENCES

Banerjee, N., A. Rahmati, M. D. Corner, S. Rollins, and L. Zhong. 2007. Users and batteries: Interactions and adaptive energy management in mobile systems. Proceedings of the 9th International Conference on Ubiquitous Computing. Association of Computing Machinery (ACM), Innsbruck, Austria, Sept. 16–19, 2007, pp. 217–234.

Bellavista, P., A. Küpper, and S. Helal. 2008. Location-based services: Back to the future. *IEEE Pervasive Computing* 7 (2): 85–89.

Brown, L., K. A. Karasyov, V. P. Levedev, A. Y. Starikovskiy, and R. P. Stanley. 2006. Linux laptop battery life. Proceedings of the Linux Symposium Ottawa, Canada, July 19–22, 2006.

Civilis, A., C. S. Jensen, and S. Pakalnis. 2005. Techniques for efficient road-network-based tracking of moving objects. *IEEE Transactions on Knowledge and Data Engineering* 17 (5): 698–712.

Constandache, I., S. Gaonkar, M. Sayler, R. R. Choudhury, and L. Cox. 2009. EnLoc: Energy efficient localization for mobile phones. Proceedings of the IEEE INFOCOM 2009 Mini Conference. IEEE Rio de Janiero, Brazil, April 19–25, 2009.

Hayes, B.. 2008. Cloud computing. *Communications of the ACM* 51 (7): 9–11.

King, T., and M. B. Kjærgaard. 2008. Composcan: Adaptive scanning for efficient concurrent communications and positioning with 802.11. Proceedings of the 6th International Conference on Mobile Systems, Applications, and Services. Association for Computing Machinery (ACM), pp. 67-80.

Kjærgaard, M. B., 2007. A taxonomy for radio location fingerprinting. Proceedings of the Third International Symposium on Location- and Context-Awareness. Berlin/Heidelberg: Springer Publishing, pp. 139–156.

Kjærgaard, M. B., 2010. On improving the energy efficiency and robustness of position tracking for mobile devices. Proceedings of the 7th International Conference on Mobile and Ubiquitous Systems: Computing, Networking and Services. Berlin/Heidelberg: Springer Publishing.

Kjærgaard, M. B.. 2011. Minimizing the power consumption of location-based services on mobile phones. (Forthcoming) *IEEE Pervasive Computing.*

Kjærgaard, M. B., S. Bhattacharya, H. Blunck, and P. Nurmi. 2011. Energy-efficient trajectory tracking for mobile devices. Proceedings of the 9th International Conference on Mobile Systems, Applications, and Services. Association for Computing Machinery (ACM). Bethesda, MD, U.S. June 28–July 1, 2011, 307–320.

Kjærgaard, M. B., J. Langdal, T. Godsk, and T. Toftkjær. 2009. EnTracked: Energy-efficient robust position tracking for mobile devices. Proceedings of the 7th International Conference on Mobile Systems, Applications, and Services. Association for Computing Machinery (ACM), Krakow, Poland, June 22–25, 2009, pp. 221–234.

Küpper, A. and G. Treu. 2006. Efficient proximity and separation detection among mobile targets for supporting location-based community services. *Mobile Computing and Communications Review* 10 (3): 1–12.

LaMarca, A., and E. de Lara. 2008. Location systems: An introduction to the technology behind location awareness. Bonita Springs, FL: Morgan and Claypool Publishers.

Leonhardi, A., and K. Rothermel. 2001. A comparison of protocols for updating location information. *Cluster Computing* 4 (4): 355–367.

Nokia. Nokia–Energy Profiler. 2011. Online at: http:/www.nokia.com (accessed January 18, 2011).

Pys60 Community. Python for S60. 2011. Online at: http://sourceforge.net/projects/pys60 (accessed January 18, 2011).

Reddy, S., M. Mun, J. Burke, D. Estrin, M. H. Hansen, and M. B. Srivastava. 2010. Using mobile phones to determine transportation modes. *ACM Transactions on Sensor Networks* (TOSN) 6 (2).

Skyhook Wireless. 2010. Online at: http://www.skyhookwireless.com/locationapps (accessed January 18, 2011).

Tippenhauer, N. O., K. B. Rasmussen, C. Pöpper, and S. Capkun. 2009. Attacks on public WLAN-based positioning systems. Proceedings of the 7th International Conference on Mobile Systems, Applications, and Services. Association for Computing Machinery (ACM), Krakow, Poland, June 22–25, 2009, pp. 29–40.

Chapter 2

Energy Efficient Supply of Mobile Devices

Leander B. Hörmann, Philipp M. Glatz, Christian Steger, and Reinhold Weiss

Contents

2.1 Introduction

Energy efficient supply is essential to extend the battery life of mobile devices or even to enable perpetual operation. Therefore, the power needs of the mobile device have to be analyzed to adapt the supply to the device.

This chapter introduces a tier model that can be used to describe energy supply units (ESU) using normal batteries, rechargeable batteries, or energy scavenging systems (ESSs). The model structures the hardware into five tiers with special functionality: the device tier, the measurement tier, the power control and conditioning tier, the storage access tier, and the energy storage tier. Each tier is adapted to the neighboring tier to ensure high efficiency of the power flow. This model increases the comprehension of ESSs and the design can be trimmed to energy efficiency.

The most important tier respecting the energy efficiency is the power control and conditioning tier. Typically, the voltage level of the energy storage component (ESC) depends on its state-of-charge (SOC). However, most mobile devices need a specific supply voltage. Therefore, a conditioning element must be embedded to convert the battery terminal voltage to the required supply voltage. Three different voltage conversion techniques are explained in this chapter.

The supply voltage range of mobile devices depends on the embedded and active components. Therefore, component-aware dynamic voltage scaling (CADVS) can be used to save a lot of energy. The principle behind CADVS is to set the supply voltage to the lowest possible supply voltage depending on the active components. An application scenario and a prototype with six different voltage conversion circuits has been developed and implemented to evaluate CADVS. The measurement results demonstrate the correct functionality of the prototype and show the possible energy savings of 38.7 percent using CADVS in the introduced scenario.

2.2 Related Work

Mobile devices need to be as energy efficient as possible because their available energy is always limited. As the name already says, the devices suffer from the lack of wired infrastructure and each of them needs its own energy supply unit (ESU). There are mainly three possibilities to supply mobile devices. First, they can be powered by normal batteries (alkaline). After the operating time, they have to be replaced manually. Second, they can be powered by rechargeable batteries (nickel-cadmium (NiCd), nickel-metal hydride (NiMH), lithium ion (LiIon), or lithium-ion polymer (LiPo)). After the operating time, they have to be recharged manually. Third, they can be powered by energy scavenging systems (ESSs) [1]–[3]. These systems use energy scavenging devices to convert environmental energy into electrical energy. Due to the fact that *energy of the environment is generally unpredictable, discontinuous, and unstable* [4], each ESS needs an energy storage component (ESC). However, it is proved that if the environmental energy source can be modeled in a specific way, the average power consumption of the mobile device is lower than the average power output of the energy scavenging device, and a well-dimensioned ESC is used, then the mobile device can operate forever [5]. This is called perpetual operation. Typical ESCs are rechargeable batteries and double layer capacitors (DLCs). The size and weight of a mobile device are limited in order to keep it mobile. Therefore, the size and weight of the ESCs are also limited and an efficient usage of its capacity is important.

To extend the operating time or to enable a perpetual operation of the mobile devices, an energy-efficient and well designed ESU is essentially needed. The design of the ESU depends on the energy scavenging device and on the power needs of the device itself. The design can be supported by using a tier model of the ESS. We have introduced this model in Hörmann et al. [6] and it can be used to describe an ESU using batteries as well as an ESU using energy scavenging. It consists of five different tiers: the device tier, the measurement tier, the power control and conditioning tier, the storage access tier, and the energy storage tier. Each tier consists of one or more elements, which are necessary to fulfill its functionality. The overall efficiency of the ESU can be optimized by adapting the tiers and the components of the tiers to each other.

Table 2.1 shows six different types of ESCs that can be used to supply mobile devices. The difference between the start voltage and the end voltage is the operating range of the ESC [7–9]. The DLC values are obtained from [7].

Most mobile devices consist of one or more electronic components, which include a digital CMOS (complementary metal oxide semiconductor) circuit. This type of circuit has a static and a dynamic power consumption [10]. The reasons for the static power consumption are leakage and bias currents. It can be neglected in most systems with a power consumption of more than 1 mW. The calculation of the dynamic power consumption is shown in equation (2.1).

$$P_{Dynamic} = C \cdot f \cdot V_{Supply}^2 \tag{2.1}$$

It is assumed that the capacities of the single gates of the CMOS circuit are merged to a common switching capacity *C*. It can be seen that the dynamic power consumption has linear dependency on the clock frequency f and it has a quadratic dependency on the supply voltage V_{Supply}. Therefore, much energy

Table 2.1 Six Different Types of ESCs and their Characteristics for Supplying Mobile Devices

Battery Type	*Start Voltage*	*End Voltage*	*Average Discharge Voltage*	*Recharge-able*	*Energy Density*	*Overcharge Tolerance*
Alkaline	1.50 V	0.7 V	≈ 1.20 V	no	≈ 145 Wh/kg	–
NiMH	1.38 V	0.8 V	≈ 1.25 V	yes	≈ 75 Wh/kg	low
NiCd	1.48 V	0.8 V	≈ 1.26 V	yes	≈ 35 Wh/kg	moderate
LiIon	4.10 V	2.5 V	≈ 3.76 V	yes	100–158 Wh/kg	very low
LiPo	4.10 V	2.8 V	≈ 3.80 V	yes	136–190 Wh/kg	very low
DLC [7]	2.70 V	0.0 V	≈ 1.35 V	yes	up to 5 Wh/kg	very low

Source: D. Linden and T. B. Reddy. 2002. *Handbook of batteries*, 3rd ed. New York: McGraw-Hill; C. Kompis and S. Aliwell. 2008. Energy harvesting technologies to enable remote and wireless sensing. Online at: http://host.quid5.net/ koumpis/pubs/pdf/energyharvesting08.pdf. With permission.

can be saved if the supply voltage of mobile devices is reduced to the minimal possible supply voltage. This is the main idea of dynamic voltage scaling (DVS) [11]–[14]. It is used to adapt the voltage of a processor on its clock frequency. The clock frequency depends on the current workload of the processor. Therefore, the overall power consumption of the processor can be significantly reduced during times with low workload. In Powell, Barth, and Lach [15], they have implemented an MSP430 microcontroller platform that is able to adapt the clock frequency and the supply voltage of the MSP430 to its workload. However, they did not consider the leakage current or the low power states (sleep states) of the MSP430.

The unused components of the hardware can be switched off completely in order to save energy. The controller of the mobile device switches the power supply of the components, e.g., sensors, communication module, and display. We have described this idea in Hörmann et al. [16].

The active power supply switching of the components and the minimization of the supply voltage results in CADVS. We have introduced this low-power idea in Hörmann et al. [17].

The rest of the chapter is organized as follows. Section 2.3 introduces the tier model and describes each tier in detail. Section 2.4 shows the principles of the energy-efficient supply of mobile devices. Three different voltage conversion techniques and CADVS are described. Section 2.5 discusses the consequences on the software of mobile devices. Section 2.6 introduces a scenario that is used to evaluate CADVS and it presents the measurement results. Finally, Section 2.7 concludes this chapter.

2.3 Tier Model for Energy Scavenging Systems

This section describes the tier model and each of its five tiers in detail. A generic structure of the model is shown in Figure 2.1.

Each tier provides a special functionality and is interacting with the bordering tiers. Here, interacting means the flow of power. The two possible directions of power flow are from an outer tier toward an inner tier and vice versa. The first one represents the input power into the ESS provided by an energy scavenging device. The second one represents the output power of the ESS that is used to supply a mobile device. The basic idea of the tiers is to encapsulate special functionalities that are necessary to provide a safe power flow through the model. These encapsulated functionalities form single elements of a tier. Examples for the functionalities are measuring, switching, or transforming the power flow.

The input stage consists of the input measurement, the power switch, the maximum power point tracker (MPPT), and the charge element. The function of this stage is the measurement, the control of the power provided by the energy scavenging device, and the optimal and safe charging of ESC. The output stage consists

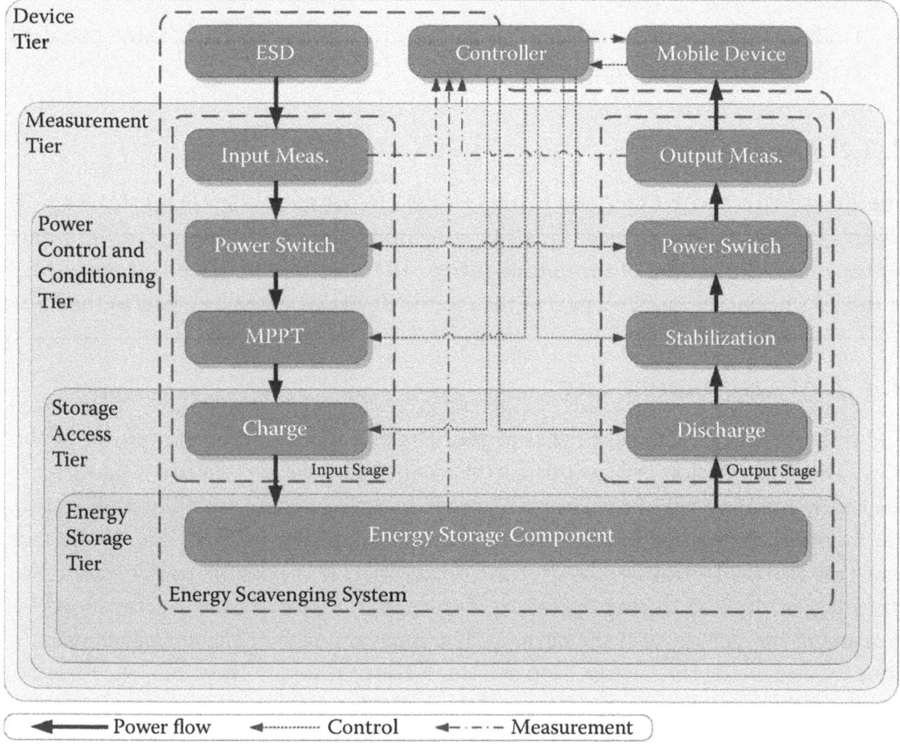

Figure 2.1 Tier model of an energy scavenging system.

of the discharge element, the voltage converter, the power switch, and the output measurement. The function of this stage is the safe discharge of the ESC, the conversion of the output voltage to a constant level, and the control and measurement of the output power.

The dotted arrows from the controller to some elements represent the control possibilities that can be used by the controller to influence the power flow through the model. Thus, the efficiency of the system can be enhanced. For example, the operating point of the energy scavenging device can be adjusted by varying the operating point of the MPPT. The dash-dotted arrows from some elements to the controller represent the flow of measurement information about the power flow or state-of-charge (SOC) of the ESC. The measurement elements measure the input and the output power that is flowing through them. The SOC of the ESC can be measured by the controller. These control and measurement information lines are not always necessary. There are two special types of ESS that should be mentioned. First, if no element can be controlled by the controller, the behavior of the ESS cannot be changed. Therefore, it is a fixed ESS. However, the system has to be designed well to operate correctly. Second, if no measurement information is provided to the controller, it cannot determine the current state of the ESS. Therefore, it is a blind ESS.

As already mentioned, each tier has its own functionality. Each of the five tiers is described in the following sections.

2.3.1 Device Tier

The device tier is the outer tier of the model. It contains the energy scavenging device and the mobile device. This tier is responsible for scavenging of the input power and consumption of the output power. The controller is placed on the border of the ESS because it can be a part of the mobile device or it can be a part of the ESS.

2.3.2 Measurement Tier

The task of the measurement tier is to measure the input and output power of the ESS. This information is forwarded to the controller, which can use it to forecast the future input power of the energy scavenging device. There are different forecasting methods as described in Bergonzini, Brunelli, and Benini [18]. Due to the fact that this tier is placed directly after the device tier, it is possible to determine the efficiency of the ESS during operation. Typically, the power can be determined by measuring the voltage and the current. The measurement of the current is usually done by measuring the voltage drop at a small shunt resistor. Therefore, the measuring element is usually a lossy element, i.e., the output power of such an element is lower than the input element.

If this tier is omitted, the input and output power can only be estimated. It is possible to estimate the power consumption of a mobile device by software as described in Dunkels et al. [19] and Glatz, Steger, and Weiss [20]. The input power can be estimated by using the estimation of the output power and the variation of the measured SOC of the ESC. However, this only can be a rough estimation and cannot replace the measurement tier if detailed information of the input and output power is necessary.

2.3.3 Power Control and Conditioning Tier

The power control and conditioning tier is responsible for an optimal power flow through the ESS. Power switches and conversion elements are used to control the power flow. In simple ESSs, the power switches enable the active control of the SOC of the ESC. Furthermore, it is possible to deactivate energy scavenging or the mobile device completely. The conversion element of the output stage provides a constant supply voltage to the mobile device. In more complex ESSs, this tier can be used to bypass power from the input stage to the output stage in order to unburden the ESC. Therefore, the lifetime of the ESS can be enhanced. The output voltage of the conversion element of the output stage can be varied in order to minimize the power consumption of the mobile device. Section 2.4 deals with the efficient supply of mobile devices. The MPPT is used to optimize the power output of the scavenging device. There are two possibilities for the control of the MPPT. First, it can be self-controlled by detecting the maximum power point of the energy scavenging device. Second, it is controlled by the controller. The advantage of this solution is the reduced complexity of the MPPT element, but the disadvantage is that the controller has to track the input power continuously and has to adjust the MPPT.

2.3.4 Storage Access Tier

The task of the storage access tier is to supervise and regulate the charging and discharging of the ESC. This is necessary because some types of ESCs are sensitive to overvoltage, undervoltage, or too high currents. An overcharge protection of the charge element prevents a destruction of the ESC. The importance of the overcharge protection depends on the type of the ESC as shown in Table 2.1. Some types of ESCs need a deep discharge protection because they are sensitive to undervoltage, e.g., LiIon and LiPo. This is implemented in the discharge element. The limitation of the charge and discharge current also is important in order to prevent the ESC from destruction. The value of these currents may depend on the temperature of the environment. Then, the control lines are needed to adapt the maximum allowed charge and discharge current.

2.3.5 Energy Storage Tier

The energy storage tier is the innermost tier of the model. This tier contains the ESC and is responsible for storing the energy. As already mentioned in the introduction, this energy buffer is needed to bridge times with too little input power of the energy scavenging device. In complex ESSs, different types of ESCs can be used to buffer the energy. An example of such a system is the Prometheus module [21]. It uses a DLC and a LiPo battery to buffer the energy. The primary buffer is the DLC. Due to its lower energy storage capacity, it can only buffer the energy for short periods of time. The secondary buffer is the LiPo battery. It is only discharged if the SOC of the primary buffer is too low. The advantage of this structure is that the number of charge-discharge cycles of the LiPo battery is reduced. Therefore, the lifetime of the ESS can be enhanced.

2.4 Energy Efficient Supply of Mobile Devices

Due to the fact that the available energy for mobile devices is always limited, an efficient supply of such devices is very important. Section 2.4.1 shows different voltage conversion techniques that are necessary to reduce the overall energy consumption. Section 2.4.2 describes CADVS in detail.

2.4.1 Voltage Conversion Techniques

This section focuses on the conditioning element of the output stage of the ESS. As already mentioned in the introduction, most mobile devices need a constant supply voltage for operation. This voltage should be as low as possible in order to be as efficient as possible. The variation of the ESC's terminal voltage and the supply voltage requirements of the mobile device cause the need of a voltage conversion element. There are generally three different possibilities, which are described in the following subsections.

2.4.1.1 Linear Voltage Regulator

Linear voltage regulators provide a constant output voltage by adapting the internal resistance. Therefore, the voltage difference between the battery and the supplied device drops at the voltage regulator (equation (2.2). Low dropout (LDO) regulators are a special type of linear voltage regulators that can operate with very low voltage differences.

$$V_{Battery} = V_{LDO} + V_{Device} \tag{2.2}$$

The average input current of the regulator is equal to the average input current of the supplied device (equation 2.3). No quiescent and leakage currents are considered here.

$$I_{LDO} + I_{Device} \tag{2.3}$$

Using equation (2.1), the power consumption of the whole system (the mobile device and the voltage converter) can be calculated as shown in equation (2.4).

$$P(V_{Device}) = I_{Device}(V_{Device}) \cdot V_{Battery} = C \cdot f \cdot V_{Device} \cdot V_{Battery} \tag{2.4}$$

It can be seen that the supply voltage of the device V_{Device} has a linear influence on the total power consumption of the system.

2.4.1.2 Step-Down Voltage Converter

Step-down converters, also called buck converters, use internal switching elements to convert the voltage level. Capacitors or inductors are necessary in combination with these switches. A constant output voltage is provided by changing the timing of these switches of the buck converter (pulse-width modulation). The voltage difference between the terminal of the battery and the supplied device drops at this converter (equation (2.5)).

$$V_{Battery} = V_{Buck} + V_{Device} \tag{2.5}$$

The average input current of the converter is not equal to the average output current and, thus, not equal to the average input current of the supplied device (equation (2.6)).

$$I_{Buck,Input} \neq I_{Buck,Output} = I_{Device} \tag{2.6}$$

The input and the output power of an ideal buck converter are equal. Using equation (2.1), the power consumption of the whole system (the mobile device and the voltage converter) can be calculated as shown in equation (2.7). No quiescent currents, leakage currents, and conversion efficiencies are considered here.

$$P_{Buck,Input}(V_{Device}) = P_{Buck,Output}(V_{Device}) = C \cdot f . V_{Device}^2 \tag{2.7}$$

It can be seen that the supply voltage of the device has a quadratic influence on the total power consumption of the system.

2.4.1.3 Step-Up Voltage Converter

Step-up converters, also called boost converters, are able to generate output voltages that are higher than their input voltages. They also use internal switching elements in combination with capacitors or inductors to convert the voltage level. The timing of the switches sets the output voltage of the boost converter (pulse-width modulation). The same considerations are valid for the buck converter.

Therefore, the supply voltage of the device V_{Device} has a quadratic influence on the total power consumption of the system.

2.4.2 Multiple Supply Voltages

The hardware of a mobile device consists of lots of different components. Each hardware component has its own supply voltage range. Usually, the supply voltage range of the components overlaps and a common voltage can be found to supply all components. However, sometimes the ranges of single components do not overlap with the ranges of the other ones. One solution is to supply the hardware with multiple voltages. The disadvantage of this solution is that the complexity of the hardware increases. Each voltage level needs its own voltage converter. Furthermore, a voltage level shifter circuit is needed in order to communicate between components of different supply voltage levels. However, in some cases, it is the only possible solution, e.g., if the maximal supply voltage of a microcontroller is lower than the minimal supply voltage of a sensor, which should be accessed by the microcontroller. In other situations, CADVS can be used to avoid such multiple supply voltages, e.g., if two supply voltage ranges of two sensors do not overlap, but both of them overlap with the supply voltage range of the microcontroller that wants to read the sensor information.

2.4.3 Component-Aware Dynamic Voltage Scaling

The hardware components (e.g., temperature sensor, transceiver module) of a mobile device are typically accessed by application programs to fulfill a given functionality, e.g., to track the temperature. Typically, these application programs access different hardware components over time. Therefore, these active hardware components change over time. CADVS varies the supply voltage of the mobile device according to these active hardware components. The unused components are switched off and the supply voltage is minimized. This reduces the overall power consumption in two ways. First, switching off the unused components eliminates the remaining sleep or standby power of these components. Only a very small quiescent power of the switches remains, which is typically much lower than the sleep or standby power of the components. Second, the dynamic minimization of the supply voltage reduces the power consumption of the hardware components according to equation (2.1).

The information of the active components can be stored on the list L. This list depends on the software running on the mobile device and varies over time. Each hardware component has its own supply voltage range. Inside this range, the component has full functionality. This lowest possible supply voltage (LPSV) of each active hardware component can be stored on the list L_{LPSV}. Basically, the supply voltage of all components of the mobile device is the same using CADVS. As already mentioned, the supply voltage should be as low as possible in order to save as much energy as possible. Therefore, the minimal possible supply voltage of the mobile device can be determined as shown in equation (2.8).

$$V_{Device} = \max(L_{LPSV}) \tag{2.8}$$

The active components are varying over time and, consequently, the list L_{LPSV} and the supply voltage of the mobile device V_{Device}. changes over time too.

2.5 Consequences on Software for Mobile Devices

This section discusses the consequences of the tier model and of CADVS on software for mobile devices. The software complexity of mobile devices increases faster and faster. Therefore, an intelligent integration of new functionalities is very important to keep the software quality as high as possible.

An application program running on a mobile device or WSN (wireless sensor network) node usually does not want to know specific hardware details of the energy scavenging system. It only wants information of the SOC or the input power. Therefore, the specific hardware details should be abstracted into a device driver of the operating system. Then, it is easy for the application program to access the wanted information. Furthermore, CADVS should be applied automatically by the operating system according to the currently accessed hardware components.

2.5.1 Consequences of the Tier Model

The integration of an ESS mainly depends on its complexity. For example, a blind and fixed ESS does not provide a measurement output and cannot be controlled. Therefore, the integration of such a system into the software is very easy; no integration is necessary.

Basically, it is useful to encapsulate the software related to the ESS into an independent software module. The access of higher software layers (e.g., application software) to the ESS software should be done via well-defined interfaces. This ensures high flexibility.

As soon as the software is able to control the charging and discharging of the ESC, the software has to be developed with best care and attention. A failure of the software can cause a destruction of the ESC.

The measurement of physical quantities of the ESS is generally harmless. The most important information is the SOC of the ESC. The information is necessary to predict the remaining operating time, which can be predicted by extrapolating the history of the SOC. Therefore, most ESSs provide this information to the mobile device. The measurement of the input and output power can be used to improve the accuracy of this prediction. Such a prediction is important for users of mobile devices and for supervisors of WSNs. In the case of a battery-powered device, the batteries can be exchanged or recharged early enough. In the case of an energy scavenging device, the energy scavenging process can be checked if something inhibits this process.

2.5.2 Consequences of CADVS

CADVS needs integration into the software of the mobile device because only the software knows which hardware components are needed. Basically, the CADVS functionality should be hidden from the application software. When accessing a specific hardware component, the operating system of the mobile device should automatically control the power supply. An exemplary integration of CADVS into the software and hardware of a mobile device is shown in Figure 2.2.

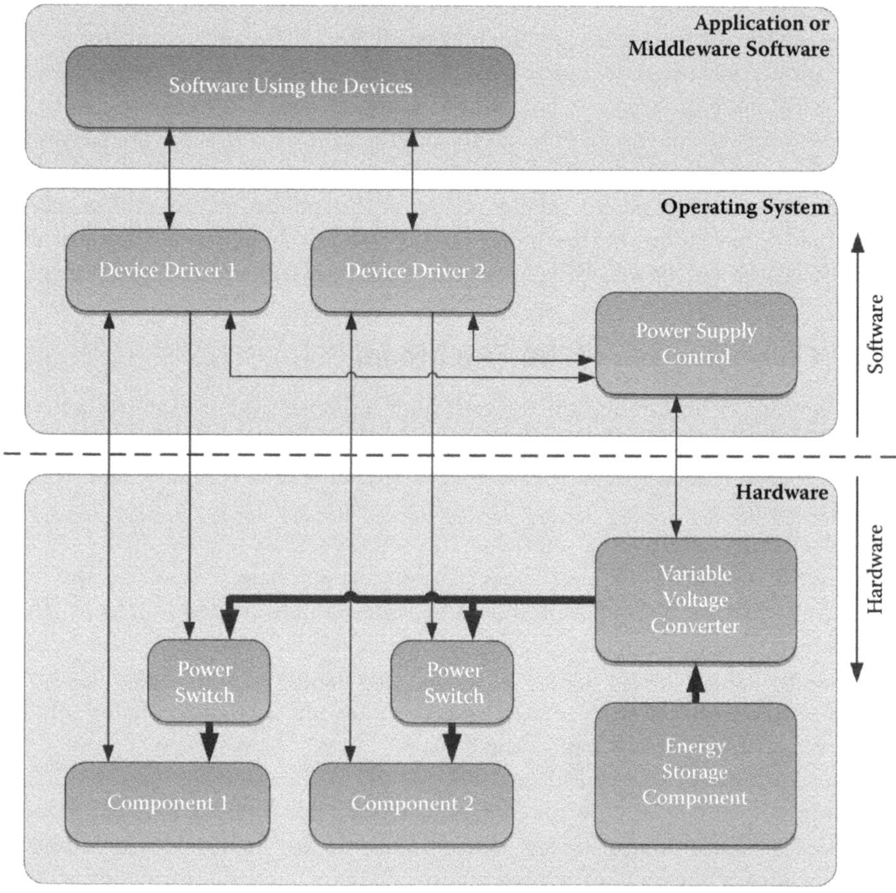

Figure 2.2 CADVS integration into software and hardware. The thick lines represent the power flow and the thin lines represent information flow.

The application software or the middleware uses the interface of the operating system to access the devices (e.g., a sensor). The device drivers are responsible for telling the power supply controller the needed supply voltage of the hardware component or device. The power supply controller sets the supply voltage to the lowest possible value. After that, the device driver switches on the power supply of the component. Then, the application software can communicate with the component via the device driver. After using the component, its power supply is switched off and the power supply controller is informed.

2.6 CADVS Example

This section shows how CADVS can be applied to a wireless sensor node. Wireless sensor networks (WSNs) are used in application areas without wired infrastructures. Precision agriculture ([22], [23]), wildlife monitoring ([24], [25]), human healthcare [26], and structural health monitoring [27] are only a few examples. Due to the lack of wired infrastructure, each sensor node needs its own power supply. ESSs can be used to extend the operating time or enable a perpetual operation of the sensor node. This chapter deals with the energy-efficient supply of such a WSN node using CADVS. Section 2.6.1 introduces the scenario and section 2.6.2 shows the measurement setup. Section 2.6.3 presents and discusses the results of the measurements. The results of this CADVS example also can be applied to other types of mobile devices.

2.6.1 Scenario

The task of the wireless sensor node is kept very simple in order to get meaningful results. It has to measure the temperature and send this information through the network. Therefore, the sensor node consists of a temperature sensor, a microcontroller, and a wireless transceiver module. Table 2.2 shows the components and their supply voltage range.

The supply range of the temperature sensor must be between 3.135 V and 3.465 V to achieve the specified accuracy [29]. To be as accurate as possible, the supply voltage during the measurement phase is set to 3.3 V.

Table 2.2 Hardware Components and their Supply Voltage Range

Hardware Component	*Name*	*Manufacturer*	*Supply Voltage Range*
Microcontroller	MSP430F1611 [28]	Texas Instruments	1.8 V to 3.6 V
Temperature sensor	TMP05B [29]	Analog Devices	3.3 V
Transceiver module	MRF24J40MB [30]	Microchip	2.4 V to 3.6 V

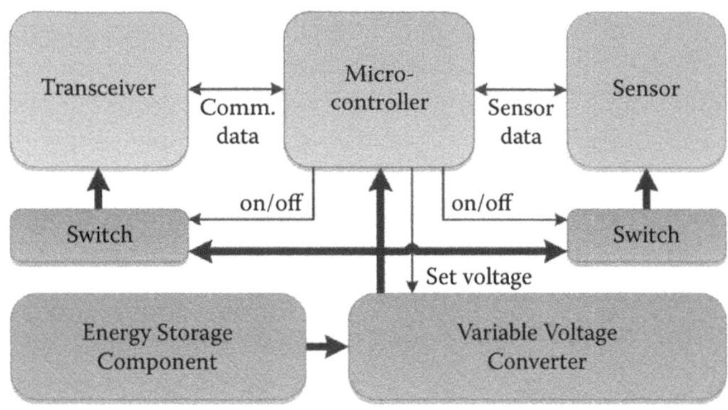

Figure 2.3 Structure of a wireless sensor node (adapted from Puccinelli and Haenggi [31]). CADVS is enabled by the switchable power supply of the components and the variable voltage converter.

To be able to apply CADVS to the sensor node, the power supply of the temperature sensor and the transceiver module has to be switchable and a variable voltage converter has to be used. The clock frequency of the microcontroller is set to 4 MHz in order to be able to use its complete voltage range. The structure of the sensor node is shown in Figure 2.3.

The temperature need not be measured continuously in this scenario. Therefore, the sensor node can be placed into a sleep state between two measurements. One measurement interval T consists of four different phases: the sleep phase, the measurement phase, the computation phase, and the communication phase. Table 2.3 shows these phases, the list of the active components during each phase, the list of the lowest

Table 2.3 The Four Phases of the Interval T and the Active Hardware Components

Phase	*Active Hardware Components L*	*Lowest Possible Supply Voltages* V_{Device}	L_{LPSV}	*Duration (%)*
Sleep	[Microcontroller]	[1.8 V]	1.8 V	85% of T
Measurement	[Microcontroller, Temp. Sensor]	[1.8 V, 3.3 V]	3.3 V	5% of T
Computation	[Microcontroller]	[1.8 V]	1.8 V	5% of T
Communication	[Microcontroller, Transceiver]	[1.8 V, 2.4 V]	2.4 V	5% of T

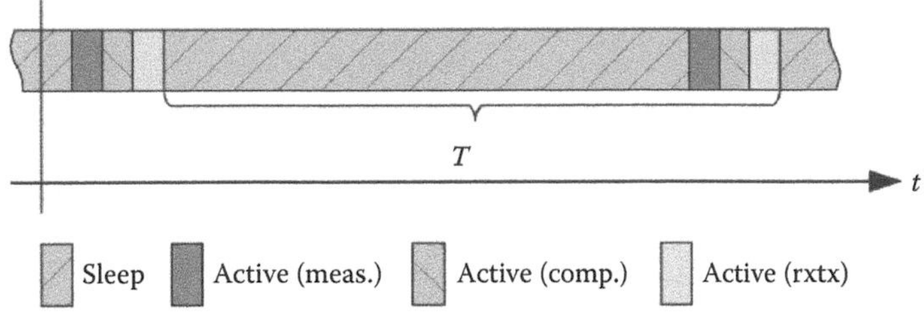

Figure 2.4 Chronological sequence of an interval *T*.

possible supply voltage, and the resulting supply voltage. Figure 2.4 shows the chronological sequence of an interval *T*. The interval is repeated as long as the sensor node is supplied.

The duty cycle is the ratio of the active time to the full interval time. In this scenario, the measurement phase, the computation phase, and the communication phase belong to the active time. The sleep phase belongs to the inactive time. Therefore, the duty cycle can be calculated according to equation (2.9).

$$Duty\ Cycle = \frac{t_{active}}{T} = \frac{t_{means} + t_{comp} + t_{comm}}{t_{means} + t_{comp} + t_{comm} + t_{sleep}} \tag{2.9}$$

Six different voltage conversion circuits have been implemented to evaluate the introduced wireless sensor node. We assume that the terminal voltage of the battery is higher than the maximal supply voltage of the sensor node. Therefore, only the linear regulators and step-down converters have been evaluated. Figure 2.5 shows the schematics of these circuits.

The first circuit, LDO1FIX, (Figure 2.5a) is implemented to show the characteristic of a bad voltage converter for this application. This linear regulator has a constant output voltage of 3.3 V and a very high quiescent current of 5 mA [32]. The supply current of the MSP430 microcontroller in full operating mode is lower than this quiescent current.

The second circuit (Figure 2.5b) is used in two different configurations. At the first configuration, LDO2FIX, the LDO regulator has a constant output voltage of 3.3 V. At the second configuration, LDO2VAR, the output voltage of the LDO regulator can be switched between 2.2 V and 3.3 V. The lower voltage (2.2 V) can be set if the lowest possible supply voltage of the mobile device or the WSN node is lower than the 2.2 V. Otherwise, the higher voltage (3.3 V) is set. Due to the fact that no smaller voltage steps are possible, CADVS can only be applied in a limited way.

The third circuit, BUCK1FIX, (Figure 2.5c) implements a step-down converter with a constant output voltage of 3.3 V.

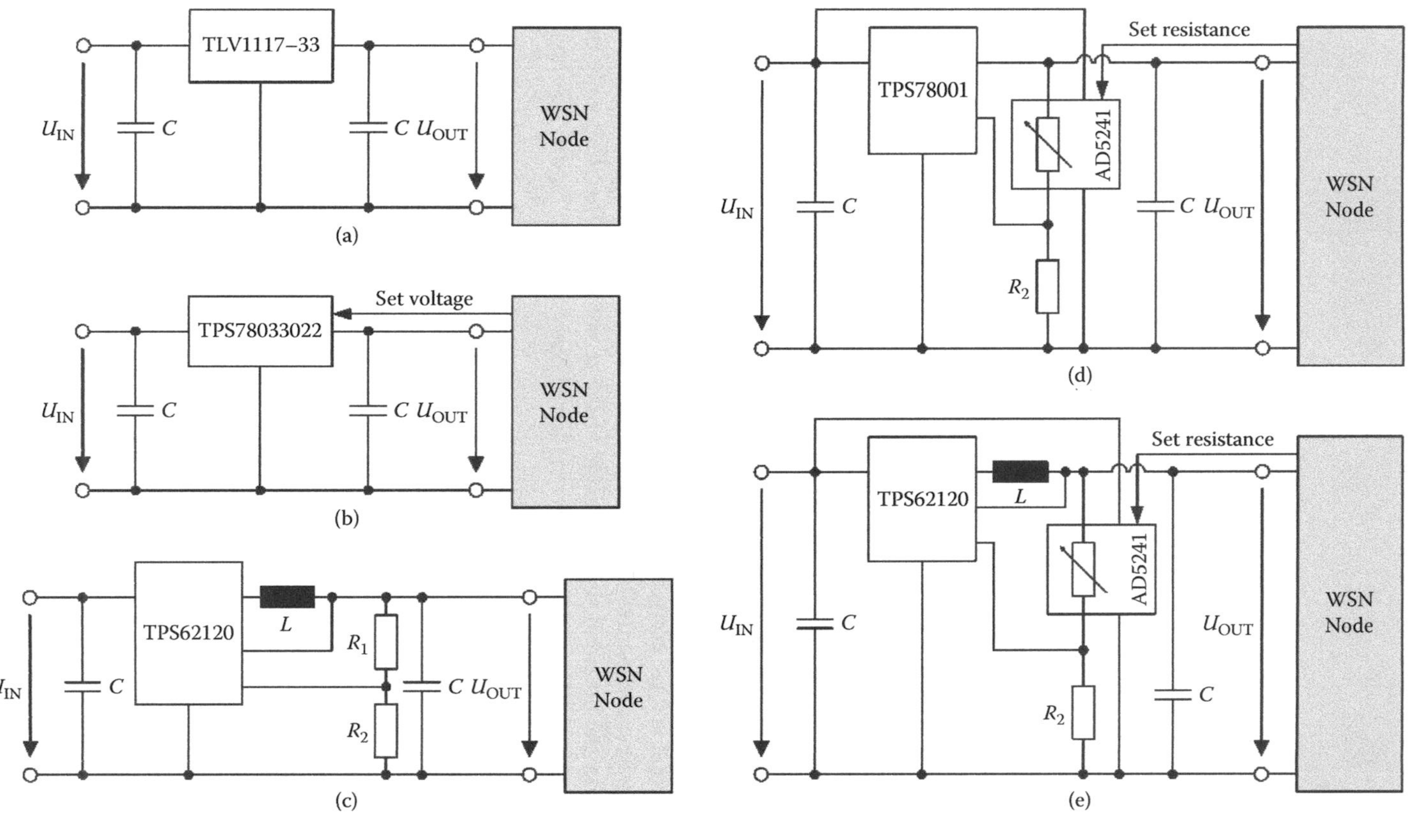

Figure 2.5 Five different voltage converter circuits to evaluate CADVS supplying a wireless sensor node.

The fourth circuit, LDO3VAR, (Figure 2.5d) enables full CADVS. The output voltage of the LDO regulator can be varied by changing the resistance of the digital potentiometer AD5241 from Analog Device. The full supply voltage range of the wireless sensor node between 1.8 V and 3.3 V can be regulated. A voltage level converter between the microcontroller and the digital potentiometer is needed in order to adjust the resistance at high voltage differences between the terminal voltage of the battery and the supply voltage of the microcontroller as described in Powell, Barth, and Lach [15].

The fifth circuit, BUCK2VAR, (Figure 2.5e) enables full CADVS as well. The output voltage of the step-down converter can be varied by changing the resistance of the digital potentiometer AD5241. The full supply voltage range of the wireless sensor node between 1.8 V and 3.3 V can be regulated. Also, this circuit needs a level shifter between the digital potentiometer and the microcontroller.

2.6.2 Measurement Setup

The measurement setup shown in Figure 2.6 is used to evaluate the six different voltage conversion circuits. This setup is similar to the measurement setup we have used to characterize ESSs in Glatz et al. [33]. The NI PXI-6221 DAQ measurement device from National Instruments is used to sample the voltages.

The values of both shunt resistors are 4 ohm. The input power of the voltage converter and of the WSN node can be calculated as shown in equation (2.10):

$$P_{conv,input} = V1 \cdot \frac{V_{R1}}{R1} \quad P_{node,input} = V1 \cdot \frac{V_{R2}}{R2} \tag{2.10}$$

The disadvantage of this setup is that the sleep current cannot be measured because it is too low. Therefore, the precise multimeter Fluke 289 is used, which can measure DC currents with a resolution of 0.01 μA and an accuracy of 0.075 percent + 20.

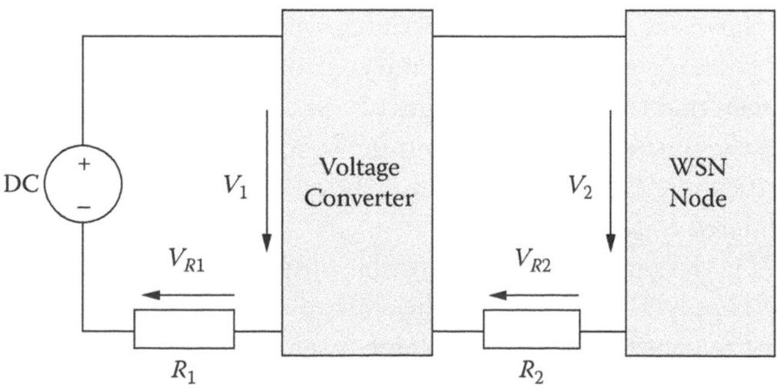

Figure 2.6 Measurement setup of the CADVS evaluation.

2.6.3 Measurement Results

This section presents and discusses the measurement results of the evaluated hardware. Three different measurements have been done to show the possible energy savings using CADVS: the measurement of the input current and power of the WSN node without a voltage converter, the measurement of the output voltages of the converters, and the measurement of the energy savings dependent on the duty cycle of the device. The duty cycle of the WSN node at this scenario has been adjusted by varying the duration of the sleep phase.

2.6.3.1 Direct Supply

The first measurement shows the input current and power of the WSN node without a voltage converter dependent on the input voltage. The input voltage is varied between 1.8 V and 3.6 V. This measurement is done for each phase of the interval.

Figure 2.7a-d shows the current and power consumption during the sleep phase, measurement phase, computation phase, and communication phase, respectively. The plots of the measurement phase, computation phase, and communication phase show a linear dependency of the supply current on the supply voltage in the allowed range. Therefore, the assumption of the introduction (equation (2.1)) is valid in these cases. The current during the sleep phase has a nonlinear dependency on the supply voltage range. However, this is caused by other effects like leakage currents and the very low supply current of the WSN node.

Furthermore, this measurement shows the high dynamic range of the supply current. It ranges from below 1 μA up to 30 mA. The voltage converter has to be selected carefully to be efficient over the whole range.

2.6.3.2 Output Voltages of the Converters

The second measurement shows the output voltage of the six different converter circuits. Figure 2.8a-d shows the output voltages of the converter circuits during the sleep phase, measurement phase, computation phase, and communication phase, respectively.

It can be seen that the LDO1FIX regulator needs a very high difference between the input and the output voltage of about 0.9 V to operate. Therefore, and because of the high quiescent current, it is recommended not to use it for this application. It is not considered in further results.

The LDO2VAR converter has two possible output voltages: 2.2 V and 3.3 V. The lower voltage is only allowed during the sleep phase and the computation phase. During the other phases, the output voltage is set to 3.3 V.

The LDO3VAR converter and the BUCK2VAR converter can set all three different supply voltages needed for this scenario: 1.8 V, 2.4 V, and 3.3 V. The output voltages of both converters are always at the minimal possible supply voltage. Only

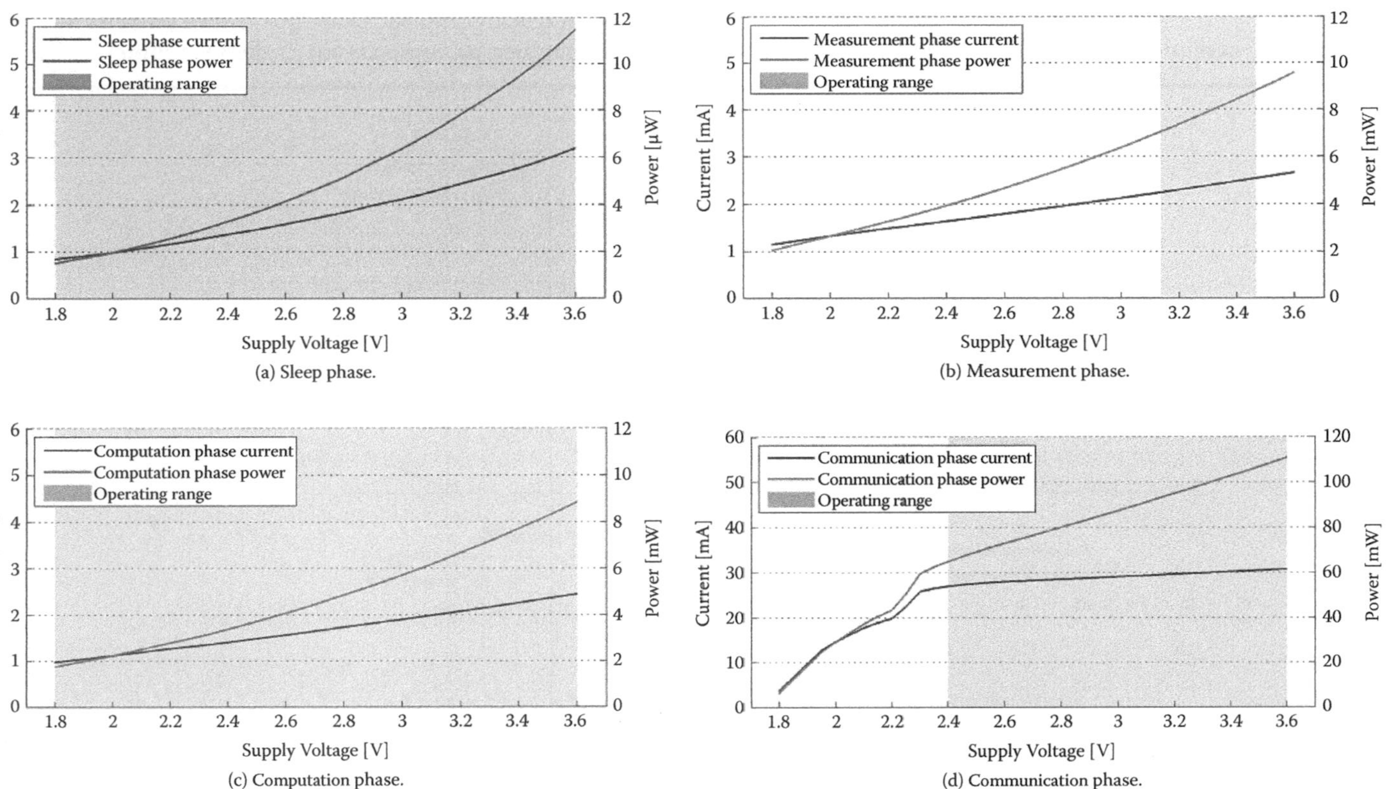

Figure 2.7 Current and power consumption of the WSN node dependent on the supply voltage of the four phases (a–d). The green area shows the allowed supply voltage range during each phase.

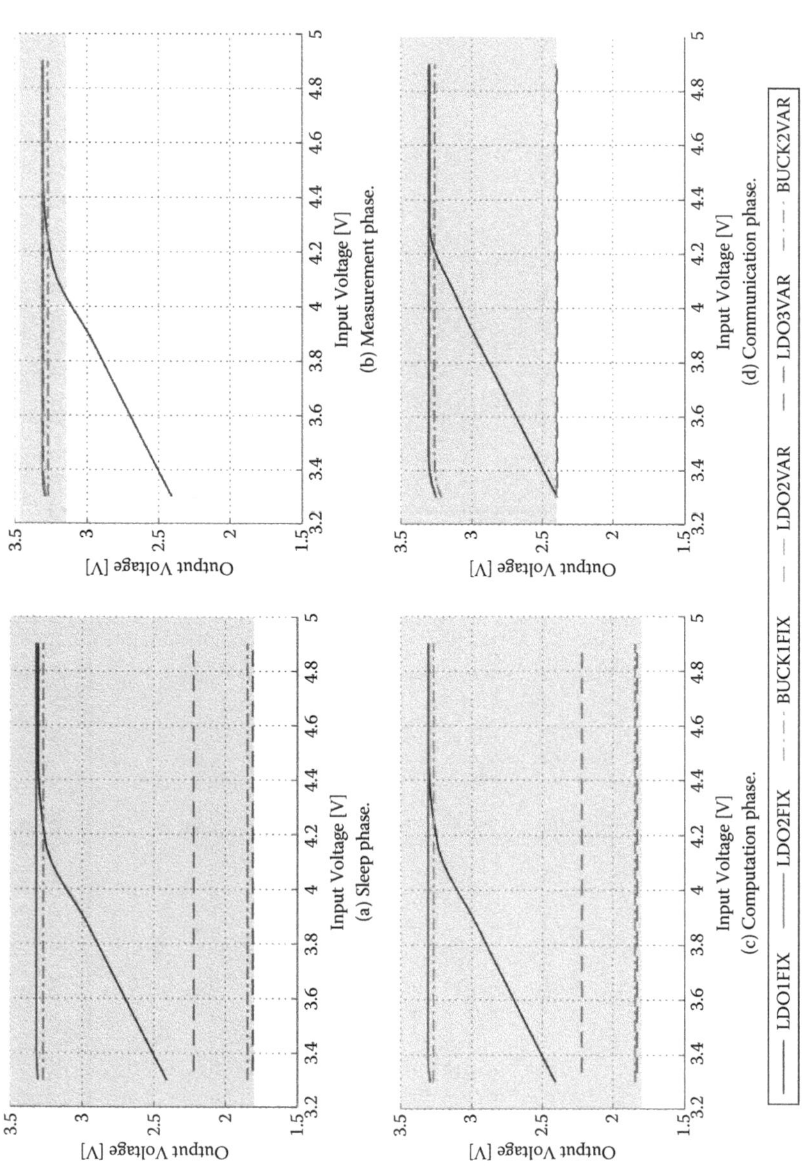

Figure 2.8 Output voltage of the six different converter circuits dependent on the input voltage of the four phases. The gray area shows the allowed supply voltage range during each phase.

during the measurement phase is it set to 3.3 V in order to enhance the accuracy of the measurement.

This measurement has shown the correct functionality of all voltage conversion circuits except LDO1FIX. The five remaining circuits are analyzed in detail.

2.6.3.4 CADVS Results

The CADVS measurement results are presented in this section. Table 2.4 shows the overall power consumption, the efficiency, and the power savings of the complete setup (voltage converter and WSN node) using five different voltage converters. Also the quiescent power consumption of each converter is listed. The averaged values refer to the introduced scenario with the interval *T*. All measurements have been done with a battery terminal voltage of 3.9 V (input voltage of the converters) and a duty cycle of 15 percent. The first part shows the power consumption of the whole system and, thus, the input power of the voltage converters during the four different phases. The second part shows the efficiency of the voltage converters. It can easily be calculated as shown in equation (2.11).

$$\eta_{Converter} = \frac{P_{Output.Converter}}{P_{Input.Converter}} \tag{2.11}$$

The third part shows the power savings compared to the LDO2FIX converter. This converter circuit is used as a reference circuit because it is a very simple solution. A direct supply of the WSN node is not possible because of the 3.9 V input voltage. Therefore, the measurement results cannot be compared to a direct supply setup.

It can be seen that the quiescent powers of the five converter circuits are very different. The quiescent powers of the LDO2FIX and the LDO2VAR are much smaller than those of the other converters. The higher quiescent powers of the BUCK1FIX and the BUCK2VAR were expected [17]. The high quiescent power of the LDO3VAR was not expected. It is caused by the high quiescent power of the digital potentiometer and its circuit.

The overall power consumption during the sleep phase ranges from 5.38 μW to 97.38 μW. During this phase, only the microcontroller is supplied. Therefore, the power consumption of the WSN node itself ranges from 2 μW at 1.8 V to 8.5 μW at 3.3 V at this phase. This is much lower than the maximum overall power consumption. Therefore, the overall power consumption mainly depends on the quiescent power using the voltage converters BUCK1FIX, LDO3VAR, and BUCK2VAR. The result is a power saving of -807.5 percent during the sleep phase of the BUCK2VAR converter.

The two other converters (LDO2FIX and LDO2VAR) have a very low quiescent current. The major difference between these two converters is the output voltage. It is 3.3 V at the LDO2FIX converter and 2.2 V at the LDO2VAR converter

Table 2.4 Power Consumption, Efficiency, and Power Savings of the Five Different Voltage Converters during the Four Phases

Voltage Converter	*LDO2FIX*	*BUCK1FIX*	*LDO2VAR*	*LDO3VAR*	*BUCK2VAR*
Quiescent power of volt. conv.	0.5 μW	73.2 μW	0.8 μW	45.7 μW	93.9 μW
Power during sleep	10.73 μW	83.13 μW	5.38 μW	49.07 μW	97.38 μW
Power during measurement	9.37 mW	8.25 mW	9.38 mW	9.78 mW	8.90 mW
Power during computation	8.44 mW	7.42 mW	4.65 mW	3.59 mW	1.63 mW
Power during communication	95.26 mW	99.45 mW	94.35 mW	81.24 mW	56.89 mW
Average power	5.71 mW	5.87 mW	5.47 mW	4.83 mW	3.50 mW
Efficiency during sleep	80.6%	10.0%	48.7%	3.2%	1.7%
Efficiency during measurement	84.9%	93.2%	84.6%	84.3%	91.9%
Efficiency during computation	85.0%	93.4%	56.7%	45.3%	103.3%
Efficiency during communication	87.9%	94.9%	88.6%	63.8%	91.6%
Average efficiency	87.4%	93.6%	86.9%	64.7%	89.8%
Power savings during sleep	0.0%	–674.7%	49.8%	–357.3%	–807.5%
Power savings during measurement	0.0%	–12.0%	–0.1%	-4.3%	5.0%
Power savings during computation	0.0%	12.1%	44.9%	57.5%	80.7%
Power savings during communication	0.0%	–4.4%	1.0%	14.7%	40.3%
Average power savings	0.0%	–2.8%	4.1%	15.3%	38.7%

Note: The input voltage of the converters is 3.9 V and the duty cycle of the WSN node 15 percent.

during the sleep phase. The result is the halving of the overall power consumption during this phase.

All in all, high quiescent powers of the voltage converters result in a very poor efficiency during the sleep phase. During all other phases, the step-down converters have the best performance. The linear regulators cannot achieve such a high efficiency because they are reducing the output voltage by using an internal resistor. The 103.3 percent efficiency of the BUCK2VAR converter is physically impossible and it is caused by measurement inaccuracies. The error of the measurement setup for each power measurement channel is about 2 percent as derived in Glatz et al. [33].

As already mentioned, during the sleep phase, the voltage converters with low quiescent power consumptions perform best (LDO2FIX and LDO2VAR). During the measurement phase, only the step-down converters were able to save power (BUCK1FIX and BUCK2VAR) because the reference circuit is also a linear voltage regulator with an output voltage of 3.3 V. Due to the fact that the WSN node must be supplied with 3.3 V during this phase, both other linear voltage regulators cannot save any power (LDO2VAR, LDO3VAR). During the computation phase, the variable linear voltage regulators (LDO2VAR, LDO3VAR) perform better than the BUCK1FIX converter due to the reduced supply voltage. Only the BUCK2VAR converter performs better because of the reduced voltage and the higher conversion efficiency. During the communication phase only the LDO3VAR and the BUCK2VAR converter were able to save a significant amount of power because only these two converters were able to reduce the supply voltage of the WSN node to 2.4 V. The bad performance of the BUCK1FIX converter was not expected and can be caused by measurement errors or bad performance of the converter at high currents. Overall, the BUCK2VAR converter performs best at the entire interval of the introduced scenario.

The overall energy savings depends mainly on the phase and on the voltage conversion circuit. Therefore, the average energy savings of the complete interval depends on its segmentation. Figure 2.9 shows the average power savings dependent on the duty cycle of the WSN node. The duration of the sleep phase has been varied in order to get different duty cycles. It can be seen that the LDO2VAR converter performs best at low duty cycles and the BUCK2VAR converter performs best at high duty cycles. The reason is the very low quiescent power of the LDO2VAR converter and the high conversion efficiency of the BUCK2VAR in combination with the adjustable supply voltage. The power savings of the LDO2VAR are nearly the same at the different input voltages. However, the power savings of the BUCK2VAR increase with increasing input voltage. Therefore, the intersection of both curves moves from a duty cycle of about 0.009 at 3.4 V input voltage to a duty cycle of 0.006 at 4.9 V. It can be seen that the optimal converter depends on the duty cycle. Therefore, the information of the expected duty cycle of the WSN node is required to select the optimal converter.

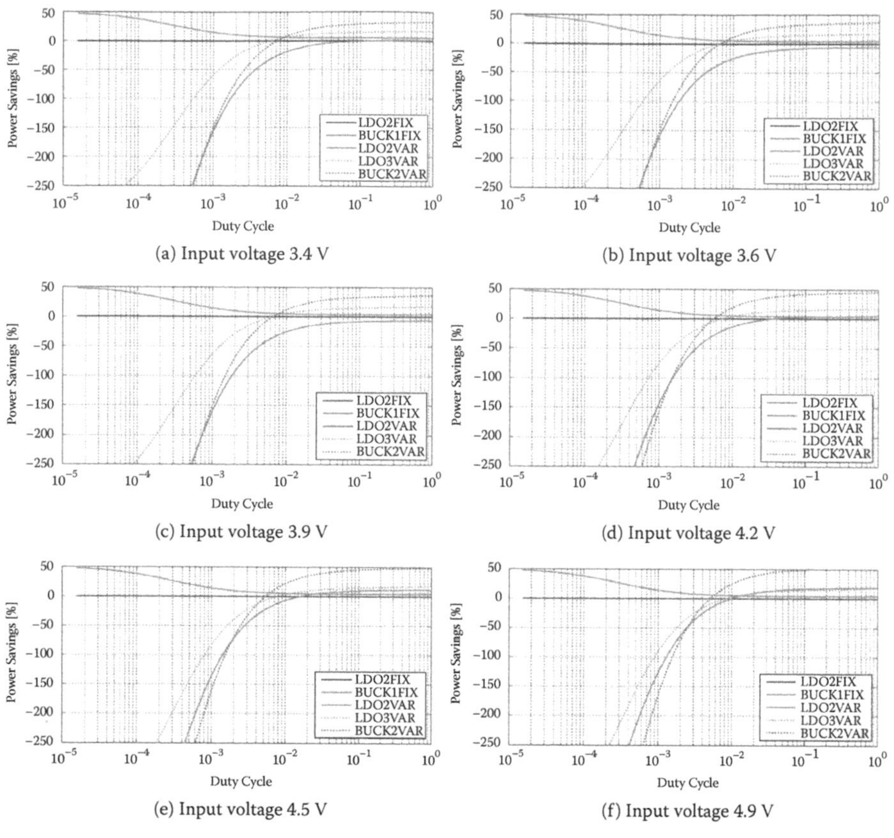

(a) Input voltage 3.4 V

(b) Input voltage 3.6 V

(c) Input voltage 3.9 V

(d) Input voltage 4.2 V

(e) Input voltage 4.5 V

(f) Input voltage 4.9 V

Figure 2.9 Power savings of the five different voltage converters dependent on the duty cycle of the WSN node. Six different input voltages of the converters have been evaluated.

An optimal solution could be to combine these two converters. They can be combined externally by using a special switching circuit controlled by itself or the WSN node, or they can be combined internally in a single chip to reduce the complexity of the supply circuit. Only the converter with better energy savings is active at a point in time. This would be the solution with the highest energy savings for WSN nodes with a high variation of the duty cycle.

These results also are valid for all other mobile devices using CADVS and a high variation of the duty cycle.

2.7 Conclusion

This chapter presented a tier model that can be used to describe ESSs. It structures the hardware into five different tiers, each with a special functionality. These tiers must be adapted to each other in order to optimize the overall efficiency. Therefore, the model can help in designing highly efficient ESSs. The second part of the chapter focuses on the power control and conditioning tier. Applying CADVS to a WSN node, it is possible to save 38.7 percent of the energy compared to a constant supply voltage using the introduced scenario. A prototype shows the correct functionality of CADVS using different voltage conversion circuits. The leakage currents of these conversion circuits have a high influence on the overall energy savings. Therefore, the duty cycle of an application influences the selection of the best voltage conversion circuit.

REFERENCES

[1]. Min, R., M. Bhardwaj, S.-H. Cho, N. Ickes, E. Shih, A. Sinha, A. Wang, and A. Chandrakasan. 2002. Energy-centric enabling technologies for wireless sensor networks. *Wireless Communications, IEEE* 9 (4): 28–39.

[2]. Rahimi, M., H. Shah, G. Sukhatme, J. Heideman, and D. Estrin. 2003. Studying the feasibility of energy harvesting in a mobile sensor network. Proceedings of the 2003 IEEE International Conference on Robotics and Automation (ICRA),Taiwan, May 12–17, Vol. 1, pp. 19–24.

[3]. Raghunathan, V., A. Kansal, J. Hsu, J. Friedman, and M. Srivastava. 2005. Design considerations for solar energy harvesting wireless embedded systems. Proceedings of the 4th International Symposium on Information Processing in Sensor Networks. Washington, D.C.: IEEE Press, p. 64.

[4]. Janek, A., C. Trummer, C. Steger, R. Weiss, J. Preishuber-Pfluegl, and M. Pistauer. 2008. Simulation based verification of energy storage architectures for higher class tags supported by energy harvesting devices. *Microprocessors and Microsystems* 32 (5-6): 330–339. Dependability and Testing of Modern Digital Systems.

[5]. Kansal, A., D. Potter, and M. B. Srivastava. 2004. Performance aware tasking for environmentally powered sensor networks. In *SIGMETRICS '04/Performance '04*: Proceedings of the Joint International Conference on Measurement and Modeling of Computer Systems. New York: ACM, pp. 223–234.

[6]. Hörmann, L. B., P. M. Glatz, C. Steger, and R. Weiss. 2011. Designing of efficient energy harvesting systems for autonomous WSNs using a tier model. IEEE 18th International Conference on Telecommunications (ICT), May 8–11, pp. 185–190.

[7]. Maxwell Technologies. Bcap0310 p270 t10–datasheet–bc power series radial d cell 310f ultracapacitor. Online at: http://www.maxwell.com/docs/DATASHEET_DCELL_POWER 1014625.PDF (accessed February 2011).

[8]. Linden, D., and T. B. Reddy. 2002. *Handbook of batteries*, 3rd ed. New York: McGraw-Hill.

[9]. Kompis, C. and S. Aliwell. 2008. Energy harvesting technologies to enable remote and wireless sensing. Online at: http://host.quid5.net/ koumpis/pubs/pdf/energyharvesting08.pdf (accessed June 2008).

[10]. Pouwelse, J., K. Langendoen, and H. Sips 2001. Dynamic voltage scaling on a low-power microprocessor. Proceedings of the 7th Annual International Conference on Mobile Computing and Networking, ser. MobiCom '01. New York: ACM, pp. 251–259.

[11]. Burd, T. D., and R. W. Brodersen. 2000. Design issues for dynamic voltage scaling. Proceedings of the 2000 International Symposium on Low Power Electronics and Design, ser. ISLPED '00. New York: ACM, pp. 9–14.

[12]. Simunic, T., L. Benini, A. Acquaviva, P. Glynn, and G. De Micheli. 2001. Dynamic voltage scaling and power management for portable systems. Proceedings of the 38th Annual Design Automation Conference, ser. DAC '01. New York: ACM, pp. 524–529.

[13]. Pouwelse, J., K. Langendoen, and H. Sips. 2001. Dynamic voltage scaling on a low-power microprocessor. Proceedings of the 7th Annual International Conference on Mobile Computing and Networking, ser. MobiCom '01. New York: ACM, pp. 251–259.

[14]. Sinha, A., and A. Chandrakasan. 2001. Dynamic power management in wireless sensor networks. *Design Test of Computers, IEEE* 18 (2): 62–74.

[15]. Powell, H. C., A. T. Barth, and J. Lach. 2009. Dynamic voltage-frequency scaling in body area sensor networks using cots components. Proceedings of the Fourth International Conference on Body Area Networks, ser. BodyNets '09. Brussels: ICST (Institute for Computer Sciences, Social-Informatics and Telecommunications Engineering), pp. 15:1–15:8.

[16]. Hörmann, L. B., P. M. Glatz, C. Steger, and R. Weiss. 2010. A wireless sensor node for river monitoring using MSP430 and energy harvesting. Proceedings of the European DSP in Education and Research Conference. Dallas: Texas Instruments, pp. 140–144.

[17]. Hörmann, L. B., P. M. Glatz, C. Steger, and R. Weiss. 2011. Energy efficient supply of WSN nodes using component-aware dynamic voltage scaling. In 17th European Wireless Conference (EW), Vienna, Austria, April 27–29, pp. 147–154.

[18]. Bergonzini, C., D. Brunelli, and L. Benini. 2009. Algorithms for harvested energy prediction in batteryless wireless sensor networks. Proceedings of the 3rd IEEE International Workshop on Advances in Sensors and Interfaces, Bari, Italy, June 25–26, pp. 144–149.

[19]. Dunkels, A., F. Osterlind, N. Tsiftes, and Z. He. 2007. Software-based on-line energy estimation for sensor nodes. EmNets '07: Proceedings of the 4th Workshop on Embedded Networked Sensors. New York: ACM, pp. 28–32.

[20]. Glatz, P. M., C. Steger, and R. Weiss. 2010. Tospie2: Tiny operating system plug-in for energy estimation. IPSN '10: Proceedings of the 9th ACM/IEEE International Conference on Information Processing in Sensor Networks. New York: ACM, pp. 410–411.

[21]. Jiang, X., J. Polastre, and D. Culler. 2005. Perpetual environmentally powered sensor networks. Proceedings of the Fourth International Symposium on Information Processing in Sensor Networks, Los Angeles, April 25–27, pp. 463–468.

[22]. Langendoen, K., A. Baggio, and O. Visser. 2006. Murphy loves potatoes: Experiences from a pilot sensor network deployment in precision agriculture. Proceedings of the 20th International Symposium on Parallel and Distributed Processing (IPDPS), Rhodes Island, Greece, April 25–29, p. 8.

[23]. Watthanawisuth, N., A. Tuantranont, and T. Kerdcharoen. 2009. Microclimate real-time monitoring based on zigbee sensor network. *Sensors, IEEE,* October: pp. 1814–1818.

[24]. Juang, P., H. Oki, Y. Wang, M. Martonosi, L. S. Peh, and D. Rubenstein. 2002. Energy-efficient computing for wildlife tracking: design tradeoffs and early experiences with zebranet. Proceedings of the 10th International Conference on Architectural Support for Programming Languages and Operating Systems (ASPLOS-X). New York: ACM, pp. 96–107.

[25]. Lindgren, A., C. Mascolo, M. Lonergan, and B. McConnell. 2008. Seal-2-seal: A delay-tolerant protocol for contact logging in wildlife monitoring sensor networks. Proceedings of the 5th IEEE International Conference on Mobile Ad Hoc and Sensor Systems, Atlanta, GA, Sept. 29–Oct. 2, pp. 321–327.

[26]. Lorincz, K., B.-R. Chen, G. W. Challen, A. R. Chowdhury, S. Patel, P. Bonato, and M. Welsh. 2009. Mercury: A wearable sensor network platform for high-fidelity motion analysis. Proceedings of the 7th ACM Conference on Embedded Networked Sensor Systems (SenSys 2009). New York: ACM, pp. 183–196.

[27]. Xu, N., S. Rangwala, K. K. Chintalapudi, D. Ganesan, A. Broad, R. Govindan, and D. Estrin. 2004. A wireless sensor network for structural monitoring. Proceedings of the 2nd International Conference on Embedded Networked Sensor Systems. New York: ACM, pp. 13–24.

[28]. Texas Instruments. 2009. Msp430f15x, msp430f16x, msp430f161x mixed signal microcontroller. Online at: www.focus-ti.com, SLAS368F

[29]. Analog Devices. 2006. ±0.5°C accurate PWM temperature sensor in 5-lead sc-70. Online at: www.analog.com, D03340 Rev.B

[30]. Microchip. Mrf24j40mb data sheet–2.4 ghz ieee std. 802.15.4 20 dbm rf transceiver module. Online at: www.microchip.com, DS70599B

[31]. Puccinelli, D., and M. Haenggi. 2005. Wireless sensor networks: Applications and challenges of ubiquitous sensing. *Circuits and Systems Magazine*, IEEE 5 (3): 19–31.

[32]. Texas Instruments. 2004. Tlv1117–Adjustable and fixed low dropout voltage regulator. Online at: www.focus-ti.com, SLVS561J

[33]. Glatz, P. M., L. B. Hörmann, C. Steger, and R. Weiss. 2010. A system for accurate characterization of wireless sensor networks with power states and energy harvesting system efficiency. Proceedings of the 8th IEEE International Conference on Pervasive Computing and Communications Workshops (PERCOM), March 29–April 2, pp. 468–473.

Chapter 3

Energy Cost of Software Applications on Portable Wireless Devices

Rajesh Palit, Ajit Singh, and Kshirasagar Naik

Contents

3.1 Introduction

Researchers have been exploring various architectural, hardware, software, and system-level optimization techniques to efficiently use the limited battery energy of portable wireless devices. The underlying objective is to maximize the amount of work that a device can perform before the battery runs out. It is not simply the amount of time that a device remains active before the battery runs out; maximizing the amount of work is a solution to the problem of extending the battery lifetime subject to system performance constraints. Various energy management strategies have been investigated at different levels of design methodology, starting from silicon at the bottom to application design at the top, with communication protocols and operating system in between. In this chapter, we discuss the issues related to energy management strategies in wireless portable devices, discuss some prior work, and then describe a model to estimate the energy cost of an application running on a portable wireless device.

3.2 Portable Wireless Devices

With advances in wireless communication technologies and hardware miniaturization, even small portable wireless devices have a large amount of communication bandwidth and computing power. These devices include cellular phones, personal digital assistants (PDAs), and other handheld devices. Typical components of state-of-art smartphones are given in Figure 3.1. The devices have low power consuming RISC (reduced instruction set computer) microprocessors limited RAM above 100 MB and pluggable flash memory in the gigabyte range. The power supply is usually equipped with a 3.7 V lithium ion battery ranging from 800 to 1,500 milliampere hour (mAh). They have Half-size VGA (video graphic array) or Quarter-size VGA color display, sometimes with touch screen. Some smartphones also have a camera, GPS (Global Positioning System) receiver, and other advanced hardware modules.

There are only a few operating systems (OSs) used in the mobile phones. As shown in Figure 3.2, according to 2010 market data for the fourth quarter, Google's OS Android owns about 33 percent of the market share, and 31 percent of the mobile devices in the market use Symbian OS. Apple's IOS is used in the iPhone, which shares 16 percent of the market. BlackBerry uses its own proprietary OS, which covers 14 percent of the cell phone market. The other OSs are Windows Mobile OS, Palm garnet OS, and some Linux-based OS and such as LiMo and Mobilinux.

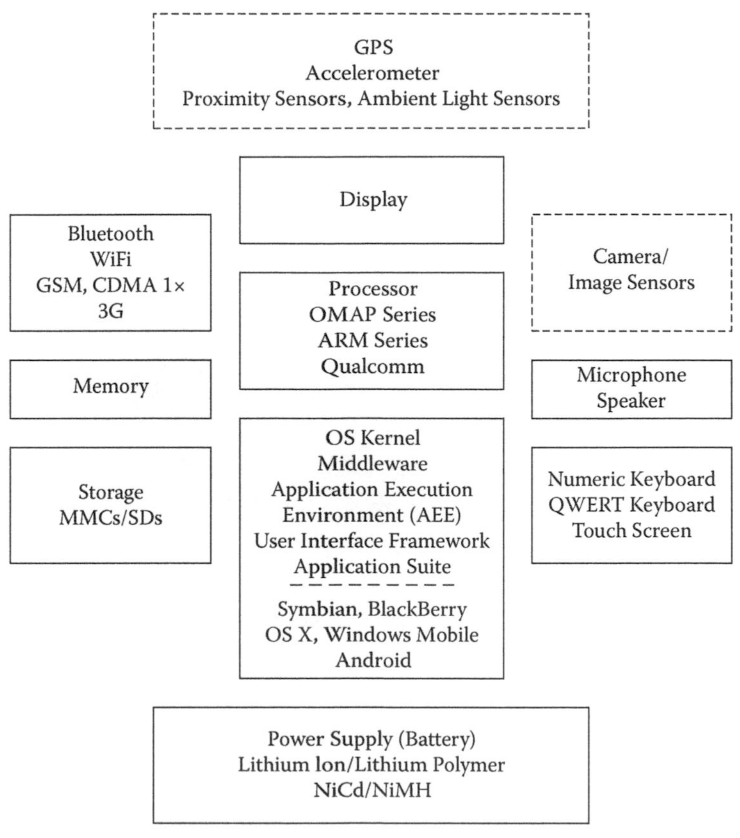

Figure 3.1 Components of a portable wireless device.

The basic mobile phones provide the users with applications such as voice calling, contacts, notes, calendar (to-do list), calculator, and simple games. However, smartphones are very feature rich, and they tend to have the tiny version of all kinds of applications that a personal computer has. Smartphone applications include a browser to surf the Internet, ability to listen and watch online multimedia, an instant messenger, GPS-assisted map applications, a VoIP (Voice over Internet Protocol) client, gadgets for weather and stock market forecasts, and so on. The current trend is to provide increasingly higher data rate and run heavier applications on these increasingly smaller wireless devices.

However, unlike the exponential growth in computing and communication technologies, in terms of speed and packaging density, growth in battery technology is a mere 5 to 10 percent increase per year in energy density [Rao03, Powers95, Naik01]. Web-based applications account for most of the energy consumption in wireless devices, and the application segment is ever-growing. With users desiring to run heavier applications, battery life is seen to pose the ultimate constraint on

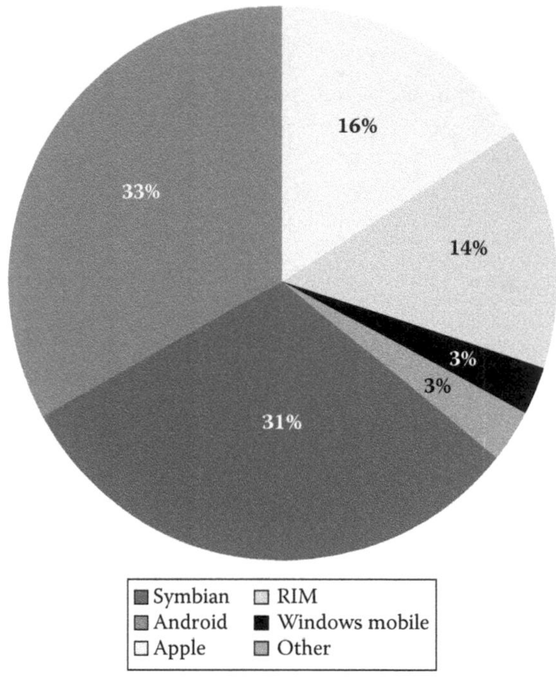

Figure 3.2 Market share of mobile OSs in 2010.

continued use of devices. For example, to support real-time multimedia applications over wireless links, batteries of a handset can only last half of the usual time [Marek02]. Thus, designers must look for energy-efficient solutions in all areas of system design, namely hardware, software, communication, architecture, and applications.

One way to achieve this goal is by using low-power-consuming hardware components as well as to take advantage of the idle states of a hardware component. All of the hardware components in a system do not remain active all of the time. From time to time, they become idle and to take advantage of this phenomenon, energy-aware hardware components can step into a low power state by temporarily reducing their speed and functionality. The duration of being in a low power state of the components also can be increased by using energy-aware applications, and by having collaboration among the OS and the applications [Creus07, Palit08a]. However, switching of modes of a component incurs overhead in terms of time and energy. If there was no overhead, energy management would be trivial: shutting down a device whenever it is idle. A component should step into low power mode only if the saved energy can justify the overhead. But a component can not switch modes by itself and software must make the decision. The set of rules that dictates when to switch into a low power state or how to increase the use of a low power state

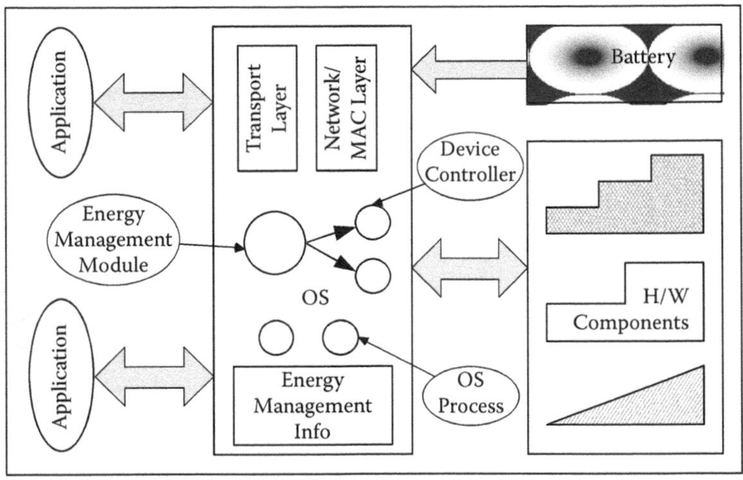

Figure 3.3 Application framework in battery-driven portable devices.

of a component is part of the energy management strategy. The ultimate objective to reduce the energy consumption of a portable wireless device consists of hardware components with energy saving features as well as software techniques.

As shown in Figure 3.3, OSs work as an interface between the hardware components and user applications in portable devices. They manage and abstract the hardware components from the applications. The OS has exclusive access to the hardware components and it needs the functionality to operate the components in a low energy state [Lu02]. Applications cannot act on power-saving mechanisms without proper coordination with the OS. They require an interface with the operating system to gain the available information about the device's energy-saving features. And, this way, an application can take advantage of those features. On the other hand, each application is required to tell the OS what its needs are and the OS notifies each application whenever there is a change in the state of the system relevant to energy management decisions. It is purely a collaborative effort, and applications and their interaction with the underlying operating system play a crucial role in energy management. As a result, the energy cost of each of the applications and how they use the hardware components first need to be investigated. Then a system-level policy is needed to facilitate the power management strategy with a view to achieving energy efficiency [Creus07].

In this chapter, we describe a model to compute the energy cost of an application. The model considers the energy costs of the computing and communication components of a device. The energy costs of other components, such as display, can be considered as fixed cost while a device is in operation. Of course, if a user keeps a device inactive for a long time, the power-saving feature in display can further reduce energy consumption. The basic cost estimation idea can be combined with the concept of an operational profile [Musa93] of a system to estimate the energy cost at a high level. Intuitively, an operational profile of a system consists of three parts: (1) the fraction of the time a system is actually running and consuming a significant amount of energy, (2) the software systems the device is executing, and (3) the set of basic activities and their frequencies of occurrences for each of the executing software systems. With the ability to model the energy cost of applications, we can estimate the battery lifetime of a device by incorporating the usage pattern of a particular user.

The rest of this chapter is organized as follows. In section 3.3, we review the related work. We present the energy cost model in section 3.4, and, in section 3.5, we discuss how to obtain the cost model. In section 3.6, we add our chapter conclusions.

3.3 Related Work

In this section, we discuss some substantial research work that has been reported in the literature, such as these based on portable devices' battery models, battery sensing parameters, and battery lifetime estimation. This is followed by a discussion of energy-aware software implementation strategies. Then, we describe some specific tools for evaluating the power consumption of software applications.

3.3.1 Sensing of Smart Battery

Rao, Vrudhula, and Rakhmatov present the discharging patterns of lithium ion batteries for different current drawing rates in [Rao03]. The discharge behavior is sensitive to numerous factors including the discharge rates, temperature, and the number of charge–recharge cycles. Thus, the battery deviates from the behavior of an ideal energy source. As the discharge rate increases, battery capacity decreases. The output voltage of the battery changes over the discharge duration. However, when the current drawn rate is well below the maximum current drawn rate, the capacity and voltage remain almost the same as the original capacity.

The authors present several battery models that capture the battery behavior and they claim that these mathematical models are capable of showing battery discharge characteristics in sufficient detail. This fact lets the designers formulate an optimization strategy to pull out maximum charge from the battery. State-of-charge (SoC) or the remaining energy in the battery is very important in making energy management decisions. Casas and Casas in [Casas05] discuss three mechanisms to sense the SoC. These techniques are based on voltage, current, and impedance, which are relatively easy to measure in a battery. Panigrahi et al. [Panigrahi01] suggest an alternative battery model that is based on rate capacity effect and recovery effect. They claim that their model estimates the battery lifetime as well as the delivered energy. However, for these sensing and measuring techniques, we need to make sure that the techniques themselves do not consume a significant amount of battery energy. Based on the battery discharging patterns and behavior, there are a couple of battery-driven power management strategies proposed in the literature [Benini00, Lahiri02].

3.3.2 Software Strategies for Energy Management

Lorch et al. [Lorch98] present a very comprehensive discussion about what should be meant by energy management in portable devices and what are the software strategies that can be considered to address the energy management problem. They mention that four things must be considered to evaluate a power management strategy:

1. How much it reduces the power consumption of a hardware component.
2. What percentage of total system power, on average, is due to that component.
3. How much it changes the power consumption of other components.
4. How it affects the battery capacity through its changes in power consumption.

They classify the software issues into three groups: transition, load change, and adaptation. Based on this concept, they discuss ways to achieve energy efficiency in secondary storage, processor, WLAN (wireless local area network), and other components. Their classification is extremely insightful and effective.

Naik and Wei [Naik01] propose a static energy-saving strategy based on the cost of executing different instructions on a processor. They did a study on how a variety of algorithm design and implementation techniques affect energy consumption. As the study is independent of a particular processor or a system, their study helps in implementing energy-conscious software applications in the design phase.

Flinn and Satyanarayanan [Flinn99] claim that a collaborative attempt between the operating system and applications can be used to meet user-specified goals for battery duration. They show how applications can dynamically modify their behavior to conserve energy. They advocate for dynamic tradeoffs rather than static tradeoffs in hardware and software design. According to the study, at runtime, more accurate knowledge of energy supply and demand allows better decisions to be made in resolving the tension between energy conservation and usability.

3.3.3 Tools for Analyzing Software Applications

Flinn and Satyanarayanan describe the design and implementation of PowerScope in [Flinn99]. This is a tool for profiling energy usage by applications. PowerScope maps energy consumption to program structure in much the same way that CPU (central processing unit) profilers map processor cycles to specific processes and procedures. PowerScope combines the hardware instrumentation to measure current level with kernel software support to perform statistical sampling of system activity. Postprocessing software maps the sample data to program structure and produces a profile of energy usage by process and procedure. Shnayder et al. [Shnayder04] present PowerTOSSIM, a scalable simulation environment for wireless sensor networks that provides an accurate, per-node estimate of power consumption. PowerTOSSIM is an extension of TOSSIM, an event-driven simulation environment for TinyOS applications. In PowerTOSSIM, TinyOS components corresponding to specific hardware peripherals are instrumented to obtain a trace of each device's activity during the simulation run. Banerjee and Agu [Banerjee05] and Dick, Lakshminarayana, and Jha [Dick00] also present frameworks for analyzing power consumption of embedded operating systems.

3.3.4 System Level Energy Management

Hardware manufacturers typically provide power-saving features in their components, and there are techniques to address the software energy efficiency issue. But it is difficult to maximize energy saving for all system components together. For example, power saving on CPUs can have a negative effect on power saving on memory. Complex interaction among resources and applications affect the total power consumption. As a result, there is a need for a system-wide, energy-saving policy. Unsal and Koren [Unsal03] survey the system-level, power aware techniques. Their work provides good understanding of power aware design for energy constraint devices. Creus and Niska provide a policy-based approach for system-level power management targeting mobile devices [Creus07]. They present a concept of application collaboration to select the optimum resource allocation. This is a system-wide, dynamic power management policy where the applications are allowed to communicate their intent to the system through a set of APIs (application program interfaces).

3.3.5 Miscellaneous Work

3.3.5.1 μSleep Technique

Brakmo, Wallach, and Viredaz [Brakmo04] propose the μSleep technique for energy reduction on handheld devices. They claim that the technique is most effective when the handheld's processor is lightly loaded, such as when the user is reading a document or looking at a Web page. The basic idea is that when possible, rather than using the processor's idle mode, it tries to put the processor in sleep mode for short periods (less than one second). To enhance the perception that the system is on, an image is maintained on the display and activity is resumed as a result of external events, such as touch screen and button activity. This is easily implementable in any OS and the OS can put the device in μSleep when no user activity is detected for a while. They have implemented μSleep on a prototype pocket computer where it has reduced energy consumption by up to 60 percent. A similar kind of technique has been proposed by Shih, Bahl, and Sinclair [Shih02]. They have shown that the lifetime of an iPAQ can be increased up to 115 percent by utilizing a wake-on-wireless technique. They need to introduce a separate low-energy channel to implement their technique.

3.3.5.2 Power-Aware Buffering

Ling and Chen [Ling07] did a study on the impact of buffering schemes on wireless sensor networks. They present theoretical analysis of the power consumption of fixed-size and fixed-interval buffering schemes. Sensor nodes' wireless radio and memory modules, as well as the data arrival rate, were considered in the analysis. Experimental results indicate that, under the same circumstances, the optimal fixed-size buffering scheme outperforms the optimal fixed-interval scheme in terms of overall power conservation. The benefit of a power-aware buffering lies in its ability to exploit the predictable idleness of the workload, and in its ability to amortize the radio wake-up energy to reduce the power consumption. This very idea can be extended to a wireless mobile device, especially when it plays multimedia contents. Their theoretical results, at least, will provide guidance for the determination of optimal buffer size based on actual power parameters of communication component and memory size, and on the data packet arrival rate.

3.3.5.3 Impact of Data Compression

Xu et al. [Xu03] investigated the use of data compression to reduce the battery power consumed by handheld devices when downloading data from proxy servers over a WLAN. The objective is to make a trade-off between the communication energy and the computing energy to perform decompression. They experimented with three lossless compression schemes using Compaq iPAQ 3650 in a wireless LAN environment. Results show that, from the battery-saving perspective, the gzip compression software (based on LZ77) is far superior to bzip2 (based on BWT). They also present an energy model to estimate the energy consumption for the compressed downloading. With this model, they are able to reduce the energy cost of gzip by interleaving communication with computation and by using a block-by-block selective scheme. They used a threshold file size below which the file is not to be compressed before transferring.

This study can well be extended for specialized compression schemes for video, music data, and for uploading of multimedia data as well. In fact, when carefully chosen and applied, data compression schemes can effectively reduce the energy cost on handheld devices in a wireless environment. However, when carelessly chosen, a data compression scheme can cause a significant energy loss instead of savings. The overall idea can further lead to choosing the encryption/decryption techniques for a portable wireless device.

3.3.5.4 Energy-Efficient Schedulers

Lee, Rosenbery, and Chong [Lee06] formulated a downlink scheduling optimization problem aimed at saving energy and proposed two heuristic scheduling policies to solve it. They consider a generic wireless system composed of an access point (AP) and several stations that offer a power-saving mode (PSM) to its users. The authors show how the length of the beacon period (BP) has a significant impact on the energy and the delay performance of wireless stations. For each of the scheduling policies, they derive a simple approximate formula for the length of the BP that minimizes the energy consumption. When the maximum allowable average packet delay is given as a QoS requirement, they define a method to find the length of the BP for the two schedulers. By adjusting the BP, the idle period of the communication part can be adjusted and it helps the device to be in sleep state for a longer period of time. The results of this paper show that a fine-tuning of the length of the BP, as well as well-designed scheduling disciplines, are essential to reducing the energy consumption of mobile stations.

3.3.5.5 GUI Design

Researchers mostly concentrate on processing or communication intensive applications rather than interactive applications, which are now dominant in mobile devices. The modern devices use graphical user interfaces (GUIs) to handle human–computer interaction. Vallerio, Zhong, and Jha [Vallerio06] are the first to explore how GUIs can be designed to improve system energy efficiency. They investigated how GUI design approaches should be changed to improve system energy efficiency and provide specific suggestions to mobile computer designers to enable them to develop more energy-efficient systems. They show that energy-efficient GUI (E^2GUI) design techniques can improve the average system energy of three benchmarks (text-viewer, personnel viewer, and calculator) by 26.9, 45.2, and 16.4 percent, respectively. Thus, using E^2GUI design techniques can contribute to prolonging the battery lifetime of mobile computers.

We can divide the research work discussed above into several categories. Improvement of battery capacity, energy efficient coding, and low energy hardware device technology are independent fields in nature, and there are always opportunities for constant improvement in these fields. Another category that we always forget to mention is user behavior. A user of a device can play an important role in saving energy in a device. For example, one can keep the brightness of screen as low as possible. However, users generally lack knowledge about power consumption of each component, and are unwilling to make frequent energy management decisions. The class of work related to analyzing software applications and estimating energy consumption is important for emulating an application without running it in the actual device.

The recent trend of research is to develop a system-wide, energy-saving policy (as described in [Creus07]), so that all the applications running on a device can act uniformly and cooperate with the OS to save energy. A framework between the OS and software applications can facilitate such a policy. OS supplies some system information, such as processor energy cost, data packet transmission/reception costs, disk read/write costs, etc., to an application so that the application can adapt to the system. On the other hand, the application provides the OS with the information, such as delay tolerance of a data packet, time-interval between two consecutive data packets, or disk read/write so that the OS can make decisions to set the hardware component in the right energy consumption state.

3.4 Energy Consumption Model

A schematic diagram of energy consumption structure in wireless devices is given in Figure 3.4. Users, software applications, operating systems, and hardware components are the four key players in a portable device system. The hardware components are powered by the battery, and users run different types of software applications on a device. The operating system sits in between software applications and hardware components, and it manages how and when a hardware component is used according to the demand from applications. If the energy cost of using a component is known, and we have a model to capture the usage patterns of different components by an application, the energy consumption of that application can be estimated. A formal energy consumption model for a portable device is presented in [Palit08b], and we discuss a practical approach to estimate the energy consumption of an application.

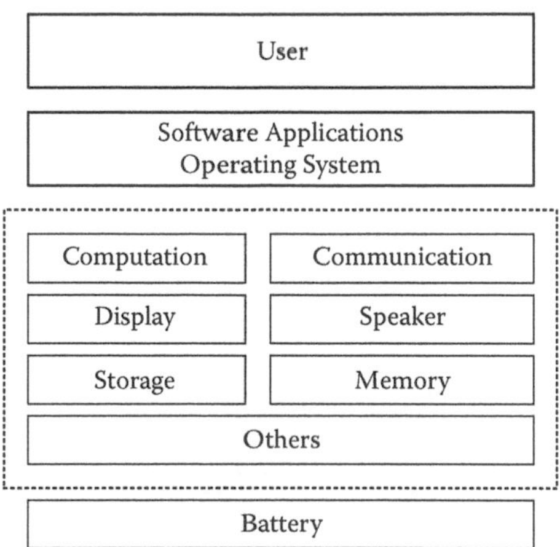

Figure 3.4 Energy consumption structure of a portable device.

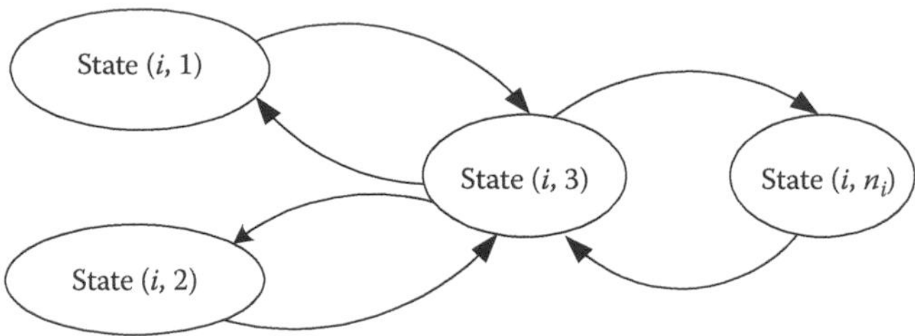

Figure 3.5 Finite State Machine for a component (*i*).

Finite State Machine (FSM) models (Figure 3.5) are widely used to describe the dynamic behavior of hardware components [Chow96, Desai03]. When a hardware component remains in a state for a certain length of time, the component draws a constant amount of power. And, if the supply voltage of a portable device remains constant, the consumed power is directly proportional to current. We use the idea of state residence time to denote a period of time in which a hardware component remains in that state. Under the constant power assumption, given the current level l of a state and the state residence time Δt, the energy consumption of a component is $v \times l \times \Delta t$, where v is the supply voltage (Figure 3.6).

Suppose that there are n numbers of hardware components in a portable device. Each of the components has a fixed number of predefined states. Let component i have n_i states. Now state $S_{i,j}$ means the jth state of component i, where $1 \leq j \leq n_i$. Suppose component i switches to state j, $n_{i,j}$ number of times in a given time period T and it stays $\delta t_{i,j,k}$ amount of time each time it enters $S_{i,j}$, where $1 \leq k \leq n_{i,j}$. From the above model definition and description, we can now derive the formula of energy consumption by the device in time period T. In equation (3.2), $\Delta t_{ij} (= \sum_{k=1}^{n_{i,j}} \delta t_k)$ is

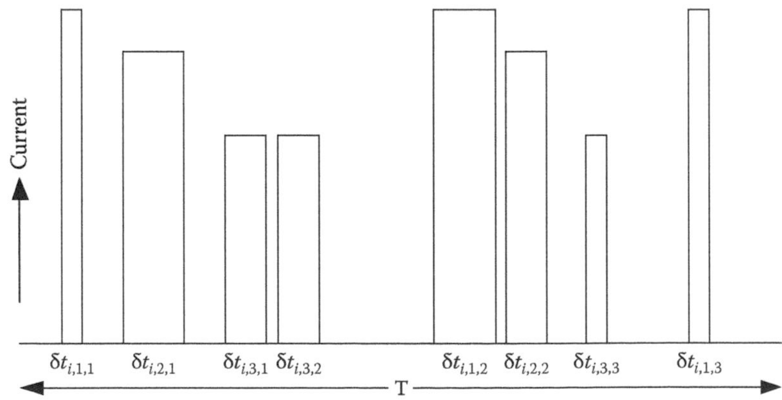

Figure 3.6 Instantaneous state residence times of component, *i*.

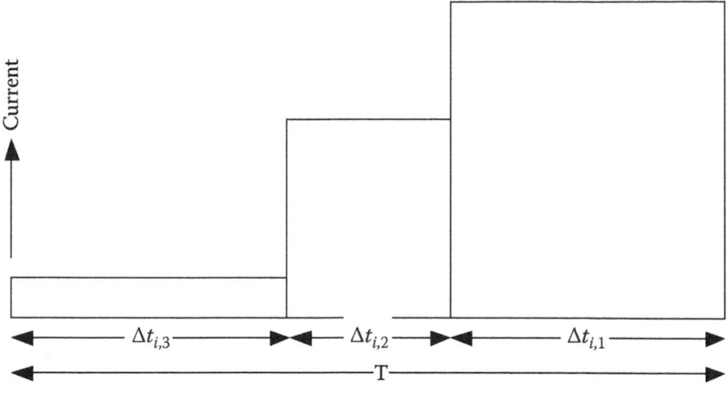

Figure 3.7 Average state residence times of component, *i* over time, *T*.

the total amount of time that component *I* stays in state *j* within time period *T*. In equation (3.3), current($S_{i,j}$) is the fixed current consumption in constant voltage supply *v*.

$$Energy(T) = \sum_{i=1}^{n}\sum_{j=1}^{n_i} Power(S_{i,j})\sum_{k=1}^{n_{i,j}} \delta t_{i,j,k} \;\ldots\;\ldots\;\ldots(1)$$

$$= \sum_{i=1}^{n}\sum_{j=1}^{n_i} Power(S_{i,j})\Delta t_{i,j} \;\ldots\;\ldots\;\ldots(2)$$

$$= v\sum_{i=1}^{n}\sum_{j=1}^{n_i} current(S_{i,j})\Delta t_{i,j} \;\ldots\;\ldots\;\ldots(3)$$

At the component level (Figure 3.7), once we know the cost of each state of the components and the state residence times, we can express the total energy consumption using equation (3.1) or equation (3.2). But, from the application level point of view, when we want to estimate the energy consumption by an application, we need to know the states and residence time of each state of all components used by that particular task. However, it is not easy to get the component states and residence times for a given task.

With each portable device there is an OS that manages and controls the hardware components. The OS schedules different tasks to access different components of the system at different times. However, there are a number of tasks running simultaneously on a system. And, the particular time when a task gets access to a component or how long it continues to access the component cannot be a known priory. Moreover, the length of time that a task uses a component at one time is very transient, which is in the millisecond range. Therefore, it is very difficult to know the timing and to get the residence time of each state mentioned in the model. If we use a monitor program to watch a given task and keep records of transient residence times, the monitor program itself consumes the component's time and system energy. Consequently, we may not get accurate values. On the other hand, evaluating the energy cost of each single state is also challenging because it is difficult to keep a device in one simple state. Therefore, we have to measure the cost of a compound state and then we need to calculate the cost of a single state by separating out the cost of other states.

3.5 Determination of Model Parameters

To estimate the energy consumption of a software application on a device, we need to know how or in what proportion the application utilizes different hardware components of the device over a given time period, *T*. For that, the total state residence times of each state of all the hardware components for that application are required. The corresponding current consumptions of each of the states also are essential to find the energy consumption. In this section, we discuss how to get state residence times and indicate how to find the energy cost of each of the states.

3.5.1 Estimation of State Residence Time

In a multitasking environment, a software application has different power-consuming states, such as active, idle, sleep, or dormant states like the hardware components. In sleep or dormant state, applications reside in the storage and are ready to launch. There is some cost for attaching the storage with the system. In an idle state, applications reside in the memory, and users do not actively interact with them. They perform some routine work, such as contacting the server, status update, etc. In the active state, a user interacts with the application and it is the highest energy-consuming state of the application, hence, we are interested in this state.

In fact, this is the most critical or complex part of estimating energy consumption. Thousands of lines of codes, numerous features, and usage scenarios make the process complex, and it is not possible to exactly figure out the residence times of an application except it is not very simple and straightforward. By executing an application on an actual device, the usage profile of the application can be obtained from the operating system. The same information also can be found from PowerSpy, PowerScope, or PowerTOSSIM simulators [Banerjee05, Flinn99, Shnayder04]. Table 3.1 shows a template for calculating state residence for an application. Using this table, the actual residence times can be calculated for given time, *T*.

3.5.2 Estimation of the Costs

Manufacturers of the hardware components are a good source of getting energy costs of each of the states of the components. However, these data sometimes are not ready to use. We need to get the energy cost when everything is integrated

Table 3.1 State Residence Time of Different Components for an Application

	Application State	
	Active	*Idle*
State of Components	**Usage of Component**	**Usage of Component (%)**
S(1,1)	$X_{1,1}$	$Y_{1,1}$
S(1,2)	$X_{1,2}$	$Y_{1,2}$
S(2,1)	$X_{2,1}$	$Y_{2,1}$
S(2,3)	$X_{2,3}$	$Y_{2,3}$
...	...	...

with each other and works together. Sometimes we are interested in getting costs for some simple task, such as sending or receiving a data packet or writing a file to storage. An experiment setup is given in Figure 3.8 that is suitable for conducting experiments for evaluating energy costs of different hardware states.

The battery connectors of a target device are connected to a high precision power supply that supplies current with constant voltage and measures the supplied current. A computer is connected to the power supply to set the required voltage and current limit on the power supply and read back the current measurement data. A wireless access point and a Web server are a part of the setup to accomplish necessary communication costs. With a cellular network access module (SIM), the device also can be connected to the EDGE/3G network, and facilitate corresponding communication costs over 3G. The accuracy and resolution of the power supply and measurement unit should be very high as the changes occur in minute intervals.

3.5.3 Cost of Processing and Communications

The energy cost of the processor's idle and sleep states can be evaluated by measuring the current from the power supply by keeping the processor in corresponding states. However, to measure the cost of processing when it is fully active, we need to keep the processor busy by executing computationally intensive programs. Figure 3.9 shows the current consumptions of a system at different levels of CPU utilization. Three benchmark programs, namely LINPACK, Whetstone, and Dhrystone packages, were used in the experiments. These packages are widely used to benchmark computer systems, and computationally complex enough to load the processor almost 100 percent. These programs were executed with different pause intervals to make the processor have some idle time in between two consecutive executions. As a result, the overall processor utilization became less with varying pause intervals.

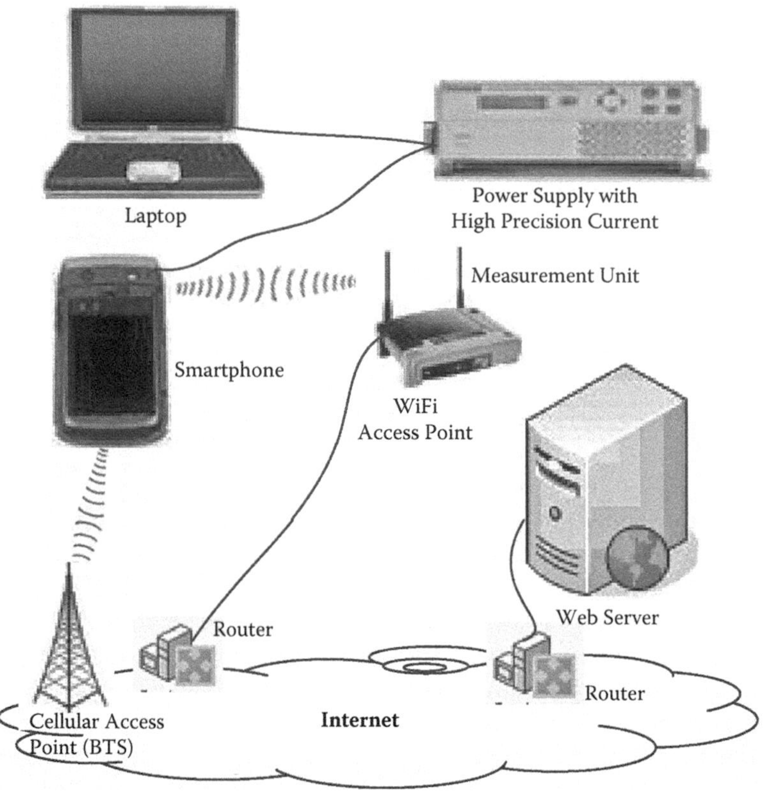

Figure 3.8 Experimental setup.

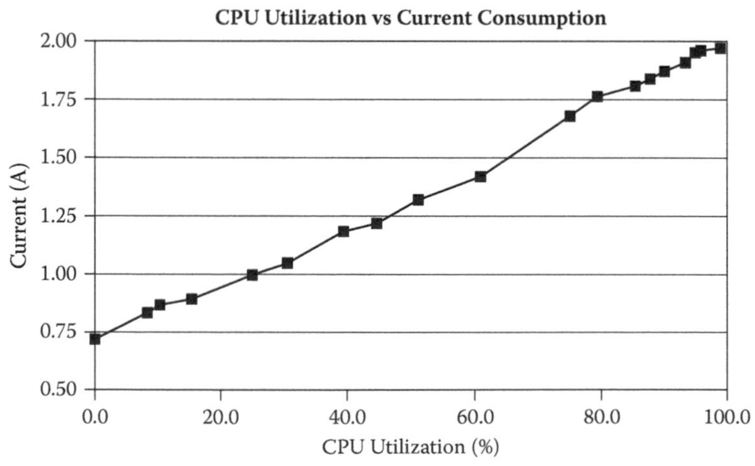

Figure 3.9 Current consumption of processor at different utilization levels.

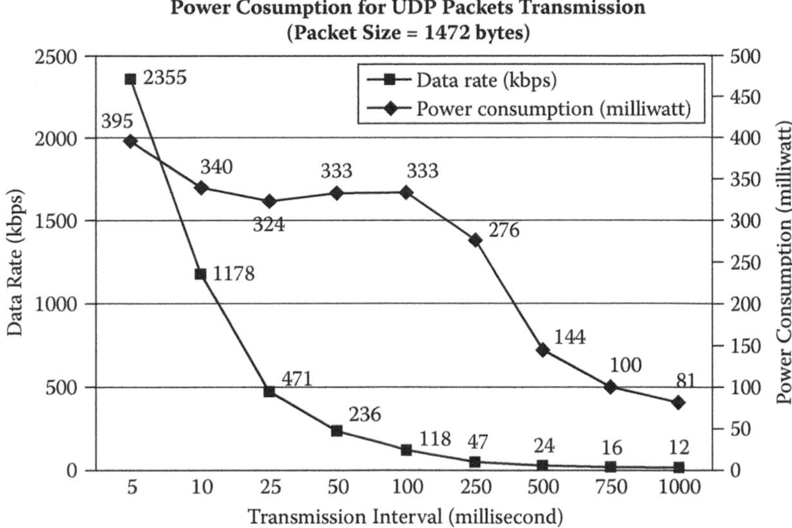

Figure 3.10 Current consumption and data rates for different transmission intervals.

There is a linear relationship in power consumption for different CPU utilization levels; however, the relationship for sending data packets at different intervals is not linear (Figure 3.10). Using the information from Figure 3.10, a suitable data rate might be chosen to save energy. The power consumption for a fixed data rate with varying packet sizes is also very interesting.

As shown in Figure 3.11, there is a dip in average current consumption when the packet size is 1,400 bytes. It shows that when the packet size is smaller, the packet generation rate from the application increases and the number of lower layer packets (MAC packets) increases accordingly. However, the default MTU (maximum transmission unit) size is 1,500 bytes in 802.11, and when the application level packet size is below 1,500 bytes, one MAC level MTU can accommodate one full application level packet and the lower level overheads associated with it. When the packet size increases further, packet generation rate decreases, but one MTU is not enough to contain an application level packet, and, therefore, more lower level packets are needed. We see the reflection of this fact in the graph. The average current consumption goes high when the packet size becomes more than MTU size.

The information of data transmission or reception in terms of byte per second is not enough for estimating the energy cost of an application. We need to know the application and lower layer packet sizes for this purpose. By observing the energy

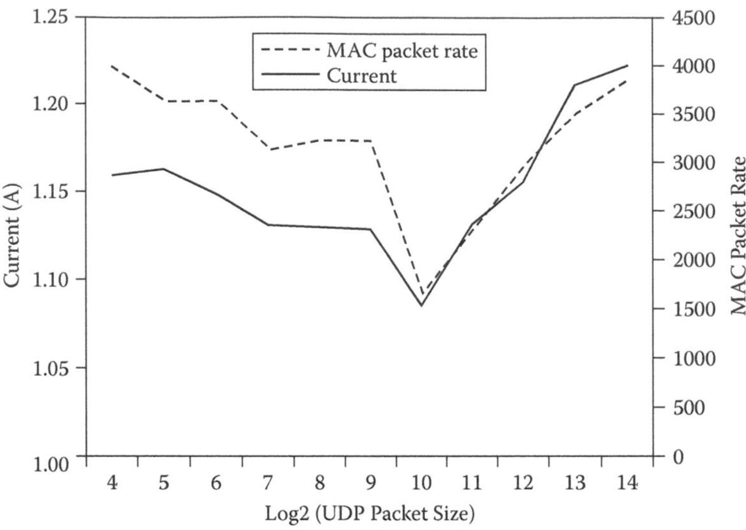

Figure 3.11 Current consumption and MAC packet rates for different packet sizes for a fixed data rate.

consumption, delay, and data rate chart, we can select a suitable packet size in order to improve the performance of the system.

3.5.4 Another Approach

Measuring energy consumption of an application on a particular device would have been a very simple task of running the application on that device for different use-case scenarios, and getting the average energy consumptions. However, we are interested in knowing the analysis of how an application consumes energy and uses the hardware components. Therefore, the designers of the software applications and manufacturers of the devices can benefit from the analysis for future design and implementations.

Figure 3.12 shows another model for evaluating energy costs of software applications. It has two parts. On the left side, developers of software applications specify the performance of their program by executing it on a simulator or emulator. They provide the utilization of the different hardware components (CPU, memory, storage) for different subtasks, such as processing, read/write on storage, and send/receive messages. They also mention the corresponding energy costs on the simulator. On the other side, manufacturers provide a list of cost mapping functions that translate the energy cost of the simulator to their particular devices. Thus, the energy cost of an application can be estimated for different devices. Using this framework, the performances of different applications for a specific purpose (e.g., two video playing applications) can be compared on the simulator, and also the performances of different devices for a specific application can be evaluated.

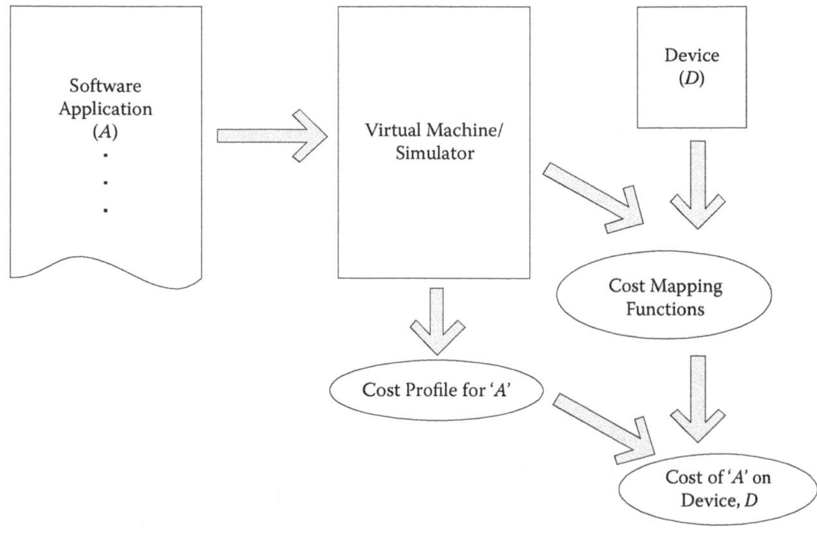

Figure 3.12 An alternative approach for evaluating energy consumption.

3.6 Conclusion

As the demand for Internet access and computation power of portable wireless devices is increasing day by day, the constrained battery energy in these devices becomes the weakest link in supporting heavy applications. The classical end-to-end argument suggests that functions placed at low levels of a system may be redundant or of little value when compared with the cost of providing them at that low level. Software applications reside at the top layer, and thus energy-saving measures are most effective at this level. The described model can easily be extended to model the other component-related cost of an application, such as secondary storage.

REFERENCES

Banerjee05 Kutty S. Banerjee and Emmanual Agu. Powerspy: Fine-grained software energy profiling for mobile devices. In Proceedings of the International Conference on Wireless Networks, Communications and Mobile Computing, vol. 2, pp. 1136–1141, June 2005.

Benini00 Luca Benini, Giuliano Castelli, Alberto Macii, Enrico Macii, and Riccardo Scarsi. Battery-driven dynamic power management of portable systems. In ISSS '00: Proceedings of the 13th International Symposium on System Synthesis, pp. 25–30, Washington, D.C.: IEEE Computer Society, 2000.

Brakmo04 L. S. Brakmo, D. A. Wallach, and M. A. Viredaz. µsleep: A technique for reducing energy consumption in handheld devices. In Proceedings of MobiSys 2004, Boston, MA, June 6–9, pp. 48–56.

Casas05 Roberto Casas and Oscar Casas. Battery sensing for energy-aware system design. *IEEE Computer* 38 (11): 48–54, November 2005.

Chow96 S.-H. Chow, Y.-C. Ho, and T. Hwang. Battery modeling for energy aware system design. *ACM Transactions on Design Automation of Electronic Systems* 1 (3): 315–340, 1996.

Creus07 G. B. i Creus and P. Niska. System-level power management for mobile devices. In 7th IEEE International Conference on Computer and Information Technology, pp. 799–804, Fukushima, Japan, October 16–19, 2007.

Desai03 M. P. Desai, H. Narayanan, and S. B. Patkar. The realization of finite state machines by decomposition and the principal lattice of partitions of a submodular function. *Discrete Applied Mathematics* 131 (2): 299–310, September 2003.

Dick00 Robert P. Dick, Ganesh Lakshminarayana, and Niraj K. Jha. Power analysis of embedded operating systems. In Proceedings of ACM/IEEE Design Automation Conference, pp. 312–315, Los Angeles, CA, June 5–9, 2000.

Flinn99 Jason Flinn and M. Satyanarayanan. Powerscope: A tool for profiling the energy usage of mobile applications. In WMCSA '99: Proceedings of the Second IEEE Workshop on Mobile Computer Systems and Applications, p. 2, Washington, D.C. IEEE Computer Society, 1999.

Lahiri02 Kanishka Lahiri, Sujit Dey, Debashis Panigrahi, and Anand Raghunathan. Battery-driven system design: A new frontier in low power design. In ASP-DAC '02: Proceedings of the 2002 Conference on Asia South Pacific Design Automation/VLSI Design, p. 261, Washington, D.C.: IEEE Computer Society, 2002.

Lee06 Jeongjoon Lee, Catherine Rosenberg, and Edwin K. P. Chong. Energy efficient schedulers in wireless networks: Design and optimization. *Mobile Networks and Applications* 11 (3): 377–389, 2006.

Ling07 Yibei Ling and Chung-Ming Chen. Energy saving via power-aware buffering in wireless sensor networks. In Proceedings of the IEEE INFOCOM 26th International Conference, pp. 2411–2415, Anchorage, AK, May 6–12, 2007.

Lorch98 Jacob R. Lorch and Alan J. Smith. Software strategies for portable computer energy management. *IEEE Personal Communications Magazine,* 5(3), 60–73, June 1998.

Lu02 Yung Hsiang Lu, Luca Benini, and Giovanni De Micheli. Power-aware operating systems for interactive systems. *IEEE Transactions on VLSI* 10: 119–134, 2002.

Marek02 S. Marek. Battling the battery drain. *Wireless Internet Magazine*, January, 2002.

Musa93 J. D. Musa. Operational profiles in software reliability engineering. *IEEE Software* 10 (2): 14–32, 1993.

Naik01 K. Naik and D. S. L. Wei. Software implementation strategies for power-conscious systems. *Mobile Networks and Applications* 6 (3): 291–305, 2001.

Palit08a Rajesh Palit, Kshirasagar Naik, and Ajit Singh. Estimating the energy cost of communication on portable wireless devices. In 1*st IFIP Wireless Days*, pp. 346–353, November 2008.

Palit08b Rajesh Palit, Ajit Singh, and Kshirasagar Naik. Modeling the energy cost of applications on portable wireless devices. In Proceedings of the 11th International Symposium on Modeling, Analysis and Simulation of Wireless and Mobile Systems (MSWiM), pp. 346–353, Vancouver, Canada, October 2008.

Panigrahi01 Debashis Panigrahi, Sujit Dey, Ramesh Rao, Kanishka Lahiri, Carla Chiasserini, and Anand Raghunathan. Battery life estimation of mobile embedded systems. 14th International Conference on VLSI Design, 0: 57, Bangalore, India, January 3–7, 2001.

Powers95 R. Powers. Batteries of low electronics. Proceedings of IEEE, 83 (4), April 1995.

Rao03 R. Rao, S. Vrudhula, and D. Rakhmatov. Battery modeling for energy aware system design. *Computer* 36 (12): 77–87, 2003.

Shih02 E. Shih, P. Bahl, and M. Sinclair. Wake on wireless: An event driven energy saving strategy for battery operated devices. In Proceedings of the 8th Annual International Conference on Mobile computing and Networking (ACM MobiCom), Atlanta, GA, September 23–28, 2002.

Shnayder04 V. Shnayder, M. Hempstead, B. Rong Chen, G. W. Allen, and M. Welsh. Simulating the power consumption of large-scale sensor network applications. In Proceedings of the 2nd International Conference on Embedded Networked Sensor Systems (SenSys), Baltimore, MD, November 2004.

Unsal03 Osman S. Unsal and Israel Koren. System-level power-aware design techniques in real-time systems. In Proceedings of the IEEE, pp. 1055–1069, 2003.

Vallerio06 K. S. Vallerio, L. Zhong, and N. K. Jha. Energy-efficient graphical user interface design. *IEEE Transactions on Mobile Computing* 5 (7): 846–859, 2006.

Xu03 Rong Xu, Zhiyuan Li, Cheng Wang, and Peifeng Ni. Impact of data compression on energy consumption of wireless-networked handheld devices. In Proceedings of the 23rd International Conference on Distributed Computing Systems (ICDCS), pp. 302–311, Washington, D.C.: IEEE Computer Society, 2003.

Chapter 4

Striking a Balance between Energy Conservation and QoS Provision for VoIP in WiMAX Systems

Xiao-Hui Lin, Ling Liu, Hui Wang,
Jing Liu, and Yu-Kwong Kwok

Contents

4.1 Introduction

Energy conservation is a critical issue in the emerging standard IEEE 802.16e/m WiMAX supporting mobility. To reduce the energy consumption in VoIP (Voice over Internet Protocol) transmission, a feasible approach is to let the mobile station (MS) enter sleep mode during the speech silence period, while remaining active during the talk-spurt period. However, VoIP service has tight QoS (quality of service) constraints: average delay and loss rate. Therefore, how to tune the sleep window size while meeting the QoS requirements is a challenging issue in the system design.

In this chapter, we examine the energy performance tradeoff for the sleep window adjustment in simplex and duplex VoIP transmissions. Specifically, we give a mathematical model to analyze the performance of the sleep scheme, with the natural talking speed, sleep window size, and energy expenditure taken into consideration. Guided by this model, we locate the optimal window adjustment parameters for the sleep scheme. Extensive simulation results have validated the analytical model, and indicated that, compared with the traditional scheme, the optimized scheme can achieve as much as 90 percent reduction in energy dissipation during the silence period and prolong battery lifetime for about 20 to 30 percent, thus striking a balance between energy conservation and QoS provision.

4.2 WiMAX: Low-Cost Broad Bandwidth for Consumers

The ever-growing demand on high-speed and ubiquitous wireless Internet access has spurred the standard of WiMAX (Worldwide Interoperability for Microwave Access), which aims at providing broad bandwidth at low cost for residential and business areas [1–7]. For years of amending and enhancing efforts by the IEEE 802.16 Working Group, WiMAX has the potential to support high-speed and high-capacity wireless access for both fixed and mobile users, providing mobile Internet, mobile VoIP, and multimedia data services. Due to the high energy consumption involved in these services by the communication and computation units, how to intelligently manage the energy is a hot research topic. Additionally, the mobility support in WiMAX also requires that the carried battery should sustain power as long as possible. To effectively utilize the limited power, sleep mode operation is specified in the standard [11–16].

Sleep mode is a state in which a mobile station (MS) conducts prenegotiated periods of absence from the serving base station (BS) air interface [17, 20–21]. The periods are characterized by the unavailability of the MS, as observed from the serving BS, to DL (downlink) or UL (uplink) traffic [3, 4]. When there is no traffic between BS and MS, through the signaling exchange, the MS can enter sleep mode and power down relevant communication units to save energy. Specifically, the MS can initiate the sleep operation by sending an MOB_SLP-REQ message, which defines the requested sleep profile, to the BS. In this sleep profile, some parameters, such as initial-sleep window, final-sleep window base, final-sleep window exponent, listening window, and start frame number for first sleep window, are included. On receiving the MOB_SLP-REQ message, the BS may comply with this profile as recommended and respond by sending an MOB_SLP-RSP message back to the MS. When MS receives the MOB_SLP-RSP message, it can enter the sleep mode. Meanwhile, during the sleep period, there may be downlink traffic addressed to the MS that is temporally unavailable to the BS. Therefore, the BS must buffer the incoming traffic for the MS. To probe the downlink traffic, from time to time, the MS must wake up in the listening window specified in the profile and receive the MOB_TRF-IND message sent from the BS. If the traffic indication flag in the message is negative, which means no incoming traffic, the MS can again enter the sleep state. On the other hand, if the flag is positive, the sleep mode is deactivated, and the MS must enter active mode to receive packets. The sleep mode also can be deactivated immediately by the MS when there is outgoing traffic to be transmitted through the UL [17, 20].

According to the characteristics of traffics, three power-saving classes have been defined in the standard. Power-saving class (PSC) of type I is recommended for best effort services and nonreal time traffics, such as Web browsing. Due to the bursty behavior of the traffic arrival, the sleep window in PSC-I is doubled each time when the traffic indication flag is negative, until the window size reaches the maximum, thus avoiding unnecessary listening. PSC-II is recommended for unsolicited grant services (UGS) and real-time connections, such as VoIP. To guarantee the QoS, the sleep interval is constant and adjusted according to the arrival interval of the coded VoIP packets. PSC-III is recommended for multicast connections as well as for management operations, and is not discussed in this chapter.

In dealing with energy conservation in WiMAX, much research has been done. Xiao Yang firstly studied the energy consumption and gave a novel analysis model for the IEEE 802.16e broadband wireless access network [1]. In his work, the sleep window is increased exponentially upon no arrival traffic, and the downlink traffic arrival is modeled as a Poisson process. Zhang [2] extended the work by considering both incoming and outgoing traffics, and analyzed energy consumptions with different initial/maximum window adjustments. Periodical listening operation in sleep mode is a significant source of energy wastage. To reduce excessive listening intervals, an enhanced energy-saving mechanism (EESM) for 802.16e was proposed by Xiao, Zou, and Cheng [3] to promote energy efficiency. In EESM, when MS just exits from the previous sleep-mode operation, half of the last sleep interval is adopted as the initial sleep interval in the next sleep mode operation. Simulation results show that the proposed mechanism can obtain better performance in energy conservation. Similarly, Jang and Choi [4] showed that the parameters, such as initial sleep window size and final sleep size, are adaptively adjusted according to the traffic types to achieve power saving. In reducing the energy consumption, Kim [5] also proposed an enhanced power-saving mechanism (EPSM), which adaptively decides the sizes of the initial and final sleep window sizes by taking into account the sleep duration in the previous sleep-mode operation. Additionally, it is shown that EPSM can adaptively control energy consumption and the response delay depending on the remaining energy of an operation MS. The system load and traffic properties also have influence on the energy expenditure. Peng and Wang [6] proposed an adaptive energy-saving mechanism (AESM) to adaptively save energy with the system load and traffic characteristics taken into consideration. This scheme is shown to be effective when the traffic load is light.

All the aforementioned works are various enhancements or analysis based on PSC-I, and, for the sake of analytical simplicity, the Poisson traffic model has been employed in performance evaluation. Nevertheless, Poisson assumption and PSC-I are not applicable to VoIP traffic because: (1) packet arrival in VoIP does not obey Poisson process, and (2) PSC-I is unsuitable for VoIP due to the fact that the delay and packet loss rate constraints cannot be properly handled. To satisfy the QoS requirements, PSC-II, which adopts constant sleep and listening intervals, is recommended for VoIP by the standard. However, according to a speech model recommended by the International Telecommunication Union (ITU) [8], one-way human speech consists of alternating talk-spurt and silence (pause) periods, both complying with negative exponential distribution, with mean 1.004 and 1.587 seconds, respectively. Statistically, total silence periods can amount to nearly 61 percent of the speech in time scale. Therefore, constant listening in PSC-II may incur unnecessary wakeup during the silence period, and, in turn, may expend more energy.

To tackle this problem, a novel hybrid scheme is proposed by Hyun-Ho and Dong-Ho Cho [7]. Specifically, in the hybrid scheme, during the talk-spurt period, fixed listening and sleep intervals are adopted to fit the constant bit rate arrival of the coded voice packets. While in the silence period, exponential sleep window adjustment is used to probe the traffic arrival, thus, reducing unnecessary listening. This scheme is proved to be effective in that the energy consumption can be reduced up to 20 percent [7]. However, there exist several problems that still remain unsolved:

1. The work by Hyun-Ho and Dong-Ho Cho [7] only considers a duplex conversational scenario, and the scheme proposed cannot be applied directly to downlink simplex VoIP traffic without the consideration of the influence of sleep window adjustment on the QoS performance. Given the delay and loss rate constraints, how does one select the sleep window parameters such that the tough QoS requirements can be satisfied? A large sleep window can lead to severe distortion in the voice quality, while a small window can incur more energy consumption. We must strike a balance between QoS and energy expenditure.
2. Moreover, when analyzing the stream delay and drop probability for duplex conversational scenario, the proposed hybrid scheme does not differentiate the wakeup mechanisms for the DL and UL traffics, and ignores the fact that the sleep mode also can be deactivated by the UL traffic at the MS [2, 20]. Therefore, in the performance analysis by Hyun-Ho and Dong-Ho Cho [7], under whatever circumstances, the DL traffic has to be buffered at the BS until the end of the sleep interval. This, inevitably, can lead to analytical errors in the delay and loss rate for the DL traffic.
3. The natural human conversational speed can vary with languages, genders, and personalities. It significantly influences the performance and parameters selected in a power conservation scheme. To maximize energy efficiency, we should take this factor into consideration when adjusting the sleep window size.

In this chapter, based on the hybrid method [7], we try to solve the above problems and strive to find the optimal sleep window parameters for both the simplex and duplex VoIP traffics under the prescribed QoS constraints. In section 4.3, we present the background of the human speech model and the basic idea of the hybrid power saving scheme. To evaluate the performance of the scheme, in section 4.4, we give the mathematical analysis model and derive the delay, loss rate, and number of invalid wakeups. This is followed by the analysis results validated by the simulation experiment in section 4.5. Guided by these results, we also obtain the optimal window adjustment parameters. Finally, we conclude the chapter in section 4.6.

4.3 Human Speech Model and the Hybrid Scheme

4.3.1 Scenario 1: One-Way Voice Communication

One-way voice communication is a simplex transmission of speech from the base station to the mobile user, and only the downlink voice is supported. Normally, it is adopted in many voice broadcast scenarios, such as talk shows, news broadcasting, oral presentations, etc.

In one-way voice communication, the entire speech duration can be divided into adjacent talk-spurt and silence periods, alternating in the time scale. The voice system switches between the talk state and silence state as illustrated in Figure 4.1. According to the human speech model, both the probability density function (PDF) for silence and talk-spurt durations with hangover in monologue can be approximated by two exponential functions [8–9]. Specifically, the cumulative distribution function (CDF) for the duration of talk-spurt can be expressed as $P(t < \tau) = 1 - e^{-\frac{\tau}{T_\lambda}}$, while the CDF for the silence period is $P(t < \tau) = 1 - e^{-\frac{\tau}{T_\lambda}}$, with $T_\lambda = 1.004, T_s = 1.587$, respectively.

4.3.2 Scenario 2: Two-Way Voice Communication

Two-way voice communication is a duplex transmission of conversational speech between two end users, and is one of the basic services supported by WiMAX systems. Based on the conversational speech model suggested by ITU recommendations [8], the two-way voice communication process can be classified into four states: (1) S_1: A - Talk, B - Silence; (2) S_2: A - Silence, B - Silence; (3) S_3: A - Talk, B - Talk; and (4) S_4: A - Silence, B - Talk. The transition among different states is shown in Figure 4.2.

According to the conversational model in ITU recommendations, the duration of each state S_i is a random variable, and the CDF can be approximated by the following exponential functions:

$$\begin{cases} ST: \quad 1 - e^{-\frac{\tau}{T_{st}}} \\ DT: \quad 1 - e^{-\frac{\tau}{T_{dt}}} \\ MS: \quad 1 - e^{-\frac{\tau}{T_{ms}}} \end{cases} \tag{4.1}$$

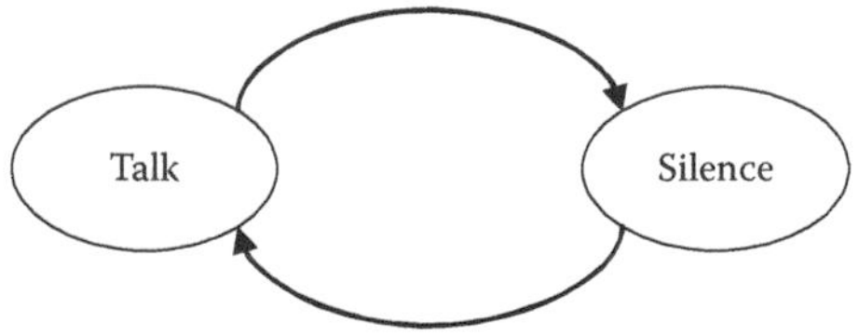

Figure 4.1 State transition in one-way voice communication.

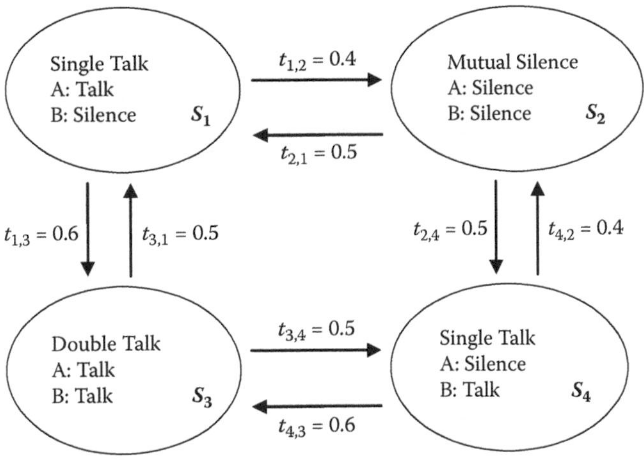

Figure 4.2 State transitions in two-way voice communication.

where T_{st}, T_{dt}, T_{ms} are the mean duration for single talk (ST), double talk (DT), and mutual silence (MS), with typical values 854 ms, 226 ms, and 456 ms, respectively. In Figure 4.2, $t_{i,j}$ is the transition probability from state S_i to state S_j.

4.3.3 Hybrid Scheme

To avoid unnecessary listening during the silence period, in hybrid scheme [7], the sleep window is doubled when the traffic indication is negative, until the window size reaches the maximum. Specifically, the evolution of sleep window is given by:

$$T_n = \begin{cases} T_1, & n = 1 \\ \min(2^{n-1}T_1, \quad T_{\max}), & n > 1 \end{cases} \tag{4.2}$$

where T_1 and $T_{\max}$ are the initial and the maximum sleep window size, respectively.

In the talk-spurt period, the listening and sleep intervals are constant values, and during the listening interval, the MS can receive the downlink VoIP traffic from the BS or transmit the uplink traffic to the BS. Therefore, the hybrid scheme is a combination of PSC-I and PSC-II. The hybrid power-saving scheme is illustrated in Figure 4.3.

4.4 Performance Analysis

VoIP service has tight QoS requirements: average delay and packet loss rate [18]. To avoid severe deterioration of the performance, these two constraints must be properly handled. In the hybrid scheme, the most significant factor influencing the performance is sleep window adjustment. A large sleep window can avoid frequent listening

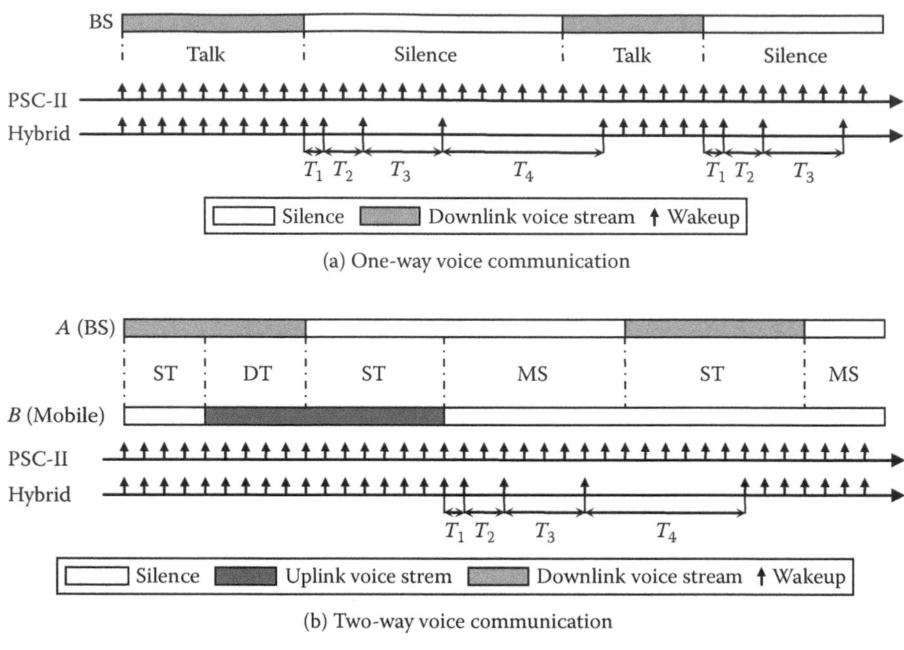

Figure 4.3 Hybrid power-saving scheme.

and, thus, save energy. However, it also can render excessive delay and loss rate. On the other hand, constant listening has better communication performance. However, it is more energy consuming and, thus, undesirable in view of energy efficiency. In system optimization, it is a waste to spend extra energy on achieving more communication performance than needed. Therefore, how to properly tune the sleep window to conserve energy while still meeting the QoS requirement is a challenging issue. To get the optimal tuning parameters, in this section, we give the mathematical analysis of the system performance. Before starting the performance analysis, Table 4.1 summarizes the notations to be used in the rest of this chapter.

4.4.1 Scenario 1: One-Way Voice Communication

4.4.1.1 Average Downlink Talk-Spurt Delay

Talk-spurt delay is defined as the period from the time the first talk-spurt packet reaches the buffer at the BS to the time the MS begins the reception of this talk-spurt packet. We assume that, in the silence period, the sleep window $T_i\ (i \geq 1)$ is adjusted according to equation (4.2). We also specify that $T_{\max} = 2^K T_1$, where K is the maximum sleep window index, which is also the maximum value that i can reach. Without loss of generality, let t_0 be the beginning time of a silence period, which is also the starting point of the first sleep window. For the MS, the wakeup

Table 4.1 Notations

Notation	*Description*
T_l	Mean talk-spurt duration for one-way voice communication
T_s	Mean silence duration for one-way voice communication
T_{st}	Mean single talk duration for two-way conversational speech
T_{dt}	Mean double talk duration for two-way conversational speech
T_{ms}	Mean mutual silence duration for two-way conversational speech
T'_s	Mean UL/DL silence duration for two-way conversational speech
T_f	Frame length
T_1	Initial sleep window size
T_{max}	Maximal sleep window size
K	Maximum sleep window index

time for checking the downlink traffic is t_1, where $i > 0$ and $t_i - t_{i-1} = T_i$. If downlink traffic arrives at time t_1 , where $t \in [t_i, t_{i+1}]$, the talk-spurt delay can be written as $t_{i+1} - t$ because MS must sleep until t_{i+1} to check the downlink traffic indication as illustrated in Figure 4.4. As the sleep window size T_i is doubled each time when the traffic indication flag is negative, t_1 can be expressed as:

$$t_i = \begin{cases} 0, & i = 0 \\ T_1 \sum_{j=0}^{i-1} 2^j, & 1 \le i \le K+1 \\ T_1 \sum_{j=0}^{K} 2^j + (i-K-1)T_{\max}, & i > K+1 \end{cases} \tag{4.3}$$

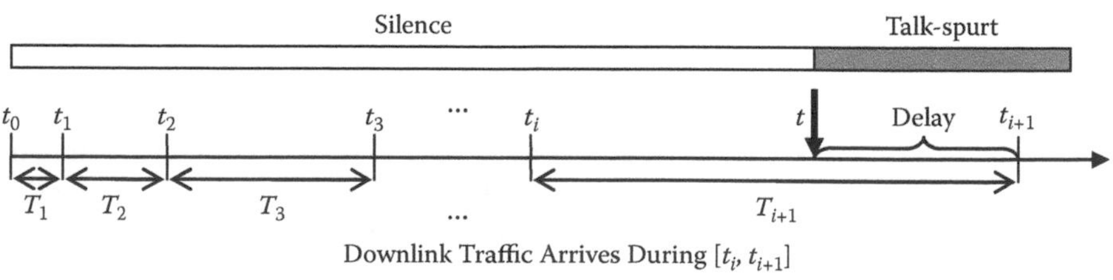

Figure 4.4 Downlink traffic arrival during the *ith* sleep window.

According to the speech model mentioned above, the PDF for the silence period can be written as $\frac{1}{T_s}e^{-\frac{t}{T_s}}$. The average delay for the talk-spurt $E[D]$ can be calculated by:

$$E[D]=\sum_{i=0}^{K}\int_{t_i}^{t_{i+1}}\frac{1}{T_s}e^{-\frac{t}{T_s}}(t_{i+1}-t)dt+\sum_{i=0}^{\infty}\int_{t_{K+1}+iT_{\max}}^{t_{K+1}+(i+1)T_{\max}}\frac{1}{T_s}e^{-\frac{t}{T_s}}(t_{K+1}+(i+1)T_{\max}-t)dt \tag{4.4}$$

The first integral in equation (4.4) can be written as $(t_{i+1}-t_i)e^{-\frac{t_i}{T_s}}+T_s(e^{-\frac{t_{i+1}}{T_s}}-e^{-\frac{t_i}{T_s}})$. The second integral in equation (4.4) can be expressed as $T_{\max}e^{-\frac{t_{K+1}+iT_{\max}}{T_s}}+T_s(e^{-\frac{t_{K+1}+(i+1)T_{\max}}{T_s}}-e^{-\frac{t_{K+1}+iT_{\max}}{T_s}})$ Combining these two integrals, $E[D]$can be rewritten as:

$$\begin{aligned}
E[D]=&\sum_{i=0}^{K}\underbrace{(t_{i+1}-t_i)}_{=T_{i+1}=2^iT_1}e^{-\frac{t_i}{T_s}}+T_s\underbrace{\sum_{i=0}^{K}(e^{-\frac{t_{i+1}}{T_s}}-e^{-\frac{t_i}{T_s}})}_{=-e^{-\frac{t_0}{T_s}}+e^{-\frac{t_{K+1}}{T_s}}=e^{-\frac{t_{K+1}}{T_s}}-1}+T_{\max}e^{-\frac{t_{K+1}}{T_s}}\sum_{i=0}^{\infty}e^{-\frac{iT_{\max}}{T_s}}\\
&+T_se^{-\frac{t_{K+1}}{T_s}}\underbrace{\sum_{i=0}^{\infty}(e^{-\frac{(i+1)T_{\max}}{T_s}}-e^{-\frac{iT_{\max}}{T_s}})}_{=-e^0=-1}
\end{aligned} \tag{4.5}$$

$$=\sum_{i=0}^{K}2^iT_1e^{-\frac{t_i}{T_s}}+T_s(e^{-\frac{t_{K+1}}{T_s}}-1)+T_{\max}e^{-\frac{t_{K+1}}{T_s}}(1-e^{-\frac{T_{\max}}{T_s}})^{-1}-T_se^{-\frac{t_{K+1}}{T_s}}$$

$$=\sum_{i=0}^{K}2^iT_1e^{-\frac{t_i}{T_s}}+T_{\max}e^{-\frac{t_{K+1}}{T_s}}(1-e^{-\frac{T_{\max}}{T_s}})^{-1}-T_s$$

where $T_1, T_{\max}$ can be calculated from equation (4.2), and t_{K+1} can be calculated from (4.3).

4.4.1.2 Loss Rate

When the MS is in a sleep state during the silence period, the incoming downlink VoIP traffic must be buffered at the BS. However, there is a delay constraint imposed on the VoIP stream because (1) buffer capacity at the BS for each MS is limited, and (2) long buffering time can render packet dropping due to delivery expiration. Therefore, we assume that buffer capacity allocated to each MS is B, i.e., the BS can only buffer B seconds' voice stream for the MS.

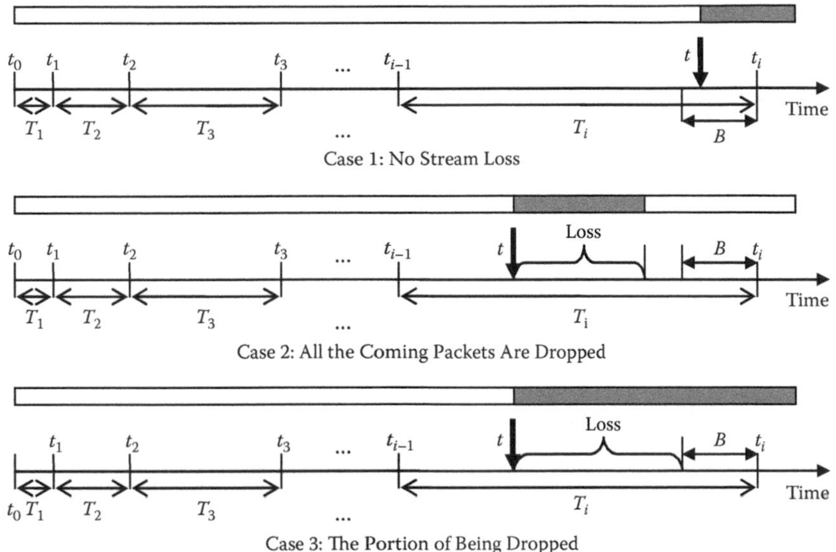

Figure 4.5 Talk-spurt arrival in various time intervals.

Again, let $t_1, t_2, t_3, t_4, \cdots$ be the downlink traffic checking points, and the starting and ending time for the talk-spurt be t and t', respectively. For the analysis of the stream loss, we have three cases, as shown in Figure 4.5.

Case 1: If $t \in [t_i - B, t_i]$, all the coming VoIP packets during this interval are buffered at the BS and there is no stream loss.

Case 2: If $t \in [t_{i-1}, t_i - B]$ and $t' \in [t_{i-1}, t_i - B]$, all the coming VoIP packets are dropped due to expiration.

Case 3: If $t \in [t_{i-1}, t_i - B]$ and $t' \in [t_i - B, \infty]$, the portion of being dropped due to expiration is $t_i - B - t$.

Let $j = \min\arg\{i \mid T_i > B\}$. For Case 2, the loss rate can be calculated as:

$$
\begin{aligned}
E(L_1) &= \sum_{i=j}^{\infty} \int_{t_{i-1}}^{t_i - B} \underbrace{\frac{1}{T_s} e^{-\frac{t}{T_s}}}_{\text{means } t \in [t_{i-1}, t_i - B]} dt \int_{t_{i-1}}^{t_i - B} \underbrace{\frac{1}{T_\lambda} e^{-\frac{t'-t}{T_\lambda}}}_{\text{means } t' \in [t_{i-1}, t_i - B]} \underbrace{(t' - t)}_{\text{lost portion}} dt' \\
&= \sum_{i=j}^{\infty} \int_{t_{i-1}}^{t_i - B} \left(\frac{T_\lambda}{T_s} e^{-\frac{t}{T_s}} - \frac{t_i - B - t + T_\lambda}{T_s} e^{-\frac{t}{T_s} - \frac{t_{i+1} - B - t}{T_\lambda}} \right) dt \\
&= \sum_{i=j}^{\infty} \left(T_\lambda e^{-\frac{t_{i-1}}{T_s}} - \frac{T_\lambda T_s^2}{(T_\lambda - T_s)^2} e^{-\frac{t_i - B}{T_s}} \right) - \frac{(T_\lambda^2 - T_\lambda T_s)(t_i - t_{i-1} - B + T_\lambda) - T_\lambda^2 T_s}{(T_\lambda - T_s)^2} e^{-\frac{(T_\lambda - T_s)t_{i-1} + T_s t_i - T_s B}{T_s T_\lambda}}
\end{aligned}
\tag{4.6}
$$

For Case 3, the loss rate can be calculated as:

$$E(L_2)=\sum_{i=j}^{\infty}\int_{t_{i-1}}^{t_i-B}\underbrace{\frac{1}{T_s}e^{-\frac{t}{T_s}}}_{\text{means } t\in[t_{i-1},t_i-B]}\underbrace{e^{-\frac{t_i-B-t}{T_\lambda}}}_{\text{means } t'-t>t_i-B-t}\underbrace{(t_i-B-t)}_{\text{lostp ortion}}dt$$

$$=\sum_{i=j}^{\infty}\left[\frac{(T_\lambda^2-T_\lambda T_s)(t_i-t_{i-1}-B)-T_\lambda^2T_s}{(T_\lambda-T_s)^2}e^{-\frac{(T_\lambda-T_s)t_{i-1}+T_st_i-T_sB}{T_sT_\lambda}}+\frac{T_sT_\lambda^2}{(T_\lambda-T_s)^2}\cdot e^{-\frac{t_i-B}{T_s}}\right] \tag{4.7}$$

Therefore, the total average loss rate $E(L)$ is the combination of $E(L_1)$ and $E(L_2)$, i.e.,

$$E(L)=E(L_1)+E(L_2)$$
$$=\sum_{i=j}^{\infty}\left[T_\lambda e^{-\frac{t_{i-1}}{T_s}}+\frac{T_\lambda T_s}{T_\lambda-T_s}\cdot e^{-\frac{t_i-B}{T_s}}-\frac{T_\lambda^2}{T_\lambda-T_s}e^{-\frac{(T_\lambda-T_s)t_{i-1}+T_st_i-T_sB}{T_sT_\lambda}}\right] \tag{4.8}$$

4.4.1.3 Average Number of Invalid Wakeups

If the MS wakes up and receives negative traffic indication, this wakeup is invalid. To reduce energy consumption due to an invalid wakeup, in the proposed hybrid scheme, the sleep window is increased exponentially. The average number of invalid wakeups $E(N_w)$ in one silence period can be expressed by:

$$E[N_w]=\sum_{i=0}^{K}i\int_{t_i}^{t_{i+1}}\frac{1}{T_s}e^{-\frac{t}{T_s}}dt+\sum_{i=0}^{\infty}(K+i+1)\int_{t_{K+1}+iT_{max}}^{t_{K+1}+(i+1)T_{max}}\frac{1}{T_s}e^{-\frac{t}{T_s}}dt$$

$$=\underbrace{\sum_{i=0}^{K}i\left(e^{-\frac{t_i}{T_s}}-e^{-\frac{t_{i+1}}{T_s}}\right)}_{=\sum_{i=1}^{K}e^{-\frac{t_i}{T_s}}-Ke^{-\frac{t_{K+1}}{T_s}}}+\underbrace{\sum_{i=0}^{\infty}(K+i+1)\left(e^{-\frac{t_{K+1}+iT_{max}}{T_s}}-e^{-\frac{t_{K+1}+(i+1)T_{max}}{T_s}}\right)}_{=-e^{-\frac{t_{K+1}}{TS}}\left(1-e^{-\frac{T_{max}}{TS}}\right)\left[K\sum_{i=0}^{\infty}e^{-\frac{iT_{max}}{T_s}}+\sum_{i=0}^{\infty}(i+1)e^{-\frac{iT_{max}}{T_s}}\right]} \tag{4.9}$$

$$=\sum_{i=1}^{K}e^{-\frac{t_i}{T_s}}+e^{-\frac{t_{K+1}}{T_s}}\left(1-e^{-\frac{T_{max}}{T_s}}\right)^{-1}$$

4.4.2 Scenario 2: Two-Way Voice Communication

In the two-way conversational speech scenario, if the MS has uplink traffic (outgoing stream) and it is still in sleep mode, the MS can deactivate the sleep mode immediately to send out the uplink traffic. Therefore, there is no delay in the uplink transmission, and we only need to focus on the downlink delay. In addition, as the MS can enter active mode before the predetermined wakeup time due to the arrival of incoming uplink traffic, the downlink delay at the BS can be shortened, and, hence, the performance analysis is different from that by Hyun-Ho and Dong-Ho Cho [7]. Before we derive the mean delay and loss rate, we must first get the distribution of silence duration of one single side T_s' in the conservational speech.

Proposition: The duration that the system stays in some state S_k can be approximately expressed by an exponential function.

Proof: Assume that the current state is S_k, and let p_k be the probability that system transits from state S_k within the observation time T_f (e.g., frame length in this chapter) to another state. Normally, when observation time is sufficiently small, we have $p_k << 1$. Thus, the probability that the system state keeps unchanged within t can be expressed by:

$$p_k(\tau < t) = p_k + (1-p_k)p_k + (1-p_k)^2 p_k + \cdots + (1-p_k)^{[t/T_f]} p_k \tag{4.10}$$

where $[t/T_f]$ denotes the largest integral number no greater than t/T_f. From equation (4.10), we get:

$$p_k(\tau < t) = 1-(1-p_k)^{[t/T_f]+1} \approx 1-(1-p_k)^{t/T_f+1} \tag{4.11}$$

Therefore, the probability density function of the time that the system remains in current state S_k is given by:

$$f_k(t) = \frac{d}{dt} p_k(\tau < t) = -\frac{(1-p_k)\ln(1-p_k)}{T_f}(1-p_k)^{t/T_f} = (1-p_k)\frac{\ln(\frac{1}{1-p_k})}{T_f} e^{-\frac{\ln(\frac{1}{1-p_k})}{T_f}t} \tag{4.12}$$

Let $\lambda_k = \ln(\frac{1}{1-p_k})/T_f$, and $1-p_k \approx 1$, then equation (4.12) can be rewritten as $f_k(t) = \lambda_k e^{-\lambda_k t}$, which is an exponential function. Moreover, the mean duration that the system stays in state S_k is λ_k^{-1}. (End of Proof)

From the above analysis, we have $p_k = 1-e^{-\lambda_k T_f}$, which is the state transition probability from the current state S_k during observation period T_f. With this

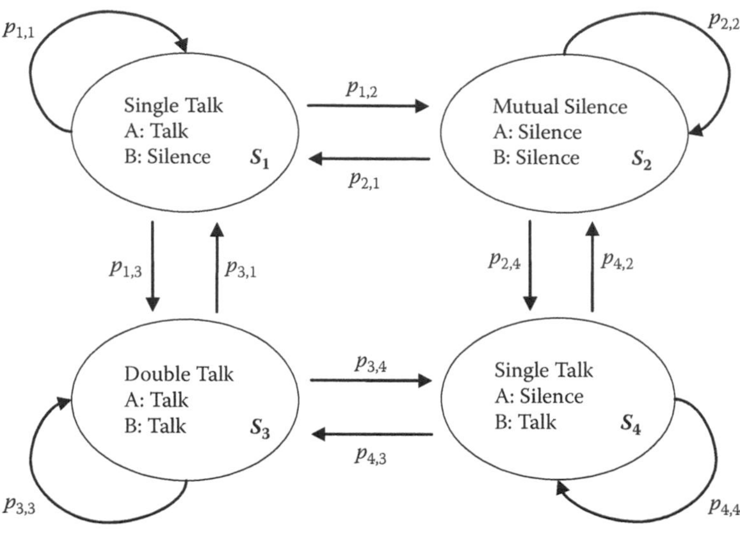

Figure 4.6 Markov model for the two-way conversational speech.

proposition, we can convert the two-way conservation model in Figure 4.2 into a 4-state Markov model as illustrated in Figure 4.6.

In Figure 4.6, the state transition probability can be expressed with matrix $\bar{P} = [p_{i,j}]_{4\times4}$, where

$$\begin{cases} p_{1,1} = 1 - p_1 = e^{-\lambda_1 T_f} = e^{-T_{st}^{-1} T_f} \\ p_{2,2} = 1 - p_2 = e^{-\lambda_2 T_f} = e^{-T_{ms}^{-1} T_f} \\ p_{3,3} = 1 - p_3 = e^{-\lambda_3 T_f} = e^{-T_{dt}^{-1} T_f} \\ p_{4,4} = p_{1,1} \\ p_{i,j} = (1 - p_{i,i}) \times t_{i,j} \quad i \neq j \end{cases} \tag{4.13}$$

Let $\Pi = [\pi_1\ \pi_2\ \pi_3\ \pi_4]$ be the stationary distribution of the four states $\bar{S} = [S_1\ S_2\ S_3\ S_4]$. Thus, we have the following equation group:

$$\begin{cases} \Pi \times \bar{P} = \Pi \\ \sum_{i=1}^{4} \pi_i = 1 \end{cases} \tag{4.14}$$

By solving equation (4.14), we get the stationary probability distribution $\{\pi_i\}$.

We assume the two-way communication sides are A and B. Now, we derive the silence duration distributions of a single side, say, side A. Let $p(A_T|A_S)$ be the transition probability that A transits from silence state to talk state in the observation period T_f. Thus, we have one step transition probability:

$$
\begin{aligned}
p(A_T|A_S) &= p(A_T B_S|A_S) + p(A_T B_T|A_S) \\
&= p(A_T B_S|A_S B_S) \times \frac{\pi(A_S B_S)}{\pi(A_S)} + p(A_T B_T|A_S B_T) \times \frac{\pi(A_S B_T)}{\pi(A_S)} \\
&= p_{2,1} \times \frac{\pi_2}{\pi_2 + \pi_4} + p_{4,3} \times \frac{\pi_4}{\pi_2 + \pi_4}
\end{aligned}
\tag{4.15}
$$

Similarly,

$$
\begin{aligned}
p(A_S|A_T) &= p(A_S B_S|A_T) + p(A_S B_T|A_T) \\
&= p(A_S B_S|A_T B_S) \times \frac{\pi(A_T B_S)}{\pi(A_T)} + p(A_S B_T|A_T B_T) \times \frac{\pi(A_T B_T)}{\pi(A_T)} \\
&= p_{1,2} \times \frac{\pi_1}{\pi_1 + \pi_3} + p_{3,4} \times \frac{\pi_3}{\pi_1 + \pi_3}
\end{aligned}
\tag{4.16}
$$

Moreover, we also have:

$$
\begin{cases}
p(A_S|A_S) = 1 - p(A_T|A_S) \\
p(A_T|A_T) = 1 - p(A_S|A_T)
\end{cases}
\tag{4.17}
$$

With the proposition and analysis above, the silence duration of one single side, say A, can be approximated by an exponential function with PDF given by $\lambda_S e^{-\lambda_S t}$. The mean silence duration $T_S' = \lambda_S^{-1} = -\frac{T_f}{\ln[1 - p(A_T|A_S)]}$.

4.4.2.1 Average Downlink Talk-Spurt Delay

We assume that the MS is in $(i + 1)$ the sleep window, with the predetermined sleep interval $[t_i, t_{i+1}]$. We further assume that there is downlink traffic arrival at the BS during this interval, and the arrival time is t. We need to consider two cases

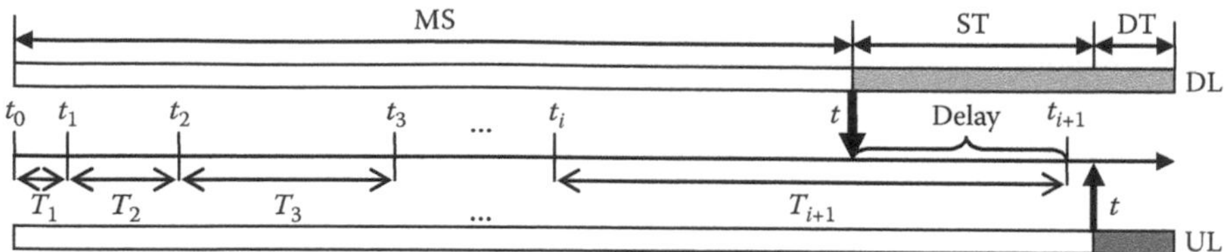

Case 1: Downlink Traffic Arrives Time $t\varepsilon$ $[t_i, t_{i+1}]$, and No Uplink Traffic During $[t, t_{i+1}]$

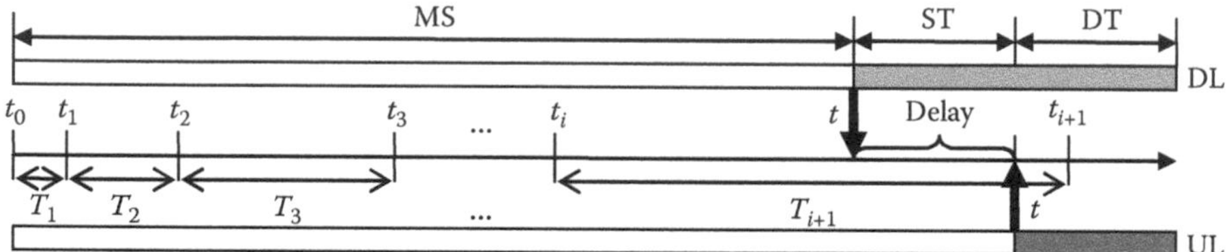

Case 2: Downlink Traffic Arrives Time $t\varepsilon$ $[t_i, t_{i+1}]$, and Uplink Traffic Arrives During $[t, t_{i+1}]$

Figure 4.7 Downlink delay in a two-way conversational speech scenario.

as illustrated in Figure 4.7. Let t and t' be the arrival time of downlink and uplink traffic, respectively.

In case 1, during $[t, ti + 1]$, there is no uplink traffic from the mobile user and we have $t \in [t_i, t_{i+1}]$ and $t' \in [t_{i+1}, \infty]$, thus, the downlink traffic has to be buffered at the BS until the MS wakes up at time t_{i+1} to check the downlink traffic indication. The downlink delay in this case can be written as $t_{i+1} - t$. Therefore, the mean delay can be expressed as:

$$E[D_1] = \sum_{i=0}^{\infty} \int_{t_i}^{t_{i+1}} \frac{1}{T_{ms}} e^{-\frac{t}{T_{ms}}} dt \int_{t_{i+1}}^{\infty} \frac{1}{T_s'} e^{-\frac{t'-t}{T_s'}} (t_{i+1} - t) dt' \tag{4.18}$$

In equation (4.18), T_s' is the mean silence duration of the uplink stream in the conservational speech, which has been derived above.

In case 2, there is uplink traffic coming during $[t, ti + 1]$, thus, MS terminates the sleep mode ahead of the prescheduled wakeup time t_{i+1}. In this case, we have $t \in [t_i, t_{i+1}]$ and $t' \in [t, t_{i+1}]$, and the downlink delay is $t' - t$. The delay can be written as:

$$E[D_2] = \sum_{i=0}^{\infty} \int_{t_i}^{t_{i+1}} \frac{1}{T_{ms}} e^{-\frac{t}{T_{ms}}} dt \int_{t}^{t_{i+1}} \frac{1}{T_s'} e^{-\frac{t'-t}{T_s'}} (t' - t) dt' \tag{4.19}$$

The mean downlink traffic delay is the sum of $E[D_1]$, and $E[D_2]$ can be expressed as:

$$
\begin{aligned}
E[D] &= E[D_1] + E[D_2] \\
&= \sum_{i=0}^{\infty}\int_{t_i}^{t_{i+1}} \frac{1}{T_{ms}} e^{-\frac{t}{T_{ms}}} dt \int_{t_{i+1}}^{\infty} \frac{1}{T_s'} e^{-\frac{t'-t}{T_s'}} (t_{i+1}-t)dt' + \sum_{i=0}^{\infty}\int_{t_i}^{t_{i+1}} \frac{1}{T_{ms}} e^{-\frac{t}{T_{ms}}} dt \int_{t}^{t_{i+1}} \frac{1}{T_s'} e^{-\frac{t'-t}{T_s'}} (t'-t)dt' \\
&= \sum_{i=0}^{\infty} \frac{T_s'}{T_{ms}} \int_{t_i}^{t_{i+1}} e^{-\frac{t}{T_{ms}}} dt - \sum_{i=0}^{\infty} \frac{T_s'}{T_{ms}} \int_{t_i}^{t_{i+1}} e^{\left(\frac{1}{T_s'}-\frac{1}{T_{ms}}\right)t - \frac{t_{i+1}}{T_s'}} dt \qquad (4.20) \\
&= T_s' - \frac{T_s'^2}{T_{ms}} \left[\sum_{i=0}^{\infty} e^{-\frac{t_{i+1}}{T_{ms}}} - \sum_{i=0}^{\infty} \left(e^{-\frac{t_i}{T_{ms}} + \frac{t_i - t_{i+1}}{T_s'}} \right) \right]
\end{aligned}
$$

4.4.2.2 Loss Rate

The base station can accommodate B seconds' VoIP downlink stream for the mobile user, i.e., the maxim buffer length at BS is B. Again, let t and t' be the arrival time of downlink and uplink traffic, respectively. Obviously, if $t \in [t_{i+1} - B, t_{i+1}]$, there is no traffic loss. Thus, with respect to stream loss, we only need to consider two cases, as shown in Figure 4.8.

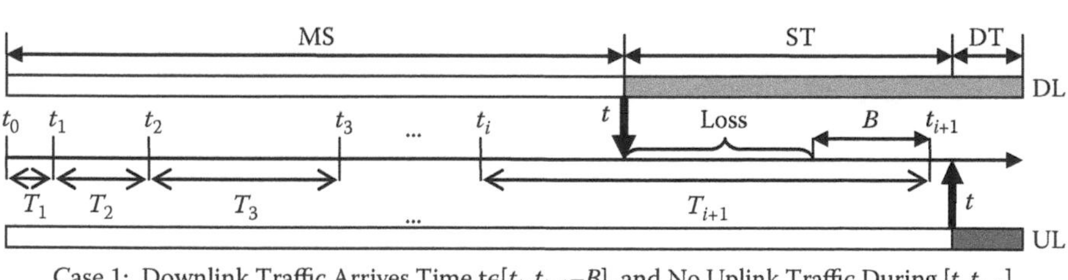

Case 1: Downlink Traffic Arrives Time te$[t_i, t_{i+1}-B]$, and No Uplink Traffic During $[t, t_{i+1}]$

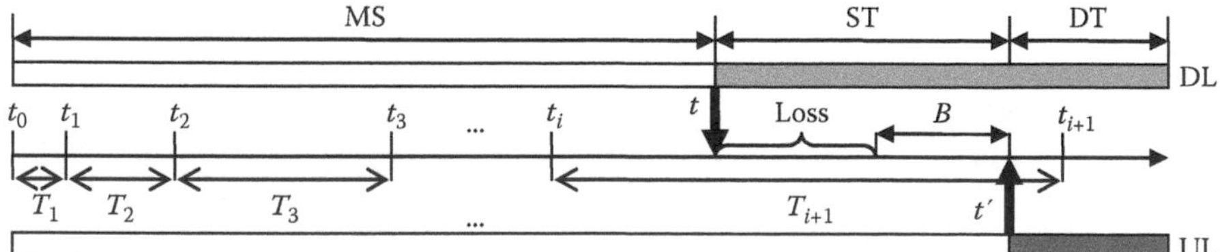

Case 2: Downlink Traffic Arrives Time te $[t_i, t_{i+1}-B]$, and Uplink Traffic Arrives During $[t, t_{i+1}]$

Figure 4.8 Downlink stream loss in a two-way conversational speech scenario.

In case 1, downlink traffic arrives during $[t_1, t_{i+1} - B]$, and there is no uplink traffic during $[t, t_{i+1}]$, i.e., $t \in [t_i, t_{i+1} - B]$ and $t' \in [t_{i+1}, \infty]$. Therefore, the amount of stream loss is $t_{i+1} - B - t$. Let $j = \min \arg\{i | T_i > B\}$. The average loss can be expressed as:

$$E[L_1] = \sum_{i=j}^{\infty} \int_{t_i}^{t_{i+1}-B} \frac{1}{T_{ms}} e^{-\frac{t}{T_{ms}}} dt \int_{t_{i+1}}^{\infty} \frac{1}{T_s'} e^{-\frac{t'-t}{T_s'}} (t_{i+1} - B - t) dt'$$

$$= \sum_{i=j}^{\infty} \frac{T_s'^2 T_{ms}}{(T_{ms} - T_s')^2} e^{\frac{(T_{ms}-T_s')(t_{i+1}-B) - T_{ms} t_{i+1}}{T_s' T_{ms}}}$$

$$- \sum_{i=j}^{\infty} \frac{T_s'}{(T_{ms} - T_s')} \left(t_{i+1} - t_i - B + \frac{T_s' T_{ms}}{T_{ms} - T_s'} \right) e^{\frac{(T_{ms}-T_s')t_i - T_{ms} t_{i+1}}{T_s' T_{ms}}} \tag{4.21}$$

In case 2, downlink traffic arrives during $[t_i, t_{i+1} - B]$, and we have outgoing uplink traffic during $[t + B, t_{i+1}]$, i.e., $t \in [t_i, t_{i+1} - B]$ and $t' \in [t + B, t_{i+1}]$. The average loss can be written as:

$$E[L_2] = \sum_{i=j}^{\infty} \int_{t_i}^{t_{i+1}-B} \frac{1}{T_{ms}} e^{-\frac{t}{T_{ms}}} dt \int_{t+B}^{t_{i+1}} \frac{1}{T_s'} e^{-\frac{t'-t}{T_s'}} (t' - B - t) dt'$$

$$= \sum_{i=j}^{\infty} \frac{2T_s'^2 T_{ms} - T_s'^3}{(T_{ms} - T_s')^2} e^{\frac{(T_{ms}-T_s')(t_{i+1}-B) - T_{ms} t_{i+1}}{T_s' T_{ms}}}$$

$$- \sum_{i=j}^{\infty} [(t - t_{i+1} + B)(T_s' T_{ms} - T_s'^2) - 2T_s'^2 T_{ms} - T_s'^3] \, e^{\frac{(T_{ms}-T_s')t_i - T_{ms} t_{i+1}}{T_s' T_{ms}}}$$

$$+ \sum_{i=j}^{\infty} T_s' e^{\frac{-B}{T_s'}} \left(e^{-\frac{t_i}{T_{ms}}} - e^{-\frac{t_{i+1}-B}{T_{ms}}} \right) \tag{4.22}$$

Totally, the mean downlink traffic loss is the sum of $E[L_1]$ and $E[L_2]$:

$$E[L] = E[L_1] + E[L_2] \tag{4.23}$$

4.4.2.3 Average Number of Invalid Wakeup

In the two-way conservational speech scenario, the sleep window size is increased exponentially during the mutual silence period. Therefore, the average number of invalid wakeup $E(N_w)$ in one mutual silence period can be calculated as:

$$E[N_w] = \sum_{i=0}^{\infty} i \int_{t_i}^{t_{i+1}} \frac{1}{T_{ms}} e^{-\frac{t}{T_{ms}}} dt$$

$$= \sum_{i=0}^{K} i \int_{t_i}^{t_{i+1}} \frac{1}{T_{ms}} e^{-\frac{t}{T_{ms}}} dt + \sum_{i=0}^{\infty} (K+i+1) \int_{t_{K+1}+iT_{\max}}^{t_{K+1}+(i+1)T_{\max}} \frac{1}{T_{ms}} e^{-\frac{t}{T_{ms}}} dt$$

$$= \sum_{i=1}^{K} e^{-\frac{t_i}{T_{ms}}} + e^{-\frac{t_{K+1}}{T_{ms}}} (1 - e^{-t_{K+1}\frac{T_{\max}}{T_{ms}}})^{-1} \tag{4.24}$$

4.5 Numerical Results and Performance Analysis

The tuning of the sleep window size significantly influences the performance of the hybrid scheme in the QoS provision and energy consumption. In this section, we change the initial and maximum sleep window, and present the numerical results on average talk-spurt delay, loss rate, and number of invalid wakeup. We hope to spend the least amount of energy, which is just sufficient to meet the QoS requirements. To validate the accuracy of performance analysis in section 4.4, we also perform extensive simulations.

According to the QoS specifications in ITU document G.114 [10], for satisfactory transmission quality, the maximum end-to-end delay of VoIP should not exceed 250 ms. The factors contributing to the end-to-end delay include speech processing delay, codec delay, and backbone transmission delay [22]. Therefore, in system design, we should leave some redundancy for these delay components and require that the average talk-spurt delay should not exceed 200 ms. In addition, to avoid severe speech quality distortion at MS, the maximum packet loss rate must be less than 10 percent.

4.5.1 Scenario 1: One-Way Voice Communication

At the transmitting side, the speech is sampled, coded, and encapsulated into IP (Internet protocol) streaming packets according to the format specifications [19]. We assume that, during the talk-spurt period, the VoIP packet is generated for every 20 ms. Accordingly, during this active duration, the MS should also wake up

Table 4.2 Simulation Parameters for Scenario 1

Parameter	*Value*
Listening Power	1.5 W
Sleep Power	0.05 W
Frame Length T_f	5 ms
Mean Talk-spurt Period T_l	1.004 s
Mean Silence Period (Base) T_s	1.587 s
Initial Sleep Window Size T_1	20 ms
Delay Tolerance	<200 ms
Loss Tolerance	<10%
Buffering Length B	100 ms
Initial Battery Energy Level	1000 Joule

every 20 ms to receive the downlink traffic. In the silence period, the sleep window is adjusted exponentially according to equation (4.2). After the sleep window, the MS wakes up to receive the downlink traffic indication broadcast from the BS. We assume that the reception of the traffic indication needs 1 ms. If negative indication is presented, the MS does not need to receive the rest frame and goes directly into the sleep state. Otherwise, if positive indication is presented, the MS should keep active for the whole frame length (5 ms) to receive its downlink packets. The physical parameters for analysis and simulation are listed in Table 4.2. We write MATLAB code to compute the performance according to the above derivations. To validate the accuracy of the derivations, we also write simulation codes with C language. All the codes can be downloaded from http://cie.szu.edu.cn/uploads/userup/63/1305361619-1230.rar

4.5.1.1 Average Talk-Spurt Delay

Figure 4.9 and Figure 4.10 are the mean downlink talk-spurt delay versus the varied maximum sleep window exponent, with the initial window exponents 2 and 4, respectively.

The initial sleep window T_1 in equation (4.2) is calculated by:

$$\textit{Initial Sleep Window} = \textit{Frame Length} \times 2^{\textit{Initial Window Exponent}} \quad (4.25)$$

The maximum/final sleep window $T_{\max}$ is calculated by:

$$\textit{Maximum Sleep Window} = \textit{Frame Length} \times 2^{\textit{Maximum Window Exponent}} \quad (4.26)$$

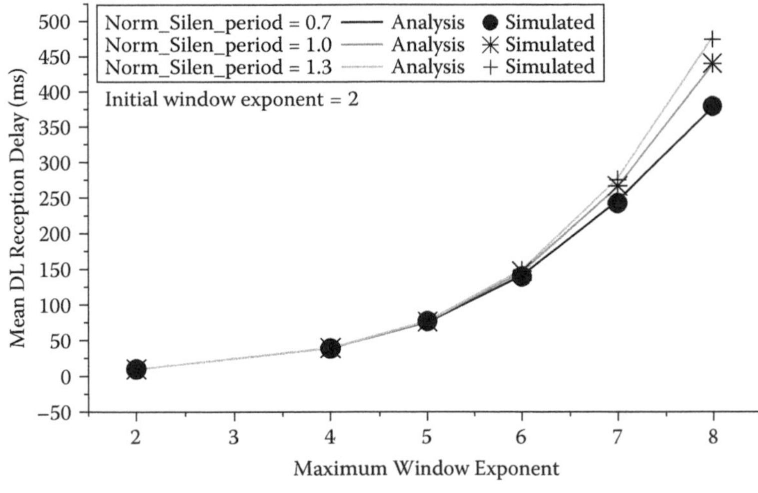

Figure 4.9 Mean talk-spurt delay versus maximum sleep window exponent (initial window exponent = 2).

Thus, K in equation (4.3) can be expressed as:

$$K = \textit{Maximum Window Exponent} - \textit{Initial Window Exponent} \quad (4.27)$$

According to the above equations, the initial sleep windows in Figure 4.9 and Figure 4.10 are 20 ms and 80 ms, respectively, while the maximum sleep window in both figures is 1,280 ms (with $K = 8$). Note that the average duration of the silence period can vary in terms of language, gender, or even individual personality. In our simulations, we change the silence duration by multiplying the mean silence period in Table 4.2 by "Norm_Silen_period," which reflects the speed of natural human speech.

In both figures, we can see that the simulated results match well with the analytical results, which validates the accuracy of the delay model in section 4.4. The mean delay increases with the maximum window. This is natural because the MS spends more time on sleeping, resulting in more delay in the traffic reception. As expected, for the same maximum window exponent, the mean delay increases with the silence period due to the increase in the interval of two consecutive talk-spurts. As the BS can buffer stream for 100 ms, to reduce unnecessary wakeup, we further increase the initial window exponent from 2 to 4, which corresponds to the first sleep window of 80 ms (exponent of 5 or above is not selected because it might render stream loss in exactly the first sleep window). It is observed that the mean delay has a slight increase when the initial sleep window is prolonged, as shown in Figure 4.10. In both figures, a valuable observation is that, if mean talk-spurt delay of less than 200 ms can be satisfied, the maximum window exponent should not exceed 6, i.e., maximum sleep window $T_{max} = 320\ ms$.

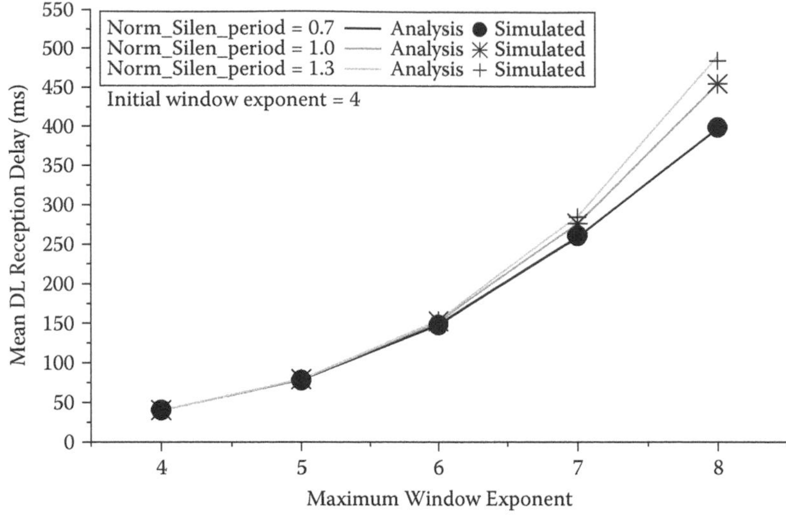

Figure 4.10 Mean talk-spurt delay versus maximum sleep window exponent (initial window exponent = 4).

4.5.1.2 Stream Loss Rate

Figure 4.11 and Figure 4.12 are the mean stream loss rate versus the maximum sleep window. Again, the simulation results fit quite well with the analytical lines. We can observe that a larger maximum window exponent can lead to more stream loss due to longer sleeping period. Similarly, the increase in the silence period can incur more loss, which is more obvious when the maximum window exponent is large. In the analysis and simulations, we have set the buffering length to 100 ms, thus no stream is lost if the maximum window exponent does not exceed 4. Comparing Figure 4.12 with Figure 4.11, we find that increasing the initial window exponent from 2 to 4 has a slight influence on the loss rate. It is illustrated in Figure 4.11 and Figure 4.12 that, to keep the loss rate less than 10 percent, the maximum window exponent should not exceed 6, or equivalently, the final sleep window is 320 ms.

4.5.1.3 Number of Invalid Wakeups Per Sleep Period

We calculate the average number of invalid wakeups according to equation (4.9) and perform simulations to validate the accuracy. These results are plotted in Figure 4.13 and Figure 4.14. From the figures, it is shown that, with the increase of the maximum window exponent, the number of invalid wakeups significantly decreases because the sleeping period is prolonged. For the same maximum window

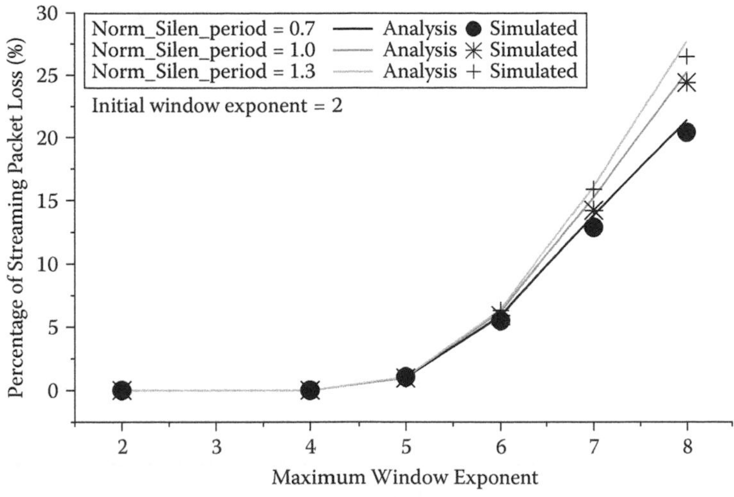

Figure 4.11 Loss rate versus maximum sleep window exponent (initial window exponent = 2).

exponent, the extension in the traffic silence period results in more invalid wakeups because the MS more frequently probes the traffic arrival. In Figure 4.14, the initial window exponent increases from 2 to 4, and it is observed that the number of invalid wakeups is considerably reduced. Thus, an initial exponent of 4 can further enhance the energy efficiency.

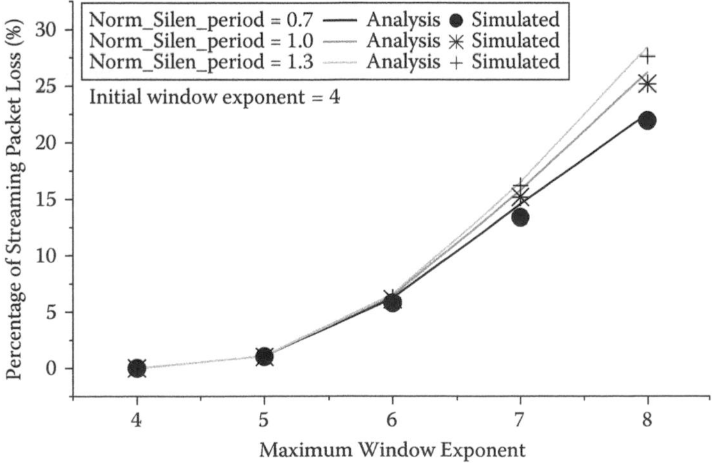

Figure 4.12 Loss rate versus maximum sleep window exponent (initial window exponent = 4).

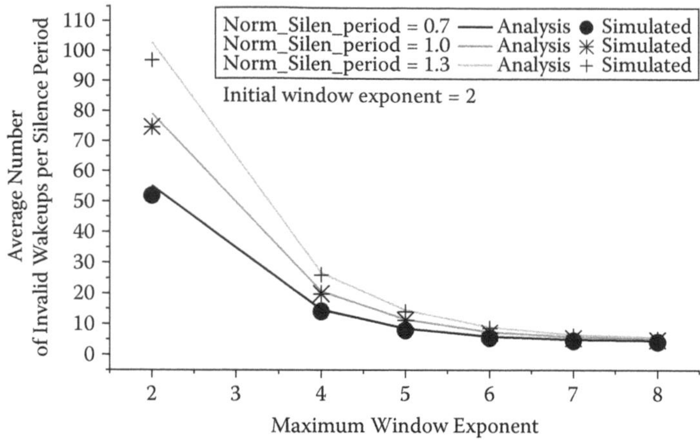

Figure 4.13 Number of invalid wakeups versus maximum sleep window exponent (initial window exponent = 2).

4.5.1.4 Energy-Saving Effects

From the above discussion, we know that, to provide satisfactory QoS for the downlink VoIP, the maximum window exponent should be set to 6. In addition, as the BS has a buffering length of 100 ms for each VoIP stream, the initial sleep window exponent can be set to 4 (corresponding to the sleep window size of 80 ms). Therefore, given the prescribed QoS requirement, to conserve the energy to the best

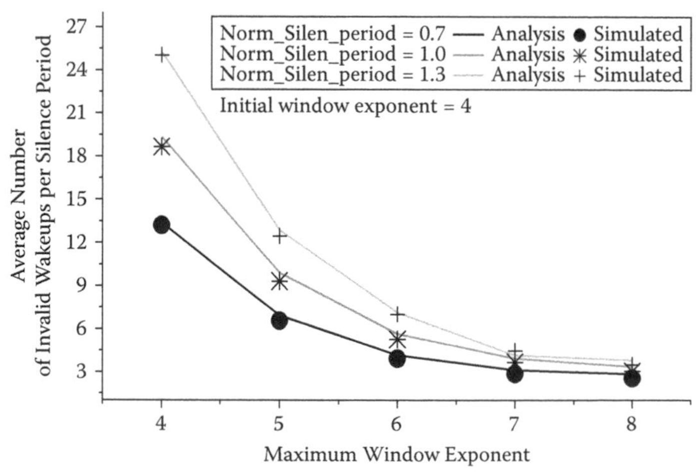

Figure 4.14 Number of invalid wakeups versus maximum sleep window exponent (initial window exponent = 4).

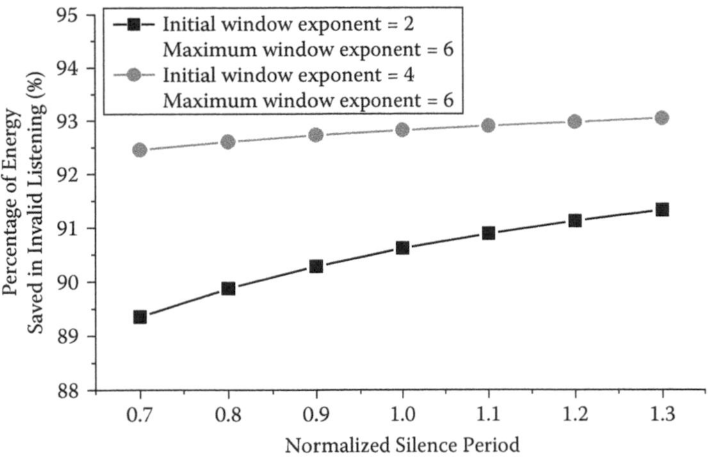

Figure 4.15 Percentage of energy saves in listening versus normalized silence period.

effort, [Initial Window Exponent = 4, Maximum Window Exponent = 6] is the optimal working parameter for the hybrid scheme.

To demonstrate the energy conserving effect, we use PSC-II as the baseline. As the packet is generated every 20 ms in the speech active period, to fit this arrival rate, the constant sleep interval in PSC-II is also set to 20 ms.

Figure 4.15 shows the percentage of energy saved for invalid listening to the downlink traffic indication in the speech silence period versus the normalized silence period. We can see that the hybrid scheme can reduce energy wasted in unnecessary wakeup for about 89 to 94 percent. The gain is more obvious with the increase of the speech silence period.

To see the effect of energy conservation more clearly, we perform simulations to test how long the battery life can be extended with the hybrid scheme. The initial battery capacity and the energy consumption in different states are listed in Table 4.2. Figure 4.16 shows the life extension versus normalized silence period. Due to the reduction in unnecessary listening in the speech silence period, compared with traditional PSC-II, the battery life in the hybrid scheme can be extended for about 14 to 28 percent. The improvement is more obvious with a longer silence period. This proves the effectiveness of the hybrid scheme.

4.5.2 Scenario 2: Two-Way Voice Communication

In the two-way conversational speech scenario, we use the model given in Figure 4.2, which is suggested by the ITU document. Some parameters listed in Table 4.3, such as T_{st} , T_{dt} , and T_{ms} are the typical values adopted by the ITU

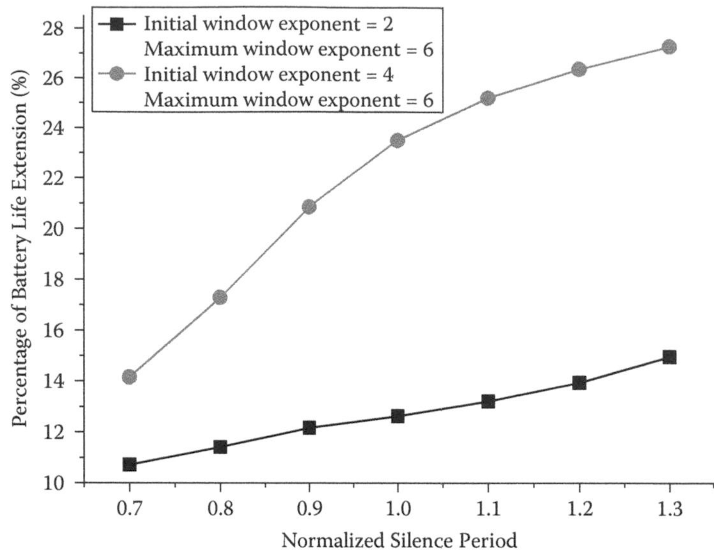

Figure 4.16 Battery life extension versus normalized silence period.

Table 4.3 Simulation parameters for scenario 3

Parameter	*Value*
Listening Power	1.5 W
Transmit Power	1.66 W
Sleep Power	0.05 W
Frame Length T_f	5 ms
Initial sleep window size T_1	20 ms
Mean single talk duration T_{st}	854 ms
Mean double talk duration T_{dt}	226 ms
Mean mutual silence duration T_{ms}	456 ms
Mean UL/DL silence duration T'_s (base)	1219.51 ms
Delay Tolerance	<200 ms
Loss Tolerance	<10%
Buffering Length B	100 ms
Initial Battery Energy Level	1000 Joule

recommendations. To have an insight into the impact of human speech speed on the performance, we also alter the mean one side silence duration (i.e., UL or DL mean silence duration) by weighting this value with "Norm_Silen_period," with its value varying from 0.7 to 1.3.

4.5.2.1 Average Downlink Delay

Figure 4.17 and Figure 4.18 are the mean DL reception delay with the maximum sleep window exponent, with the initial window exponent starting from 2 and 4, respectively. We can see that the simulated results fit quite well with the analytical values, thus validating the accuracy of theoretic analysis.

When we increase "Norm_Silen_period," the downlink reception delay also increases. This is due to the fact that the increase in the silence period can result in lower talking speed, thus leading to a larger sleep window size. As expected, a larger initial window exponent (exponent of 4, Figure 4.18) can incur some extra reception delay.

It is observed in Figure 4.17 and Figure 4.18 that, to satisfy the prescribed delay requirements in Table 4.3, the maximum window exponent is 7 when "Norm_ Silen_period" is larger than 1.0; however, when "Norm_Silen_period" is smaller than or equal to 1.0, the maximum window exponent is 9. This phenomenon is easy to understand: smaller "Norm_Silen_period" means faster talking speed, and,

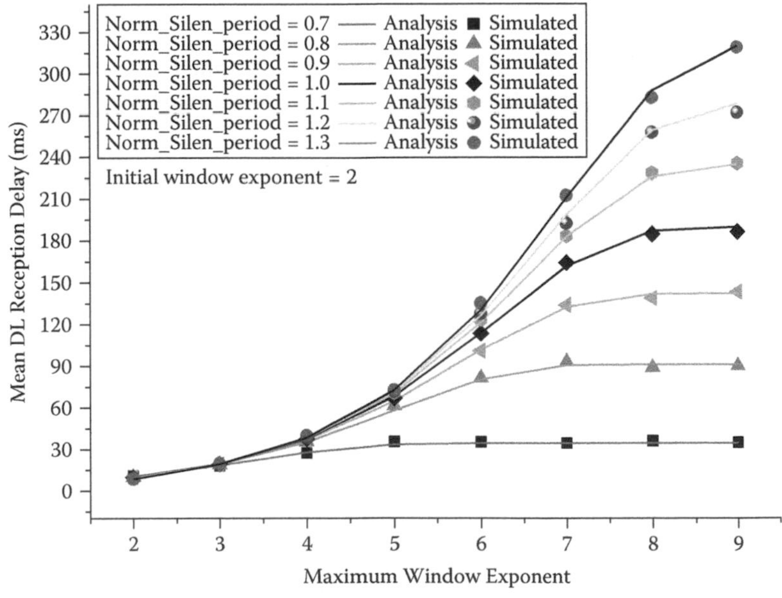

Figure 4.17 Mean DL delay versus maximum sleep window exponent (initial window exponent = 2).

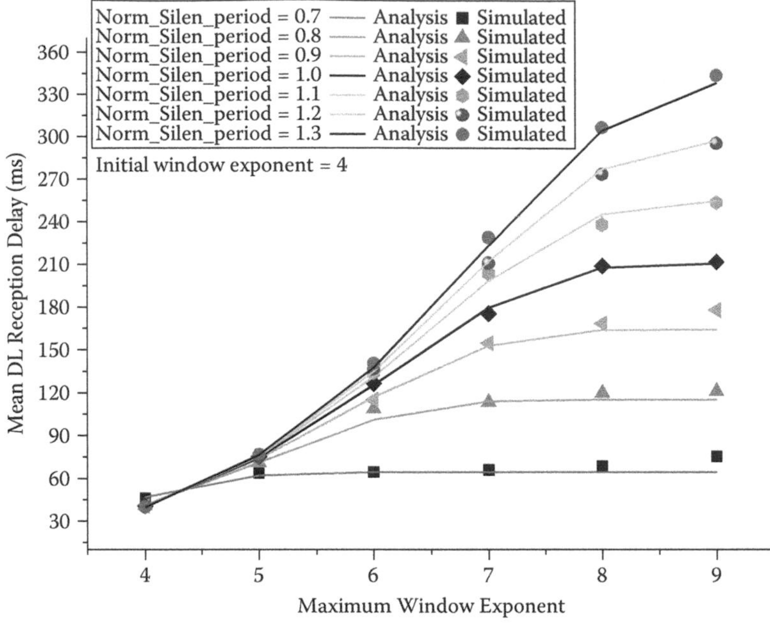

Figure 4.18 Mean DL delay versus maximum sleep window exponent (initial window exponent = 4).

hence, more frequently, the sleep mode at the mobile terminal is deactivated by the outgoing uplink stream, thus, loosing the constraint on the large window exponent.

4.5.2.2 *Stream Loss Rate*

Results of downlink stream loss versus maximum sleep window exponent are given in Figure 4.19 and Figure 4.20. From the results, it is shown that smaller value of "Norm_Silen_period" can lead to less stream loss. This is because a shorter silence interval results in higher talking speed, and, therefore, the sleep state of the mobile terminal is more often interrupted by the UL outgoing traffic, thus reducing the downlink loss due to oversleep. In both figures, it is observed that, given the prescribed loss tolerance of 10 percent in Table 4.3, for "Norm_Silen_period" larger than or equal to 1.3, the maximum window exponent allowed is 8, while for "Norm_Silen_period" smaller than 1.3, the maximum window exponent allowed is 9.

4.5.2.3 *Number of Invalid Wakeups Per Sleep Period*

Figure 4.21 and Figure 4.22 are the mean number of invalid wakeups per mutual silence period, with initial window exponent of 2 and 4, respectively. As the buffer

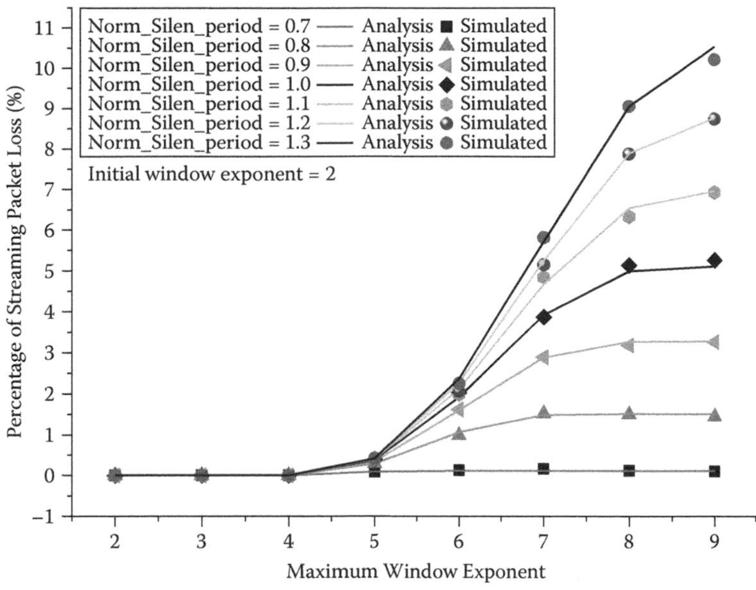

Figure 4.19 DL stream loss rate versus maximum sleep window exponent (initial window exponent = 2).

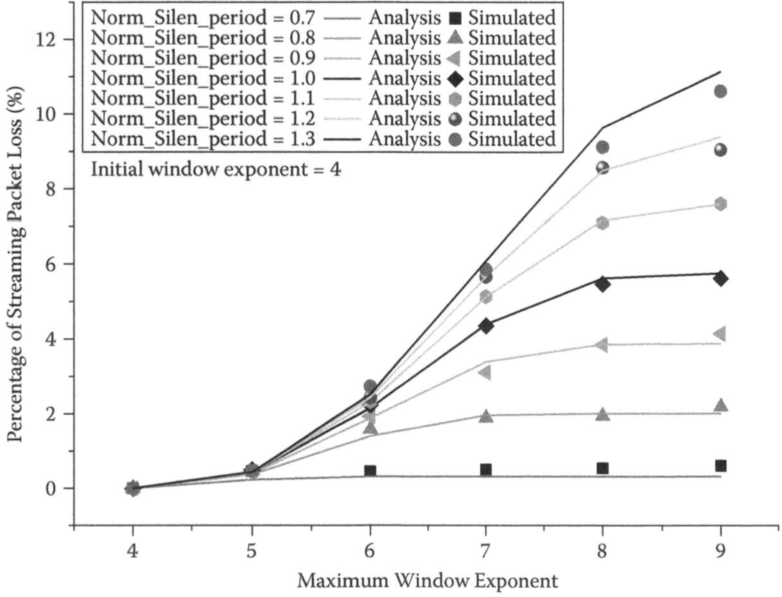

Figure 4.20 DL stream loss rate versus maximum sleep window exponent (initial window exponent = 4).

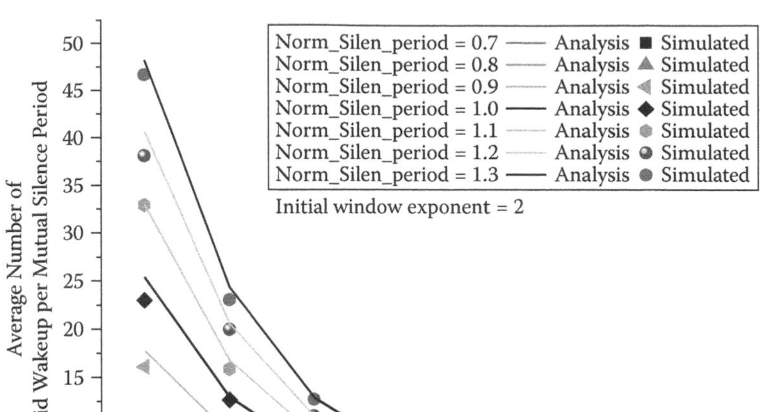

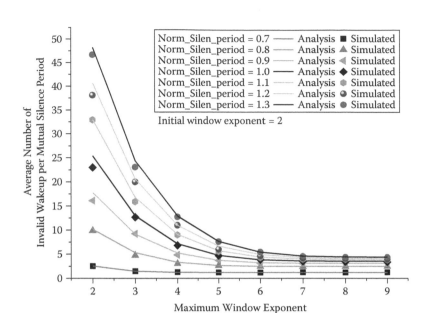

Figure 4.21 Number of invalid wakeups versus maximum sleep window exponent (initial window exponent = 2).

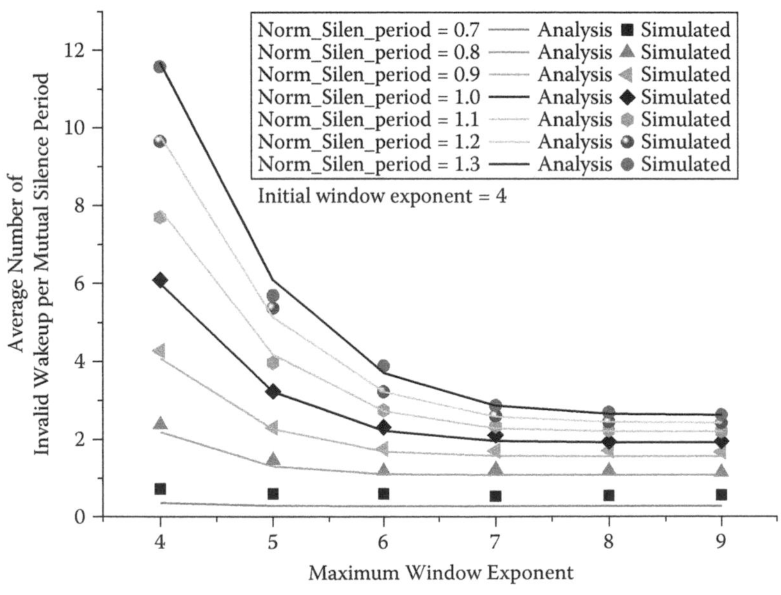

Figure 4.22 Number of invalid wakeups versus maximum sleep window exponent (initial window exponent = 4).

at the base station can temporarily accommodate 100 ms' stream for each voice connection, the initial window exponent of 4 (corresponding to window size of 80 ms) can be an optimal value for the starting window exponent. This has been confirmed in the simulated and analytical results, in that, the average number of invalid wakeups has considerably decreased in Figure 4.22.

4.5.2.4 Energy-Saving Effects

Given the prescribed QoS constraints in Table 4.3, i.e., [delay tolerance ≤200 ms, loss tolerance ≤10 percent], in the selection of maximum window exponent for the two-way conversation speech scenario, the "Norm_Silen_period" also should be taken into consideration. With the above analysis, we get the optimal value for the maximum window exponent (with both delay and loss constraints being satisfied):

$$\textit{Maximum Window Expont} = \begin{cases} 9 & \textit{Norm\_Silen\_period} \leq 1.0 \\ 7 & \textit{Norm\_Silen\_period} \geq 1.1 \end{cases} \tag{4.28}$$

Therefore, we have the optimal energy conservation policies for the two-way conservational speech scenario:

Strategy 1:

$$\begin{bmatrix} \textit{Iinitial Window Exponent}, \\ \textit{Maximum Window Exponent} \end{bmatrix} = \begin{cases} [2,9] & \textit{Norm\_Silen\_period} \leq 1.0 \\ [2,7] & \textit{Norm\_Silen\_period} \geq 1.1 \end{cases} \tag{4.29}$$

and,

Strategy 2:

$$\begin{bmatrix} \textit{Iinitial Window Exponent}, \\ \textit{Maximum Window Exponent} \end{bmatrix} = \begin{cases} [4,9] & \textit{Norm\_Silen\_period} \leq 1.0 \\ [4,7] & \textit{Norm\_Silen\_period} \geq 1.1 \end{cases} \tag{4.30}$$

The difference between strategy 1 and strategy 2 is that a different initial window exponent has been selected. Strategy 2 has exploited the buffer at the base station to cache the stream, thus reducing the number of invalid wakeups and saving energy to the best effort. However, as shown in the above results, the cost to pay is that some extra loss and delay can incur.

Again, we use PSC-II as the baseline, in which the mobile station wakes up every 20 ms to check the DL traffic or to send its outgoing UL traffic, ignoring the mutual silence period. The voice coding speed rate is 50 frames/sec during the active period. The effects of energy saving are illustrated in Figure 4.23 and

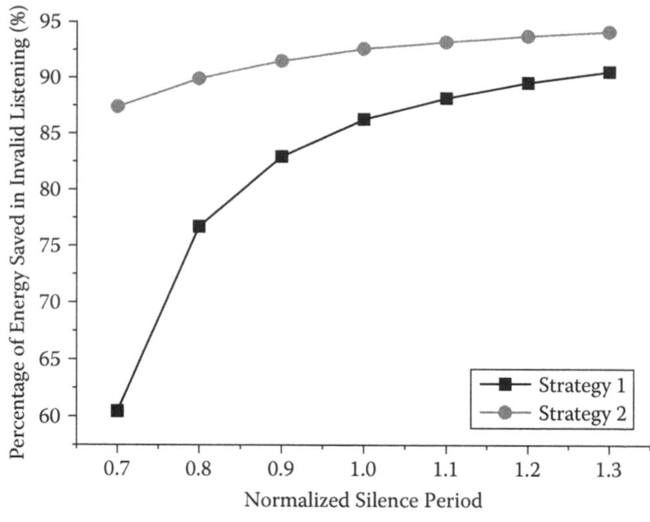

Figure 4.23 Percentage of energy saves in invalid listening versus normalized silence period.

Figure 4.24. The percentage of energy saved in invalid listening during the mutual silence period versus "Normalized Silence Period" is given in Figure 4.23. From the figure, it is illustrated that we can achieve energy reduction in invalid wakeups for about 60 to 90 percent (depending on the strategy selected and the value of "Normalized Silence Period").

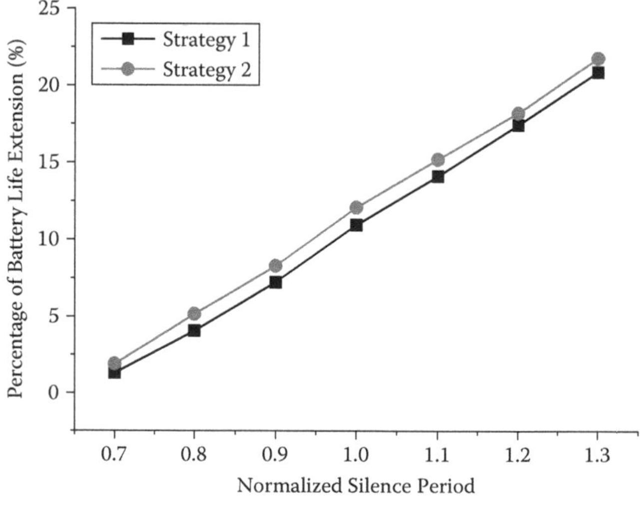

Figure 4.24 Battery life extension versus normalized silence period.

Figure 4.24 is the battery life extension versus "Normalized Silence Period" for both policies. We can observe that the larger the "Normalized Silence Period," the more obvious the energy conservation effect. On average, the battery life can be extended for about 5 to 23 percent, depending on the talking speed of the conversational speech. Compared with the corresponding values in scenario 1, the energy conservation effect is less obvious. The reason for this is that the effect of the hybrid scheme significantly relies on the portion of silence duration in the whole conservation period. In a two-way conversational speed scenario, the portion of mutual silence duration is lower than the portion of silence duration in the one-way voice scenario. Thus, the room for energy-efficiency enhancement is smaller. Even so, we can still considerably enhance energy efficiency with the hybrid scheme.

4.5.2.5 Performance Comparisons with and without Sleep Mode Deactivation

For the two-way conservational speech scenario, when there is outgoing UL voice traffic in the sleep window, the MS wakes up immediately by deactivating the current sleep mode [2, 17]. Therefore, if concurrently, there is DL traffic buffered at the BS, this DL traffic does not need to be buffered to the end of the sleep interval, and can be transmitted directly to the MS. This sleep mode deactivation function has been ignored in the analytical derivation by Hyun-Ho and Dong-Ho Cho [7], which can incur errors in performance analysis.

To evaluate the errors, we perform simulations on the power-saving schemes with and without the sleep mode deactivation. We define $Delay_{with}$ and $Loss_{with}$ as the traffic delay and loss rate with sleep deactivation function being enabled, while $Delay_{without}$ and $Loss_{without}$ are the traffic delay and loss rate with sleep deactivation function being disabled. The delay and loss errors are defined in equation (4.31) and equation (4.32) respectively.

$$Delay\ Error = \frac{Delay_{without} - Delay_{with}}{Delay_{without}} \tag{4.31}$$

$$Loss\ Error = \frac{Loss_{without} - Loss_{with}}{Loss_{without}} \tag{4.32}$$

Figure 4.25 and Figure 4.26 are the delay and loss errors, with initial window exponent of 4. From the figures, we can observe that the sleep deactivation by the UL traffic can significantly reduce the DL delay and loss. The performance gaps are more obvious when the maximum window exponent is increased. Therefore, when doing performance analysis, the influence of UL traffic on the DL delay and loss should not be neglected.

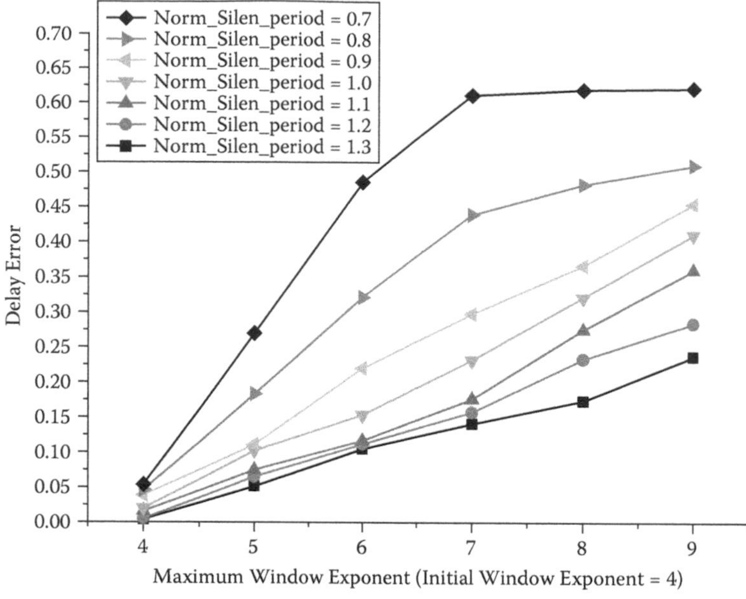

Figure 4.25 Delay error versus maximum sleep window exponent.

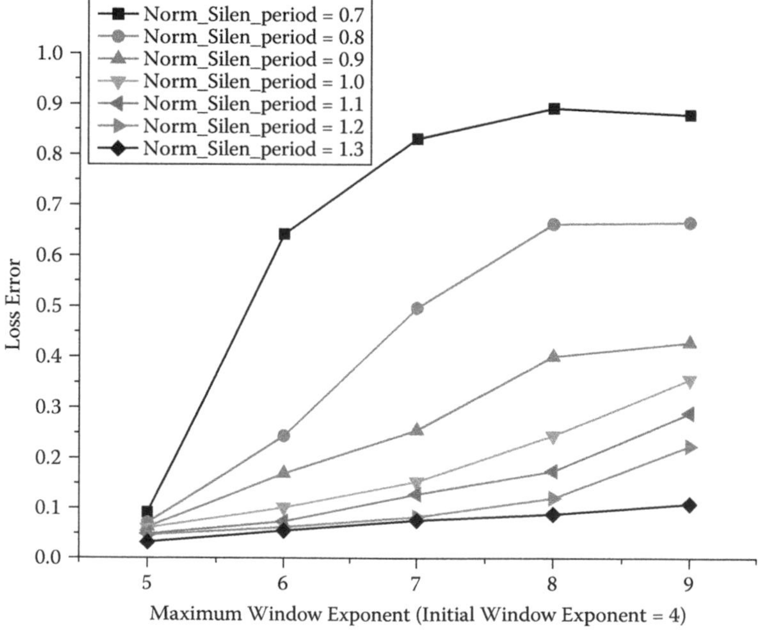

Figure 4.26 Loss error versus maximum sleep window exponent.

4.6 Conclusions

In this chapter, we have applied the hybrid energy conservation scheme to both the simplex broadcasting and duplex conversational VoIP traffics in a WiMAX system. We have also given an analysis model to evaluate the system performance. Guided by the model, we obtain the optimal window adjustment parameters under different scenarios. With these optimal working parameters, the battery life can be extended for about 10 to 30 percent (depending on the working scenarios and talking speeds), while the tough QoS requirements still can be satisfied. Therefore, by carefully tuning the sleep parameters, we can strike a proper balance between energy conservation and QoS provision. In addition, different from the prior research work, when analyzing the traffic delay and loss for a duplex conservational scenario, we have considered the effect of sleep mode deactivation by the UL traffic and, thus, our derivation model is more accurate in performance analysis.

ACKNOWLEDGEMENT

The research was jointly supported by a research grant from Natural Science Foundation of China under project numbers 60602066, 611710711, and 60773203, and grants from the Foundation of Shenzhen City under project numbers JC200903120069A, SG200810220145A, JC201005250035A, and JC201005250047A.

REFERENCES

[1] Yang Xiao, "Energy Saving Mechanism in the IEEE 802.16e Wireless MAN," *IEEE Communication Letter*, vol. 9, no. 7, pp. 595–597, July 2005.

[2] Yan Zhang and Masayuki Fujise, "Energy Management in the IEEE 802.16e MAC," *IEEE Communication Letter*, vol. 10, no. 4, pp. 311–313, Apr. 2006.

[3] Junfeng Xiao, Shihong Zou, and Shiduan Cheng, "An Enhanced Energy Saving Mechanism in IEEE 802.16e," Proceedings of GLOBECOM 2006, pp. 463–467, San Francisco, November 2006.

[4] Jaehyuk Jang and Sunghyun Choi, "Adaptive Power Saving Strategies for IEEE 802.16e Mobile Broadband Wireless Access," Proceedings of Asia-Pacific Conference on Communications, pp. 1–5, Aug. 2006.

[5] Min-Gon Kim and Minho Kang, "Enhanced Power-Saving Mechanism to Maximize Operational Efficiency in IEEE 802.16e Systems," *IEEE Transactions on Wireless Communications*, vol. 8, no. 9, pp. 4710–4719, Sept. 2009.

[6] Mugen Peng and Wenbo Wang, "An Adaptive Energy Saving Mechanism in the Wireless Packet Access Network," Proceedings of Wireless Communications and Networking Conference (WCNC) 2008, pp. 1536–1540, Las Vegas, NV, Mar. 2008.

[7] Hyun-Ho and Dong-Ho Cho, "Hybrid Power Saving Mechanism for VoIP Services with Silence Suppression in IEEE 802.16e Systems," *IEEE Communication Letter*, vol. 11, no. 5, pp. 455–457, May 2005.

[8] International Telecommunication Union–Telecommunication Standardization Sector (ITU-T) Recommendation P.59 (1993), Artificial Conversational Speech, Geneva, Switzerland.
[9] H.-H. Lee and C.-K. Un, "A Study of On-Off Characteristics of Conversational Speech," *IEEE Transactions on Communications*, vol. 34, no. 6, pp. 630–637, June 1986.
[10] International Telecommunication Union–Telecommunication Standardization Sector (ITU-T) Recommendation G.114 (2003), International Telephone Connections and Circuits—General Recommendations on the Transmission Quality for an Entire International Telephone Connection, Geneva, Switzerland.
[11] A. Bacioccola, C. Cicconetti, and E. Mingozzi, "IEEE 802.16: History, Status and Future Trends," *Computer Communications*, vol. 33, no. 2. pp. 113–123, Feb. 2010.
[12] Yu-Kwong Kwok and Vincent Kin Nang Lau, Wireless Internet and Mobile Computing: Interoperability and Performance. New York: John Wiley & Sons, September 2007.
[13] Yang Yang, Honglin Hu, Jing Xu, and Guoqlang Mao, "Relay Technologies for WiMAX and LTE-advanced Mobile Systems," *IEEE Communication Magazine*, vol. 47, no. 10, pp. 100–105, Oct. 2009.
[14] Sassan Ahmadi, "An Overview of Next Generation Mobile WiMAX Technology," *IEEE Communication Magazine*, vol. 47, no. 6, pp. 84–98, June 2009.
[15] Woonsub Kim, "Mobile WiMAX, the Leader of the Mobile Internet Era," *IEEE Communication Magazine,* vol. 47, no. 6, pp. 10–12, June 2009.
[16] K. Etemad, "Overview of Mobile WiMAX Technology and Evolution," *IEEE Communication Magazine*, vol. 46, no. 16, pp. 31–40, June 2008.
[17] IEEE 802.16e-2005, "Amendment for Physical and Medium Access Control Layers for Combined Fixed and Mobile Operation in Licensed Bands," New York: IEEE, Feb. 2006.
[18] C. Cicconetti, L. Lenzini, E. Mingozzi, and C. Eklund, "Quality of Service Support in IEEE 802.16 Networks, " *IEEE Networks*, vol. 20, no. 2, pp. 50–55, Mar. 2006.
[19] A. Esmailpour and N. Nasser, "Packet Scheduling Scheme with Quality of Service Support for Mobile WiMAX Networks," Proceedings of Local Computer Networks (LCN) Conference 2009, pp. 1040–1045, Zürich, Switzerland, Oct. 2009.
[20] IEEE 802.16, "IEEE Standard for Local and Metropolitan Area Networks—Part 16: Air Interface for Fixed Broadband Wireless Access Systems," May 2009.
[21] Chia-Chuan Chuang and Shang-Juh Kao, "Discrete-Time Modeling for Performance Analysis of Real-Time Services in IEEE 802.16 Networks," vol. 33, no. 16, pp. 1928–1936, Oct. 2010.
[22] F. Hou, P.-H. Ho, and X. S. Shen, "An Efficient Delay Constrained Scheduling Scheme for IEEE 802.16 Networks," *ACM/Wireless Networks*, vol. 15, no. 7, pp. 831–844, May 2009.

Chapter 5

QoE-Based Energy Conservation for VoIP Applications in WLAN

Adlen Ksentini and Yassine Hadjadj-Aoul

Contents

5.1 Introduction

The multiplications of multimedia-based services coupled with the recent proliferation of mobile devices, with built-in audio and video capabilities, have stimulated the interest in multimedia transmission over mobile communication systems. VoIP (Voice over Internet Protocol) over WLAN (wireless local area network) or

VoWLAN (voice over wireless local area network), particularly, remains a great concern as it is considered as a possible "killer application." This is favored by (1) the low cost call allowed by VoIP regarding 3G connections, and (2) the high data rate provided by WLAN. Supporting such services in WLAN-enabled mobile devices results, however, in significant energy consumption that severely reduces the operation time. In fact, a WLAN card exhibits high-energy consumption compared to cellular communications, such as 3G. For instance, the specifications of the Apple iPhone [1] list a talk time of 14 hours over the cellular network when the WLAN interface is off. When both the cellular and WLAN interface are on, the talk time is reduced to eight hours. With very light Web browsing and e-mail access over the WLAN (i.e., once every hour), the talk time is reduced to six hours. Even with end devices like laptops, at least 15 to 20 percent of the total energy capacity is consumed by an active WLAN interface. Reducing the energy consumed by the interface during VoIP calls is, thus, a critical step toward extending the operating lifetime of these wireless devices.

Usually, WLAN radios conserve energy by staying in the sleep mode. With real-time applications like VoIP, it is not clear how such energy can be saved by this approach since packets delayed above a threshold are lost. Indeed, a longer sleep period increases the energy conservation, but increases, at the same time, packet loss due to late receptions, which leads to a user's reduced quality of experience (QoE). On the other hand, reducing the sleep period decreases the packet loss, but increases the energy consumption [2]. Many papers in the literature addressed the energy conservation problem in WLAN. We categorize these papers into two groups: (1) single-layer based solutions, and (2) cross-layer based solutions. The first class reduces energy consumption by considering the particularity of the WLAN, in term of media access control (MAC) parameters. This class optimizes the energy consumption without considering the traffic characteristics. These solutions are, thus, more appropriate for nonreal-time applications. The second class, meanwhile, uses a cross-layer paradigm to optimize energy consumption. The solutions based on this class reflect the applications' requirements in term of quality of service (QoS) (such as end-to-end delays and packet loss rate) when deriving the sleep periods for the wireless mobiles. Solutions belonging to this class are more appropriate for VoWLAN where a trade-off is achieved between reducing energy and maintaining either end-to-end delays or packet loss lower than a certain threshold. However, it is well known that the QoS metric does not reflect user QoE. While QoS is sometimes used to produce a "general customer satisfaction" ratio, QoE is subjective and relates to the actual perceived quality of a particular service by the users. The relationship between a performance-based QoS parameter has a resulting effect on the end users' QoE because a high network performance is required to meet QoE objectives.

In this chapter, we address the problem of energy conservation in VoWLAN applications through a cross-layer solution. We introduce a novel approach that exploits user QoE in order to optimize mobile energy consumption. In other words, according to user QoE (obtained at run-time), we propose to derive the sleep periods that maximize power conservation while maintaining users' QoE above a certain threshold. To measure user QoE, we use a pseudo subjective quality assessment (PSQA) tool for VoIP (noted by PSQA-VoIP), [3] which reproduces the International Telecommunication Union-Telecom's [4] (ITU-T) perceptual evaluation of speech quality (PESQ) [5] mimics in order to compute QoE, in terms of mean opinion score (MOS). Unlike PESQ, PSQA-VoIP does not need the reference audio sequence when deriving the QoE value.

This chapter is organized as follows section 5.2 presents related work on energy conservation for VoWLAN. In section 5.3, we will present the concept of PSQA for VoIP. Section 5.4 introduces our cross-layer solution that uses QoE feedbacks to optimize energy consumption, namely a QoE-based energy conservation mechanism for VoIP applications (ECVA). We will present in section 5.5 some results obtained through simulations, which demonstrate the advantages of our proposal against the 802.11 standard solutions. Finally, we end with our conclusion in section 5.6.

5.2 Background and Related Work on Energy Conservation in WLANs

5.2.1 Background

Research activities on power management in mobile terminals demonstrate the efficiency and the limits of the hardware platforms in reducing energy consumption. In fact, a large part of the energy consumption is a direct consequence of the communication process [2]. Thus, the wireless station's battery lifetime can be significantly enhanced through designing energy-efficient communication protocols.

The main idea for saving energy in WLAN is to allow the wireless interface to go to sleep as much as possible by decreasing the idle time durations (i.e., time spent in the idle mode). However, one of the main issues associated with this solution is to compute exactly when the interface should go to the sleep mode and when it should wake up. This is not an easy task because the station has to be active for packets reception. Thus, the sleeping period must be tuned optimally in order to make the wireless interface ready when packets are coming. Solutions for saving energy in WLAN can be broadly classified into two distinct classes. The first class represents single layer-based solutions, and the second class represents approaches addressing different layers at the same time (e.g., cross layer-based solutions). Single layer-based schemes consider trade-offs existing at a predefined layer. Most of the solutions of this class are mainly based on MAC layer enhancements to reduce energy consumption. The standardization activities actively address this issue in WLAN. In fact, the IEEE 802.11 group early on defined the power save mode

(PSM) [6]. More recently, the 802.11e group has introduced two other modes, namely, unscheduled power save delivery (UPSD) and scheduled power save delivery (SPSD) [7]. In PSM, the wireless station is allowed to transit to a lower power sleep state if it is not involved in sending or receiving packets. Here, the access point (AP) has to be notified of the decision made by the wireless station in order to buffer packets destined for this station. Thus, periodically, the AP informs the associated stations through the beacon frame, and, specially, by enabling the traffic indication map (TIM) field, if they have buffered packets destined for them. Figure 5.1 depicts a use case describing the PSM with IEEE 802.11 terminals. During the PSM, stations are in one of these two states, namely, the awake state and sleep state. During the awake state, wireless stations are fully powered. In contrast, during the sleep state, wireless stations are not able to transmit or receive packets, which induces very low power consumption. When a station finds out that there are pending packets in the AP, it asks for these packets through power save (PS) polls. Thus, the station stays in an awake state until buffered packets are received. Otherwise, it goes to the sleep state [1].

Aimed at addressing QoS issues that appear when the PSD mode is used, the 802.11e group included two other modes: UPSD and SPSD. In UPSD, stations decide on their own when to awake to request the frames buffered at the AP. This mode takes advantage of the fact that a station with data to transmit awakes from the sleep period to send its packets, and, hence, can receive packets without having to wait for notification in the next beacon sent from the AP. SPSD, meanwhile, is the centralized mechanism where the AP determines for the station's schedules to awake and receive frames that are buffered at the AP.

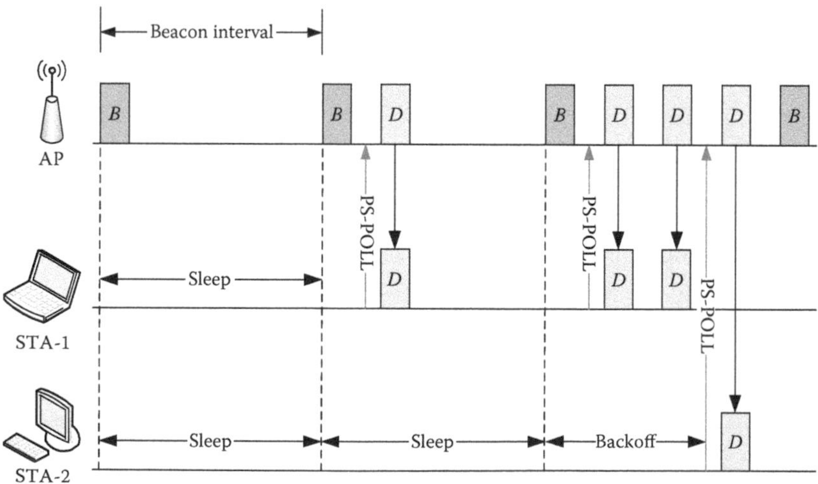

Figure 5.1 PSM mode in WLAN.

5.2.2 Related Work

Besides the standardization activities, there are tremendous works on energy conservation for 802.11 networks belonging to the first class. Based on Zhao and Huang,[8] we can further differentiate works in this class regarding which MAC parameters are considered for saving energy. Thus, we can organize these works into three main categories:

1. **Conserving energy during contention:** In this category, the main idea is to reduce the contention period (i.e., waiting time before transmitting frames) by adapting the contention window (CW), in order to reduce the energy consumption.
2. **Reducing power consumption at the transmission or the retransmission phases:** In this category, the main objective is to reduce the transmission time, by (1) compressing packets, and (2) using the highest data rate allowed by the physical layer, which reduces the energy consumption.
3. **Eliminating contentions, IFS (interframe space), and acknowledgements:** As in the first category, the main idea is to reduce the waiting time before sending data frames or acknowledgements. Thus, the proposed solutions tried to reduce the IFS and to use a block of acks (acknowledgment codes), as proposed in 802.11n.

For more information about these categories, readers can refer to Zhao and Huang [8].

Works based on cross-layer solutions, meanwhile, are mainly related to the traffic type. In fact, there are solutions related to best-effort traffic, such as transmission control protocol (TCP), and solutions related to real-time traffics, such as video streaming as well as VoIP. In this chapter, we will focus on related works that addressed energy saving for VoIP traffic in WLAN (VoWLAN). For those based on best-effort traffics, reader can refer to Zhao and Huang [8].

Saving energy for VoWLAN is more critical than for best-effort traffic as the way to select the sleep/awake periods is hard to define. Indeed, a trade-off is needed between maximizing energy conservation and maintaining acceptable users' QoE. Here, users' QoE is highly degraded when the packet delays are high, which increases late packet loss. Gleeson et al. [9] tried to compute the sleep period for VoWLAN. The authors considered the VoIP packet interarrival and the MAC layer delay variation to draw the duration of the sleep period. By considering a simple 802.11 analytical model as well as a VoIP call using 64 kbps (typically, G.711 call), they found that the optimal value of the sleep period is equal to 20 ms. Gleeson et al. [9] and Zhu et al. [10] proposed to derive a dynamic sleep strategy to tune sleep and packetization intervals dynamically according to the collision probability of the WLAN. By using the collision probability, the authors tried to find out a trade-off between maximizing network capacity and energy conservation. If the collision probability is low, the station goes to sleep for a period inversely proportional to this probability. However, if this probability is high, the station is maintained in the awake period. Both works presented above are difficult to deploy outside one hop WLAN, as they used 802.11 MAC parameters to tune the sleep periods. In addition, work in Gleeson et al. [9] considered only the case of *G*.711 audio codec, while VoIP applications usually employ more efficient audio codecs, such as Internet Low Bit Rate (ILBC), [11] G.729a, etc. The work presented by Namboodiri and Gao [12] is similar to our proposition in a sense that the authors considered the end-to-end delay and the packet loss rate (both seen at the VoIP level) as the main parameters to derive the sleep period duration. This solution is not related solely to the one hop 802.11, but it covers the multihop communication, including the Internet. However, the main difference with our solution is in the parameters used to compute the sleeping period and the proposed algorithm, which is based on the control theory to derive the optimal values despite the possible variable network conditions. Further, the authors considered QoS parameters that are not very efficient to reflect the user-perceived quality, while we used an automatic QoE measuring tool that reproduces the behavior of the PESQ model without any audio reference sequence.

5.3 Quality of Experience (QoE) and Pseudo Subjective Quality Assessment (PSQA)

QoE is defined by the International Telecommunication Union–Telecommunication Standardization Sector (ITU-T) [13] as "the overall acceptability of an application or service, as perceived subjectively by the end user." QoE is different from the QoS network indicators in terms of bandwidth, loss rate, and jitter, which are not sufficient to get a precise idea about the quality of a received audio or video sequence. QoE instead focuses on the overall experience of the end user. Table 5.1 shows the MOS value regarding the user's subjective evaluation.

For several years, ITU-T's PESQ has been the reference for objective speech quality assessment. It is widely deployed in commercial products where it is used to

Table 5.1 MOS Versus Subjective Evaluation

MOS	*User Feeling*
5	Perfect. Like face-to-face conversation or radio reception.
4	Fair. Imperfections can be perceived, but sound still clear. This is (supposedly) the range for cell phones.
3	Annoying
2	Very annoying. Nearly impossible to communicate.
1	Impossible to communicate

measure user QoE. Although PESQ gives an interesting correlation with the subjective scores for VoIP, it needs both the original and the received audio sequences in order to compute the QoE score. As the original audio sequence is not available, PESQ cannot be applied in real-time situations.

In this work, we used the PSQA version for G.711 audio codec presented by Basterrech, Rubino, and Varela [3]. PSQA is a quality assessment tool, which can be classified as a hybrid tool between subjective and objective evaluation techniques. The idea is to do subjective tests for several distorted audio sequences and use the results of this evaluation to teach an RNN (random neural network) the relation between the parameters that cause the distortion and the perceived quality. The procedure consists, first, in identifying the parameters that have an impact on QoE in the given context. The parameters used in the PSQA-VoIP version are the packet loss rate (LR) and the mean burst loss size (MBLS). Here, MBLS is the number of consecutive losses in a loss period, i.e., the mean length of loss bursts in the flow. These two parameters, in fact, represent the network impairments that degrade the VoIP communication [3]. Despite the fact that jitter could be considered as a relevant parameter for VoIP quality, it can be folded into the loss rate if no particular attention is being paid to dejittering buffer sizes and algorithms. An audio sequence database is then generated by simulating these parameters after choosing a set of representative values for each of them, together with an interval for the parameter, according to the conditions under which we expect the system to work. Then, uniformly sampled subset *S* of this audio sequence database is objectively evaluated by using the PESQ tool. After statistical processing of the results, the audio sequence receives the QoE value (often, this is a MOS). It results in S configurations of the parameters and a corresponding QoE score or MOS. Then, some of the configurations are used for training the RNN and the remaining ones are used for validation and not shown to RNN during training. This PSQA version takes advantage of the PESQ, in terms of high correlation with the human subjective score, while it can be used in real time fashion, as there is no need for the reference audio sequence. Thus, it easily could be used to compute in real time the user QoE in the context of VoWLAN application.

For more information on PSQA-VoIP, readers can refer to Basterrech, Rubino, and Varela [3].

5.4 ECVA: A QoE-Based Energy Conservation Mechanism for VoIP Applications

WLAN radios conserve energy by staying in sleep mode. With real-time applications like VoIP, it is not clear how such energy can be saved by this approach since packets delayed above a threshold are lost. In this section, we will present our proposition to reduce energy consumption for VoIP conversation while maintaining good QoE. Through the proposed algorithm, which is named ECVA, we try to allow a controlled trade-off between the QoE of the call and energy conserved. Based on the minimum QoE that users can tolerate in their conversations, our proposition adjusts dynamically how aggressively the interface tries to stay in the sleep state. This enables our algorithm to maximize energy consumption while targeting a specified level of application quality. Like in Zhu et al.[10] and Namboodiri and Gao,[12] we propose that, after successfully receiving/transmitting incoming/outgoing packets, wireless stations have to enable the sleep period. This period is derived as follows:

$$\begin{aligned} Sleep_k &= Sleep_{k-1} \\ &+k_p\left(1+\frac{T}{T_i}+k_p\frac{T_d}{T}\right)(Q_k-Q_{ref}) \\ &-k_p\left(1+\frac{2T_d}{T}\right)(Q_{k-1}-Q_{ref}) \\ &+k_p\frac{T_d}{T}(Q_{k-2}-Q_{ref}) \end{aligned}$$

where $Sleep_i$, k_p, T, Q_i, and Q_{ref} represent respectively the measured sleep period (which is the function of the beacon) at the time i, the proportional gain, the time index, the measured QoE (in terms of MOS) at the time i, the reference QoE (in terms of MOS), which represents the targeted QoE value. The parameters T_i, and T_d, which depend on the proportional gain k_p, the integral gain k_i and the derivative gain k_d, are equal to $\frac{k_p}{k_i}$ and $\frac{k_d}{k_p}$, respectively.

Indeed, the sleep period formula is derived after applying the effective discrete proportional integral derivative (PID) controller that monitors the perceived quality of experience at the mobile station (STA) level. We recall that perceived QoE is obtained using PSQA-VoIP, which tracks the packet loss rate as well as the MBLS at the decoder level, and returns the MOS perceived by the user. Each time the station has to compute the sleep period, our algorithm calls PSQA-VoIP. Figure 5.2 depicts the proposed approach through the controlled system point of view. The proposed controller computes the difference between the targeted QoE value and the measured QoE to derive the amount of time the wireless mobile can go to the sleep mode.

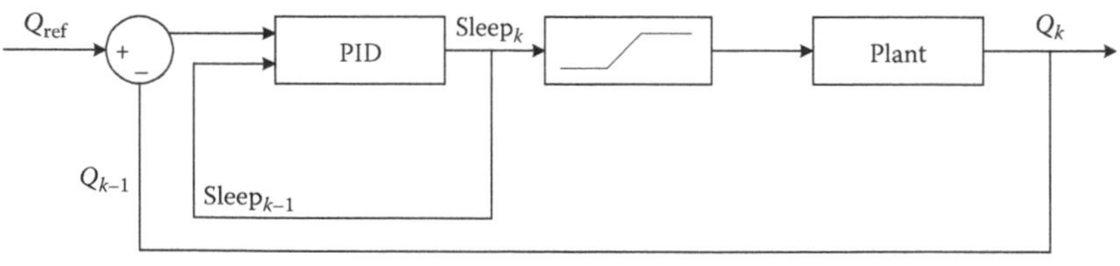

Figure 5.2 Adaptive feedback control mechanism design.

The calculated sleep period depends, then, on the three actions: (1) the proportional action to converge toward the desired value, (2) the derivative action to accelerate the convergence by using the QoE variation value, and (3) the integral action, which allows eliminating the steady state error. The energy conservation is achieved by tuning the sleep period in such a way that the QoE is maintained at acceptable levels. Thus, the main parameter that controls the system is the Q_{ref} defined as the target QoE. The higher this value, the lower the duration of the sleep period. In contrast, the lower the Q_{ref} is, the higher will be the sleep period duration because the system will try to increase the energy conservation. Typical Q_{ref} values could be 3.5.

Since the sleep period is a function of the number of beacons, the real value calculated with the equation above is approximated to the nearest integer value. Besides, to avoid late losses of packets, the application constraints are added in the computation of the sleep period, which is represented by the saturation block added at the PID module output. The lowest value the sleep period can take is fixed to 3, which corresponds to a sleep period of 200 ms as the beacon period is fixed to 100 ms.

Note that the plant in Figure 5.2 represents the simulated system, which comprises all the chain of transmission.

5.5 Performance Evaluation

Having described in detail our QoE-based energy conservation algorithm for VoIP calls, we now direct our focus to its performance evaluation using the network simulator (NS2) [14]. To support WLAN power management, we used the extension proposed by NEC Labs, which supports the legacy power management functions defined with IEEE 802.11 [15]. To demonstrate the effectiveness of the proposed solution in terms of energy conservation, we compare it to the IEEE 802.11 standard. The conducted simulations were run during 900s: a duration long enough to ensure that the system had reached its stable state. We simulated a VoIP communication of two mobiles, including a wireless interface, and connected to the same access point (AP). The VoIP call is based on G.711 codec, with 64 kbps of data rate. We fixed the max delay tolerated at the decoder side (when a packet is considered too late to be decoded) to 250 ms, which is the typical value recommended by the ITU-T. The simulation parameters are defined in Table 5.2.

Two main metrics were used to evaluate the performance of the proposed mechanism. The first metric represents the sleep period, which is a function of the number of beacons. Indeed, as explained above, the sleep period indicates how often clients serviced by the access point should check for buffered data on the access point. Thus, this metric can be translated directly into saved energy. The second metric represents the QoE (MOS), which reflects the impact of the proposed approach on the perceived quality.

We can clearly see, from Figure 5.3, that the sleep period, which was initialized to one beacon time, remains static during all the simulation for the classical

Table 5.2 Parameters Settings

Parameter	*User Value*
Q_{ref}	3.5
k_p	0.0062
k_i	0.00312
k_d	0.0051
Simulation duration	900 sec
Beacon interval	0.1 sec
Physical rate	11 Mbps
txPower	0.660 W
rxPower	0.395 W
idlePower	0.035 W
sleepPower	0.001 W

Note: Most of the deployed WLAN interfaces support PSM.

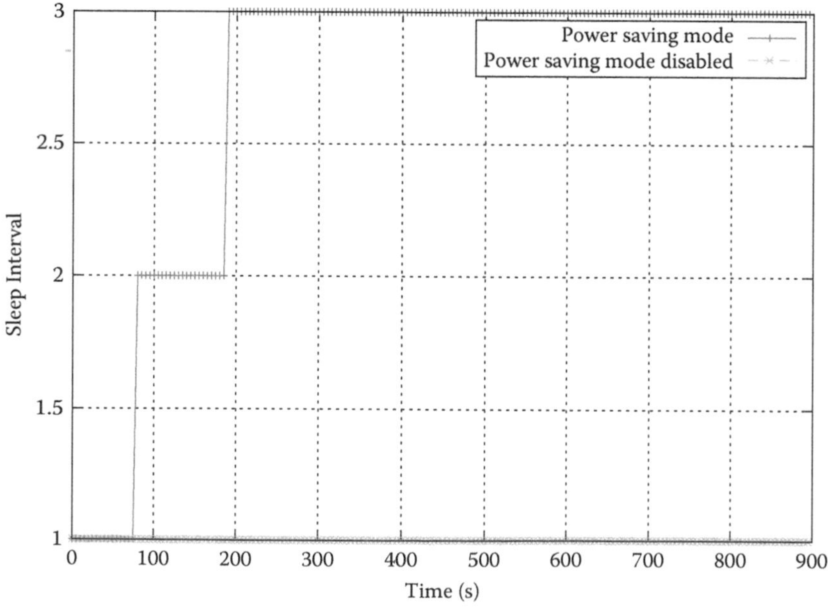

Figure 5.3 Variation of the sleep interval in function of time.

approach. Indeed, in the classical approach, no sleep periods adaptation is considered, which means that the mobile phone stays awake during the whole communication process. On the other hand, the proposed approach allows adapting the sleep period dynamically according to the VoIP's QoE. In fact, after 80 seconds of simulation, the proposed mechanism allows the mobile station to go to the sleep mode during one beacon time every two beacons, when the beacon duration is equal to 100 ms. This clearly impacts the delay, as we can see in Figure 5.4. In a second period, starting from 185 sec, the sleep period goes to three, which means that the mobile station goes to the sleep mode during two beacons' time every three beacons. This makes the end-to-end delay vary from some milliseconds to 250 ms (late packet loss threshold), while it permits increasing, considerably, the sleep period duration. Further, we can see the impact of the PID controller because the sleep period is practically constant and no high oscillation is noticed throughout the simulation. After a short duration, the controller ensures the convergence toward the Q_{ref} value (as seen in Figure 5.5). This behavior is very beneficial in terms of user-perceived quality, since it avoids the ping-pong effect of MOS.

As a direct consequence of the sleep period modification, we can see, in Figure 5.6, its impact on the energy consumption. In fact, when the sleep period goes to two, after 80 sec of simulation, we can see a slight conservation of energy compared to the classical approach. When the sleep period goes to three, we can clearly see the

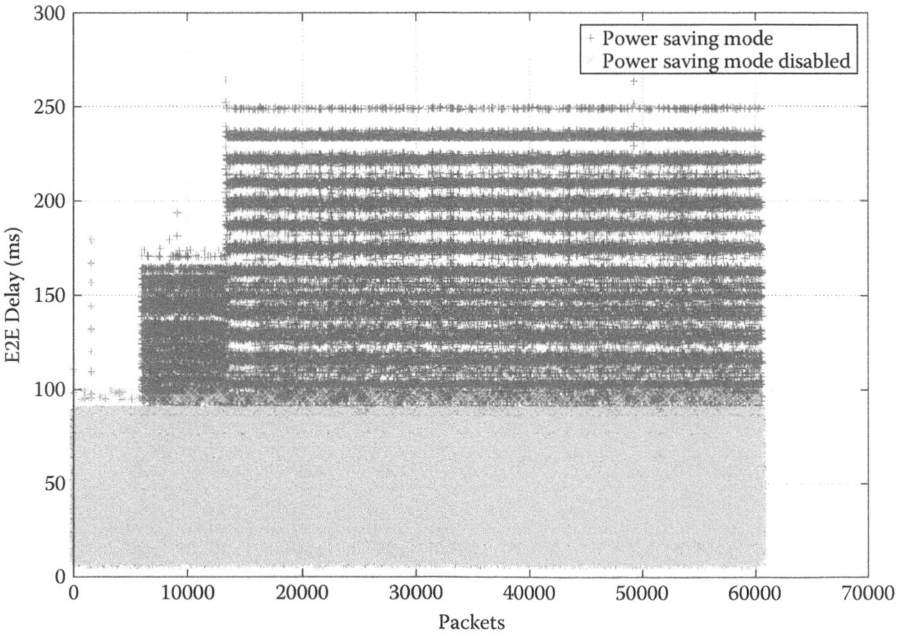

Figure 5.4 End-to-end delay.

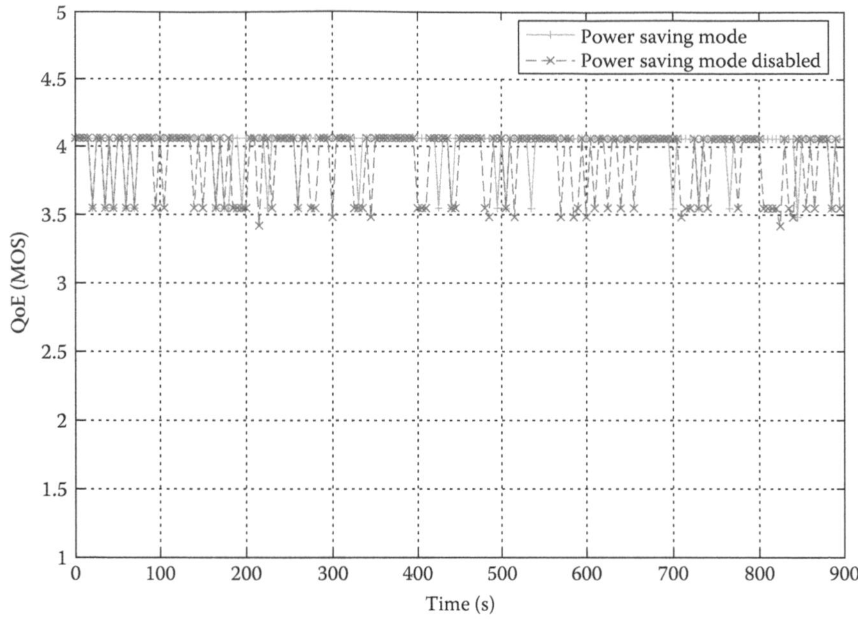

Figure 5.5 Estimated QoE when using power-saving mode versus without using power saving.

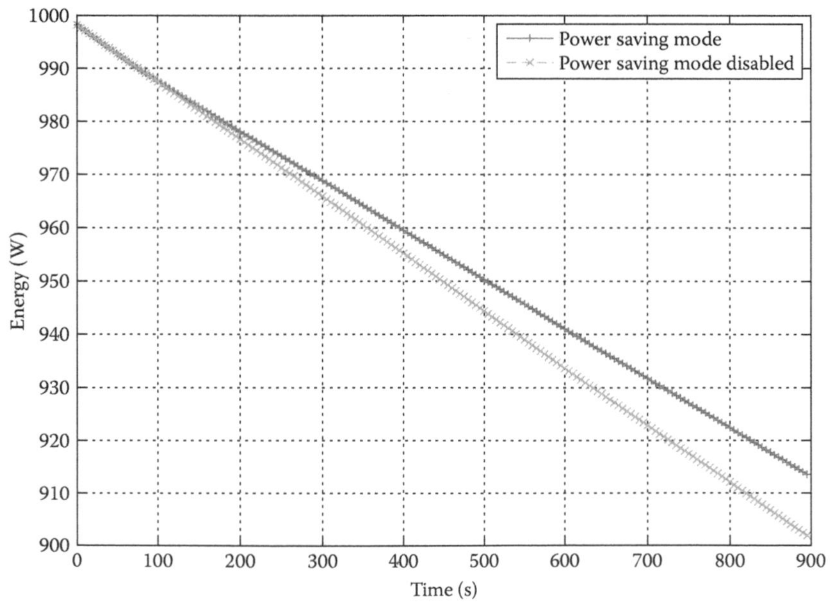

Figure 5.6 Energy consumption versus time.

augmentation of the energy conservation in the proposed approach. This represents an improvement of 15.76 percent at the depletion of the charge of the battery.

Figure 5.5 depicts the measured quality of experience of the VoIP communications in the classical and the proposed approaches. It clearly can be seen that the impact of the proposed approach on the perceived QoE is not significant. In fact, the proposed approach almost avoids packets of late losses despite the augmentation of the end-to-end delay. At the end of the simulations, the measured average QoE of the proposed approach is equal to 4.01 while the classical approach gives an average value of 3.90, which represents 2 percent of QoE loss. Nevertheless, this loss is very negligible from the user's perception. Thus, these results show clearly the advantage of using ECVA where a strict trade-off is achieved between maximizing energy saving and maintaining acceptable user QoE.

Note that the proposed algorithm allows keeping the end-to-end delay to acceptable values (see Figure 5.4), which allows maintaining the perceived quality of the communication, as we can see from Figure 5.5. The QoE difference between the two approaches is mainly due to random losses as the proposed and the classical approaches almost completely avoid packets' late losses. This clearly demonstrates the superiority of the proposed approach.

Note also that the rapidity of convergence directly depends on the values affected by the proportional, derivative, and integral gains. Indeed, small values of the gains induce a slow convergence of the algorithm and only a few oscillations in the response (i.e., in terms of QoE's variation), which allows respecting the application's constraints. On the other hand, bigger values of the gains induce a faster convergence to the desired objective (i.e., QoE equal to 3.75) with a risk of deteriorating the application's QoE.

5.6 Conclusions

In this chapter, we have addressed both the energy conservation and the QoE problems for VoWLAN. As a remedy, we have proposed a novel approach for controlling the sleep periods based on the requested quality (in term of QoE) of the VoIP flows. The proposed approach is different from the conventional approach and existing solutions as it takes into account the user QoE. The proposed scheme tracks, particularly, the VoIP quality and proposes a schedule of sleep/wakeup periods to optimize energy consumption. Extensive simulation showed that by controlling such parameters the sleep periods can be considerably increased, which augments significantly the lifetime duration of the wireless node. Future directions will address the adaptive tuning of the PID parameters (k_p, k_i, and k_d). Actually, the values of these parameters are selected heuristically. So, we choose a learning algorithm, such as neuronal network or genetic algorithm, to dynamically tune these values toward a better system stability. Furthermore, the simulation scenario used in this chapter is too simple. We thought to enhance this scenario by considering, for instance, more wireless stations in order to increase the network load, and to check the stability of our system in this case. Finally, it should be stressed out that coupling these parameters with a link scheduling approach should further enhance the performance of the proposed concept.

REFERENCES

[1] Apple iPhone Technical Specifications. Online at http://www.apple.com/iphone/specs.html

[2] G. Miao, N. Himayat, and G. Y. Li. Energy-efficient link adaptation in frequency-selective channels. *IEEE Transactions on Communications*, vol. 58, no. 2, pp. 545–554, (February 2010).

[3] S. Basterrech, G. Rubino, and M. Varela. Single-Sided Real-Time PESQ Score Estimation. In Proceedings of Measurement of Speech, Audio and Video Quality In Networks (MESAQIN 2009), Prague, Czech Republic, December 7–8, 2009.

[4] http://www.itu.int/ITU-T/

[5] International Telecommunication Union–Telecommunication Standardization Sector (ITU-T) P.862. Perceptual Evaluation of Speech Quality (PESQ): An objective method for end-to-end speech quality assessment of narrow-band telephone networks and speech codecs. Geneva, Switzerland: ITU-T (February 2001).

[6] IEEE 802.11 WG: Wireless LAN Medium Access Control (MAC) and Physical Layer (PHY) Specifications. 1999 standard (1999).

[7] IEEE 802.11e: IEEE 802.11e Wireless LAN Medium Access Control (MAC) Enhancement for Quality of Service (QoS). IEEE 802.11e standard (2009).

[8] S-L. Zhao and Ch-H. Huang. *A survey of energy efficient MAC protocols for IEEE 802.11 WLAN*. Amsterdam: Elsevier Computer Communications (2010).

[9] B. Gleeson, D. Picovici, R. Skehill, and J. Nelson. Exploring Power Saving in 802.11 VoIP Wireless Links. ACM International Wireless Communications and Mobile Computer Conference (IWCMC 2006), Vancouver, Canada, January, 2006.

[10] C. Zhu, H. Yu, X. Wang, and H-H. Chen. Improvement of Capacity and Energy Saving of VoIP over IEEE 802.11 WLANs by a Dynamic Sleep Strategy. Proceedings of the IEEE GLOBECOM (2009), Honolulu, Hawaii, November 30–December 4.

[11] Internet Low Bit Rate. Online at http://www.Ilbcfreeware.org

[12] V. Namboodiri and L. Gao. Energy-Efficient VoIP over Wireless LANs. *IEEE Trans. on Mobile Computing*, vol. 9, no. 4, pp. 566–581, (April 2010).

[13] International Telecommunication Union–Telecommunication Standardization Sector (ITU-T) SG12: Definition of Quality of Experience. COM12 LS 62 E, TD 109rev2 (PLEN/12), Geneva, Switzerland, 16–25 (January 2007).

[14] In proceedings of 4th International Workshop on Power-Save Computer Systems, PACS 2004, Portland, Oregon. December, 2004.

[15] NS-2: The Network Simulator 2. Online at http://www.isi.edu/nsnam/ns/

[16] NS-2 ext.: WLAN power management extension. Online at http://nspme.sourceforge.net/index.html

Chapter 6

Minimum Energy Multicriteria Relay Selection in Mobile Ad Hoc Networks

Komlan Egoh, Roberto Rojas-Cessa, and Swades De

Contents

6.1 Introduction

Efficient packet forwarding is essential for large-scale wireless multihop networks because of its impact on end-to-end packet success rate and energy consumption. Various forwarding schemes have been proposed for multihop wireless networks where a transmitting node selects a relay node among its neighbors using a simple criterion, such as the relaying neighbor's geographical proximity to the final destination or the energy required to transmit the packet. These forwarding schemes, where the transmitting node is the decision point, require not only that a list of all neighbors be maintained at each node, but also that the deciding node centralizes and maintains any additional decision information. Maintaining such information at all nodes in a dense and dynamic network environment and ensuring the selected relaying node is active (e.g., by wake up signals or synchronization) may be costly for the resource constrained nodes.

A majority of the proposed rules for selecting the next forwarding relay also assume a disk coverage model, wherein a node within the coverage range is considered perfectly reachable [1]; as a result, they rely solely on location metrics, such as hop-count and one-hop progress to choose the next forwarding relay node. In reality, the disk assumption does not hold from physical layer perspective; there exists a transitional region within which the geographical location and reachability are highly uncorrelated [2]–[4]. Figure 6.1 illustrates typical reachability in the case of sensor nodes equipped with Chipcon CC2024 radio transceivers. Greedy geographic forwarding, for example, has received a great deal of attention in the ad hoc networking research community. In greedy forwarding, because the goal is to reach the destination in a fewer number of hops, a transmitter typically selects a relay node that is closest to the destination, thus, also farthest away from the transmitter and, consequently, more error prone. For this reason, there has been a growing acceptance that the traditional purely greedy forwarding approach is not optimal in most practical wireless network settings where the disk assumption does not hold.

The attempts to remedy the shortcomings of greedy forwarding fall in two camps. First, a few link-aware metrics, which combine geographic location with

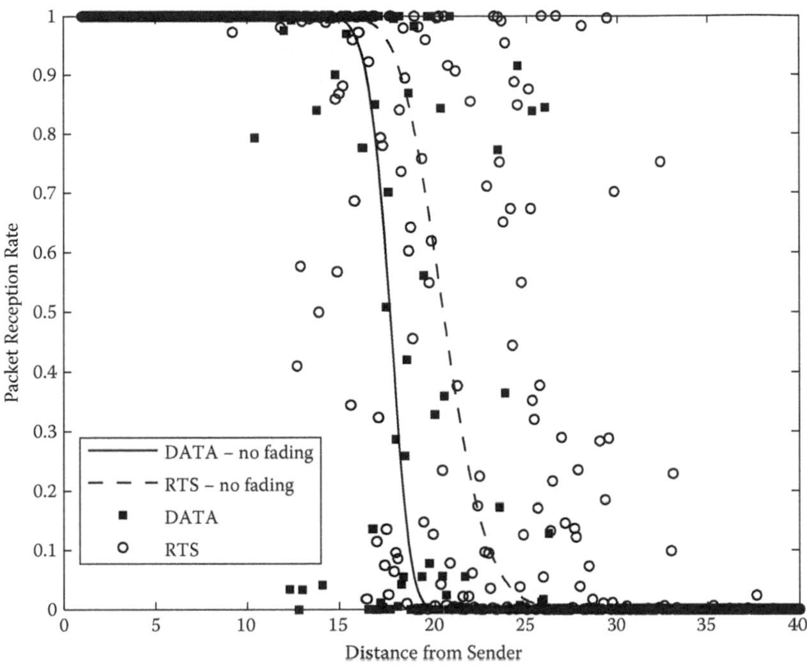

Figure 6.1 Sample of reachability in realistic wireless settings.

some measure of link quality, improve on the performance of the traditional greedy forwarding [3][5]. By incorporating a measure of link quality into the selection metric, link-aware forwarding schemes are able to avoid poor link candidates by prioritizing good links. The link-aware metrics, however, suffer additional overhead when implemented as transmitter-side selection because of added link quality information gathering (or estimation) requirements. A second group of proposals, generally termed *opportunistic*, select (or elect) the next relay node only among neighbors that have successfully received the data packet (or an initiating system packet) from the transmitting node. A number of opportunistic forwarding schemes also offer a distributed alternative to the centralized (i.e., transmitter-side) relay selection [6][7]. In these schemes, the transmitting node does not decide about selecting the next hop relaying neighbor. Instead, all neighbors contend among themselves to elect the best possible relay. Nevertheless, these distributed methods still rely on a single, location-based decision criterion and face performance degradation and higher energy consumption in lossy wireless environments.

In this chapter, we present a multicriteria distributed forwarding scheme called *receiver-side relay selection* [8][9]. The challenge in considering more than one criteria for the next hop selection lies in deciding the optimality of a particular relay candidate with respect to the other nodes. Different criteria may have conflicting consequences (e.g., longer hops versus link quality). In other words, the familiar scalar notion of optimality does not hold when multiple criteria are considered. We address this problem by formulating the forwarding task as a multicriteria, decision-making problem, and we search for the optimal trade-off among the criteria. The forwarding decision is distributed; the next hop relay at each hop is elected, opportunistically, only among neighbors who receive an initial request packet successfully. With receiver-side relay selection, information, such as received signal strength, energy scavenging sources, and actual residual energy, which are readily available at each potential relay candidate, can be combined with location information without additional information gathering overheard. We introduce a generalized multiparameter mapping function that aggregates all decision criteria into a single *criterion* used to rank the potential relay candidates. We investigate optimal rules for next hop relay as applicable to both transmitter-side selection and receiver-side selection forwarding schemes. Beyond the theoretical formulation of the generalized multicriteria-based optimum selection, as a demonstrative example of network performance evaluation, we consider the network performance based on two optimality criteria, namely one-hop progress (greediness) and packet success rate (link quality). It is shown that a suitable mapping function can be found that trades off the greediness for link quality and outperforms the reported transmitter-side, link-aware forwarding schemes. Compared to the other schemes, our distributed two-criteria optimization results show a substantially better end-to-end delay performance and a reduction of up to five times in end-to-end packet loss for the same required energy.

The rest of the chapter is presented as follows. Background information and related works are surveyed in section 6.2. In section 6.3, the basic receiver-side relay selection approach is outlined. Section 6.4 introduces the multicriteria, receiver-side relay selection and presents a general analytic framework for performance evaluation of the relaying schemes. A demonstrative example of two-criteria-based relay selection priority is also presented in section 6.5. Concluding remarks are drawn in section 6.6.

6.2 Background

6.2.1 Position-Based Forwarding

Position-based forwarding, also called *geographic routing*, is a packet forwarding scheme that relies on geographic coordinate instead of network address. Using geographic coordinate data packets from a source station (also called *network node*), they are forwarded through multiple relay nodes toward the final destination [10]. The question of *how* to select the next hop relay for optimum multihop communication has long been considered in packet radio networks [1][11–14]. Typically, the node currently holding the data packet forwards the packet to a relay node selected among its neighbors based on a predefined selection rule or metric. Many proposed position-based forwarding solutions use metrics that are the function of the locations of the final destination and the candidate relay node.

One of the first proposed rules for relay selection, named *most forward with r* (MFR) [11], uses a geographic coordinate to select, within a radius of *r*, the neighbor that offers the most progress toward the final destination. The MFR forwarding strategy is the basis of the widely adopted, greedy geographical forwarding [1][6] where the goal is to achieve maximum progress with each transmission. Greedy forwarding is a good strategy in a network with homogeneous nodes (i.e., same transmission range and without transmit power control) [12]. Reachability in these networks is often modeled by an idealized disk within which nodes are assumed to be perfectly reachable. While greedy forwarding ensures a minimum number of hops between source and destination, each hop, however, may require many retransmissions. In most practical wireless networks with unreliable wireless links, higher retransmission rates lead to diminished throughput, higher delay, and higher energy consumption.

Many attempts have been made to remedy the limitations of greedy forwarding. In Tanbourgi, Jakel, and Jondral [15], *nearest with forward progress* (NFP) improves throughput by selecting a relay candidate closer to the transmitter in order to improve throughput. A few link-aware metrics that combine geographic location with some measure of link quality improve on the performance of the traditional greedy forwarding [3][5]. By incorporating a measure of link quality into the selection metric, link-aware forwarding schemes are able to avoid poor link candidates by prioritizing a good link. The link-aware metrics, however, suffer additional overhead when implemented as transmitter-side selection because of added link quality information gathering (or estimation) requirement.

6.2.2 Opportunistic Forwarding and Distributed Schemes

Until recently, all position-related forwarding schemes proposed to make the selection of the next hop node at the transmitter-side. These schemes may work well with lightly populated, and relatively static, ad hoc networks. However, more dynamic, dense, and resource-constrained networks, such as sensor networks, require that we reconsider the question of *where* to make the next hop selection.

A second group of proposals aimed at improving the performance of greedy forwarding select the next relay node only among neighbors that have successfully received the data packet (or an initiating system packet) from the transmitting node. These proposals, generally termed *opportunistic*, also offer a distributed alternative to the centralized (i.e., transmitter-side) relay selection. Zorzi and Rao [6] and Fubler, Widmer, and Kasemann [7] have independently considered forwarding schemes in which transmitting nodes do not need to select the next hop. Rather, all eligible candidates compete among themselves to relay the packets. While Zorzi and Rao [6] considered the remaining distance to the destination-based, forwarding node selection priority criteria, it did not capture the additional media access control (MAC) contention in the selection process. Rather, it was assumed on the one hand that somehow the best relay is always elected and on the other hand that the selection process is always successful. Fubler, Widmer, and Kasemann [7] studied three possible variants of forwarding node selection aiming at reduced packet duplication where it was assumed that more than one nearly simultaneous response could be successful. Priority-dependent, MAC contention probability and the related delay in a successful relay selection process was not considered.

In these distributed schemes, the transmitting node does not decide about selecting the next hop-relaying neighbor. Instead, all neighbors contend among themselves to elect the best possible relay. Nevertheless, these distributed methods still rely on a single location-based decision criterion and face performance degradation and higher energy consumption in lossy wireless environments.

For opportunistic and distributed-forwarding schemes to work, certain networking conditions need to be ensured. Specifically, every node desiring to transmit should find with high probability at least one active neighbor in the forward direction to relay its packets. The feasibility and stability of such a network's conditions have received some attention in the research community. Yi et al. [16] considered a network of independent Bernoulli-type nodes and derived the limiting probability that every node has at least one active neighbor in highly dense networks. Zhang et al. [17] also considered the feasibility of a network of sensor nodes with independent, asynchronous duty-cycles in which transmitting nodes can simply broadcast their packets and have them relayed by available active neighbors. However, because the main focus of the authors is robustness, the approach does not discourage packet duplication and may not be optimal in dense networks.

6.3 Single-Criterion, Receiver-Side Relay Selection

6.3.1 Distributed Selection Process

We consider a network of uniformly random distributed nodes with homogeneous and circular coverage, and independent and asynchronous sleeping behavior. It is assumed that, similar to 802.11 distributed coordination function (DCF), RTS/CTS (request-to-send/clear-to-send), message exchange is done between the transmitter and a potential forwarder before the data packet forwarding. However, unlike in 802.11, the RTS message is broadcast to all local neighbors, and a forwarder's CTS response is suitably delayed to minimize the potential contention. It also is assumed that a node is aware of its geographic location or virtual (hop-count based [18]) location information of its own and the destination. In the following, we elaborate further the concepts of contention resolution, vulnerability of collision, and prioritization strategies in relation to receiver-side relay selection.

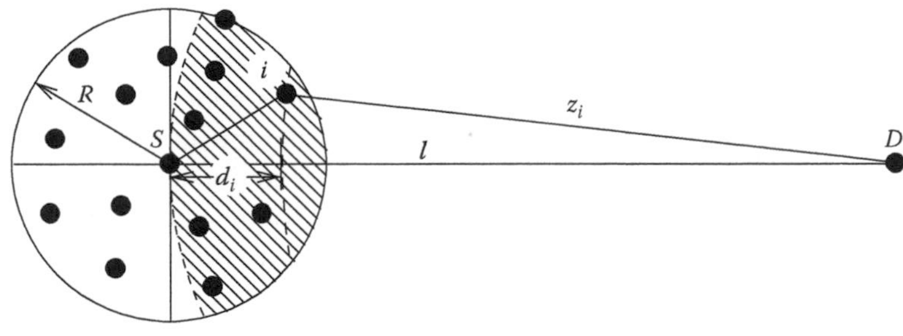

Figure 6.2 Area of contention for packet forwarding. R is the coverage range. *i* is a forwarding contender, which offers forward progress d_i towards D.

6.3.1.1 Contention Resolution

Hop-by-hop forwarding is generally done in packet radio networks based on a decision criteria to obtain the desired nodal or network-wide performance objective. In receiver-side relay selection, these decision criteria are (implicitly) used as priority measures in the distributed relay selection. Before we analyze the performance of various prioritization schemes, we first present the description of a basic selection process using time delay as contention resolution.

A node desiring to send a data packet first sends a broadcast RTS packet containing the optimality criteria and location information of itself and the final destination. After receiving the RTS packet, every eligible relay candidate i (the shaded region in Figure 6.2) schedules a reply time:

$$X_i = g(\Omega_i), \tag{6.1}$$

where Ω_i is the quality measure of node i computed based on a given criterion used by the forwarding scheme. $g(\cdot)$ is a mapping function that implements the prioritization of the selection process and its nature determines the quality of the elected relaying neighbor with respect to the set of optimality criteria and the vulnerability of the selection process to collision among two or more best candidates.

Next, every relay candidate i listens to the wireless medium between the time 0 and X_i. If no other CTS is received before time X_i, then node i considers itself the winner of the selection process and sends a CTS packet with its signature to the transmitting node. If a node overhears a CTS transmission during its waiting period, it gives up the contention, assuming that a better forwarding candidate has been found.

6.3.1.2 Vulnerability to Collision

Because of the distributed nature, receiver-side relay selection processes are vulnerable to collision. The selection process fails to elect the single next hop when two or more relay candidates schedule the same or very close RTS reply times. Note that this type of collision is different from (and additive to) regular medium access collision, such as those caused by hidden or exposed terminals. Since all forwarding schemes, independently of whether the relay selection is done at the transmitter-side or at the receiver-side, are subject to the same regular medium collision. In this work, we are not interested in quantifying this type of collision.

To quantify the collision in the selection process, assume node j schedules reply time $X_j = \min_i\{X_i\}$. Collision happens if there exists at least one node k ($k \neq j$) such that:

$$|X_i - X_k| \leq \beta, \tag{6.2}$$

where β is the collision vulnerability window, which may depend on the MAC scheme, nodes' clock precision, signal detection time, and the radio transceiver's switching time between receive and transmit modes.

Upon correctly receiving a CTS packet, the transmitter sends the data packet to the forwarder. If the transmitter receives another correct CTS packet afterwards for the same data packet (if the earlier CTS packet was unheard by some nodes in the forwarding zone), it simply discards the CTS to avoid any packet duplication. In case of any CTS message collision, all forwarding nodes give up in that contention cycle, and, as in 802.11 DCF MAC contention resolution, we assume the transmitter reinitiates the selection process by broadcasting another RTS packet after a timeout.

6.3.1.3 Prioritization Types

We categorize prioritization functions into two main groups: *purely random* and *absolute priority-based*. Below, we present definitions of the two prioritization types, assuming, without loss of generality, that the desired objective is to maximize the decision criteria Ω. For the sake of completeness, a variation of forwarding scheme, called *hybrid priority*, is also defined, which is derived from the combination of the priority-based approach and the random approach.

Definition 1—In **purely random** forwarding, no priority is given to better candidates during the selection process. If nodes j and k are two relay candidates, then the fact that node k is a better candidate than node j ($\Omega_j < \Omega_k$) does not increase the chances that k is elected over j. Formally,

$$\Pr[X_k \leq X_j | \Omega_j \leq \Omega_k] = \Pr[X_k \leq X_j]$$

Note that in this case the decision criteria Ω does not have any role in electing a relay.

Definition 2—In **absolute priority** forwarding, absolute priority is given to better candidates during the selection process. If nodes j and k are two relay candidates, then the fact that k is a better candidate than j ($\Omega_j \le \Omega_k$) guarantees that k is elected over j. That is,

$$\Pr[X_k \le X_j | \Omega_j \le \Omega_k] = 1$$

Note that absolute priority can be obtained with a function $g(\cdot)$, which is deterministic and monotonically decreasing with respect to Ω.

Definition 3—A **hybrid priority** combines absolute priority and purely random selections, and gives priority, but no guarantee to a better candidate in the selection process. If nodes j and k are two relay candidates, then the fact that k is a better candidate than j ($\Omega_j \le \Omega_k$) increases the chances, but does not guarantee that k is elected over j. That is,

$$\Pr[X_k \le X_j] < \Pr[X_k \le X_j | \Omega_j \le \Omega_k] < 1$$

However, hybrid priority will not be discussed further in this paper; it will be left as a future research extension.

6.3.2 Analytical Model

We now determine the characteristics of the prioritization function $g(\cdot)$ (see equation (6.1)) and analyze the priority-specific performance of relay selection processes. We are particularly interested in characterizing the vulnerability to collisions and the effective delay in successful relay selection processes.

It is assumed that based on the location information and set priority criteria in the RTS packet each node can have its own measure Ω_i of forwarding decision.

Two main time, delay-based contention resolution criteria that we consider are random delay and prioritized delay. In the case of random delay, a node in the forwarding zone picks up a random waiting time drawn from a given probability distribution function (PDF) before sending its CTS message. We consider uniformly random (uni_rand(t_2, t_1)) distribution of chosen waiting time in the range [t_2, t_1]. The decision criteria Ω in this case is independent of the nodes' locations (as long as they are in the forwarding region (see Figure 6.2)) and, thus, the waiting time X_i of node i, i.e., the mapping function $g(\cdot)$, is the corresponding random distribution itself.

On prioritized selection process, there could be many possible criteria, namely, residual energy, receiver signal strength, maximum per-hop progress, etc., or a combination of them. In our study, we have chosen maximum per-hop progress toward the destination as the absolute priority criteria. Accordingly, Ω_i represents the progress toward the destination d_i a relay candidate i can offer, if elected (see Figure 6.2). In this case, the mapping function $g(\Omega_i)$, which is monotonically decreasing with respect to d_i, can be expressed as:

$$X_i = g(\Omega_i) = \{a(\alpha)d_i + b(\alpha)\}^{1/\alpha} \tag{6.3}$$

where the coefficients $a(\alpha)$ and $b(\alpha)$ are chosen such that the prioritized delay of a forwarding region node remains within $[t_2, t_1]$, as in the uni_rand case. α is the shape parameter ($\alpha \neq 0$) that governs the nature of relative priority of potential relay nodes. $a(\alpha)$ and $b(\alpha)$ are given by:

$$a(\alpha) = \frac{t_2^\alpha - t_1^\alpha}{R}; \quad b(\alpha) = t_1^\alpha \tag{6.4}$$

In Figure 6.3, the priority-based mapping function $g(\cdot)$ and the corresponding mapped random variable X_i (scheduled time of node i) as a function of the shape parameter α is depicted. It is interesting to observe that, by varying α,

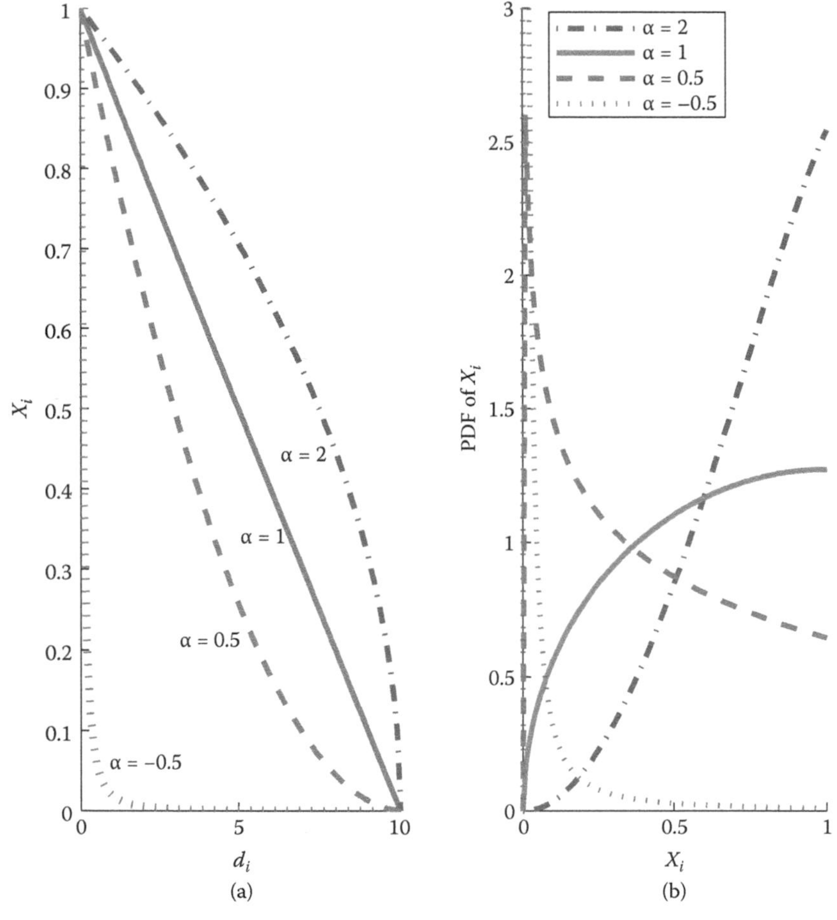

Figure 6.3 (a) Mapping functions in one-hop progress-based absolute priority as a function of α; (b) corresponding PDFs of mapped random variable X_i. $R = 10$.

relative priority of different node selection can be achieved. For example, $\alpha = 1$ corresponds to the linear mapping with respect to one-hop progress, whereas $\alpha = -1$ corresponds to the inverse mapping. What remains to be seen is the effect of different mapping on the relay selection delay and collision vulnerability, which will be discussed below.

In the following development, we first derive the generic expressions for delay and collision probability. Specific cases of relay selection strategies are taken up as examples.

6.3.2.1 Relay Selection Delay

To find the time required to successfully elect a relay, denote by $\{X_i\}_{i=1,2,\ldots}$ the set of independent and identically distributed random variables representing the scheduled reply times of all eligible relay candidates. The number of such contenders is itself another random variable C, which is Poisson distributed with parameter λ, the average number of active forward direction neighbors. The conditional duration of one election process is defined as $Y = \min_i \{X_i\}$. We are interested in the distribution ($f_Y(y)$ and $F_Y(y)$) of Y for an arbitrary prioritization scheme, i.e., for arbitrary distribution ($f_X(x)$ and $F_X(x)$) of the X_is.

If we consider $C = c$ active contending candidates, each with a scheduled time X_i $\{i = 1, 2, \ldots, c\}$, the conditional cumulative distribution function (CDF) of Y is obtained as:

$$\begin{aligned}\Pr[Y \le y|C = c] &= 1 - \Pr[Y > y|C = c]\\ &= 1 - \Pr[\min_i\{X_i\} > y]\\ &= 1 - \prod_{i=1}^{c} \Pr[X_i > y]\\ &= 1 - \{1 - F_x(y)\}^c\end{aligned}$$

Note that Y is defined only for $c \ge 1$. The unconditional CDF of Y can then be obtained by total probability:

$$\begin{aligned}\Pr[Y \le y] &= \sum_{c=1}^{\infty} \Pr[Y \le y|C = c]\frac{\Pr[C = c]}{\Pr[c \ge 1]}\\ &= \sum_{c=1}^{\infty}[1 - \{1 - F_x(y)\}^c]\frac{\frac{\lambda^c}{c!}e^{-\lambda}}{1 - e^{-\lambda}}\end{aligned}$$

After simplification, we obtain:

$$F_Y(y) = \Pr[Y \le y] = \frac{1-e^{-\lambda F_X(y)}}{1-e^{-\lambda}}$$
$$f_Y(y) = \frac{\lambda f_X(y) e^{-\lambda F_x}}{1-e^{-\lambda\lambda}} \tag{6.5}$$

Below, we apply the general result of equation (6.5) to two examples of prioritization.

Uniformly Random Forwarding—Any forwarding solution in which all relay candidates contend by setting a purely random timer falls into this category. As an illustration, we consider the uniformly distributed random time between t_2 and t_1, i.e.:

$$f_X^{rand}(x) = \frac{1}{t_2-t_1}$$
$$F_X^{rand}(x) = \frac{x-t_2}{t_2-t_1}$$

Then, closed-form expressions for the distributions with uniformly random X_i can be obtained from equation (6.5):

$$F_Y^{rand}(y) = \frac{1-e^{-\lambda\frac{y-t2}{t1-t2}}}{1-e^{-\lambda}}$$
$$f_Y^{rand}(y) = \frac{\lambda}{(t_1-t_2)(1-e^{-\lambda})} e^{-\lambda\frac{y-t2}{t1-t2}}. \tag{6.6}$$

Absolute Priority with Linear Mapping—For the absolute priority forwarding, we consider the case where scheduled time is a function of one-hop progress and linearly decreasing between t_1 and t_2. Then, from equation (6.3) and equation (6.4), setting $\alpha = 1$, $X_i = \frac{t_2-t_1}{R} d_i + t_1, 0 \le d_i \le R$, where d_i is the forward progress to the destination of candidate relay i (see Figure 6.2). The distribution of X_i can be deduced from the distribution of remaining distance [13], and can be approximately given by:

$$f_{d_i}(d) = \frac{4(l-d)}{\pi R^2} \arccos\left(\frac{(l-d)^2 + l^2 - R^2}{2l(l-d)}\right) \tag{6.7}$$

The distributions of the scheduled time are given by:

$$F_X^{lin}(x) = 1 - F_{d_i}\left(\frac{t_1-x}{t_1-t_2} R\right)$$
$$f_X^{lin}(x) = \frac{1}{|t_1-t_2|} f_{d_i}\left(\frac{t_1-x}{t_1-t_2} R\right) \tag{6.8}$$

From equation (6.6) and equation (6.8), the average delay in one relay selection attempt can be obtained in these two cases. Delay results for these two examples along with the other variants will be presented in section 6.3.3.

6.3.2.2 Election Failure Probability

As stated earlier (see equation (6.2)), the selection process fails when the first two best candidates are not at least the collision vulnerability window apart from each other. Formally, denote Y as the minimum of the set of scheduled reply times: $Y = \min_i \{X_i\}$, and Y^* as the minimum of the remaining nodes' scheduled reply times: $Y^* = \min\{\{X_i\} - Y\}$. Note that Y and Y^* can be considered identically distributed. Also, define $S_Y(y) = 1 - F_Y(y)$ as the survival function of Y, and $h(y) = \frac{f_Y(y)}{S_Y(y)}$.

Lemma 1: For a given collision vulnerability window β, *the rate of failure of the selection process is given by*

$$P_{fail} = 1 - (h \copyright S_Y)(\beta) \tag{6.9}$$

where © represents the correlation integral function defined by:

$$(h \copyright S_Y)(t) = \int_{-\infty}^{\infty} h(x) S_Y(t+x)\, dx$$

Proof: Under the condition $Y = y$, a collision occurs with the conditional probability $\Pr[Y^* \leq y + \beta \mid Y = y] = \frac{F_Y(y+\beta) - F_Y(y)}{1 - F_Y(y)}$. The unconditional probability of collision then can be obtained by:

$$\begin{aligned} P_{fail} &= \int_{t_2}^{t_1} \Pr[y \leq Y \leq y + dy] \frac{F_Y(y+\beta) - F_Y(y)}{1 - F_Y(y)} \\ &= \int_{t_2}^{t_1} f_Y(y)\, dy \frac{F_Y(y+\beta) - F_Y(y)}{1 - F_Y(y)} \\ &= \int_{t_2}^{t_1} f_Y(y) \frac{S_Y(y) - S_Y(y+\beta)}{S_Y(y)} dy \end{aligned}$$

After simplification, realizing that for $y \geq t_1$, '$S_Y(y) = 0$, and for $y \leq t_2$, $h(y) = 0$, we have,

$$P_{fail} = 1 - \int_{t_2}^{t_1} h(y) S_Y(y+\beta)\, dy$$

and, hence, the proof.

Failure probabilities for specific prioritization cases can be obtained from the respective distribution functions.

6.3.2.3 Effective Delay

With the knowledge of delay in one selection attempt and the corresponding failure probability, t. Because the distributed relay selection process can fail due to collision, repeated attempts may be required for successfully finding a next hop. To have a baseline comparison of priority-specific approaches, one needs to compute the effective delay of an eventually successful relay selection process. For this purpose, we simply assume that, in case of a collision of the RTS reply message, the sender reinitiates another selection process at the end of a fixed timeout window (t_1), and repeats this process until a CTS is received successfully.* The effective average delay before a relay is successfully found is given by:

$$D_{eff} = \frac{P_{fail}}{1 - P_{fail}} t_1 + D, \tag{6.10}$$

where P_{fail} is the collision probability given in equation (6.9), and D the average delay is one attempt, obtained from equation (6.5).

Numerical and simulation results are shown below.

6.3.3 Evaluation

To evaluate the prioritization performance in receiver-side relay selection process, we consider uni_rand(t_2, t_1) case and least remaining distance (LRD) to the destination based [13] priority forwarding. The mapping function given in equation (6.3). The range of waiting time in both cases is set in $[t_2, t_1]$. The mapping parameter α is varied to give different priority levels to different eligible nodes.

We assume independent and asynchronous sleep behavior of nodes in the network, which also gives a different realization of the set of active nodes eligible for forwarding in each transmission attempt by a sender. It also is assumed that the average number of active neighbors n of any node (also referred to as node density) is stationary. For simplicity in obtaining the analytic results, average number λ of relay candidates of a node, denoted by the shaded region in Figure 6.2, is approximated as $\lambda = \frac{n}{2}$.

* In practice, each transmission failure is followed by a binary exponential backoff, and only a finite number of reattempts are done.

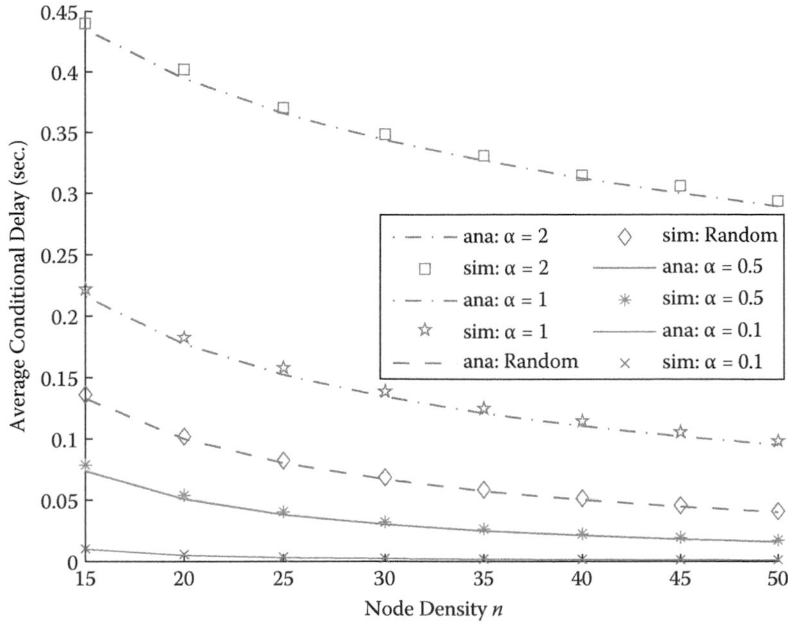

Figure 6.4 Delay versus node density in one relay selection attempt in different priority approaches.

All numerical and simulation results in this section are obtained with node transmission range $R = 10$ and a sink node at a distance $l = 100$ from the initial transmitting node. The collision vulnerability window is taken to be the change-over time of commonly available hardware ($\beta = 250\mu s$). The scheduled reply times range from $t_2 = 250\ \mu s$ to $t_1 = 1s$.

We first obtain the delay in one relay selection attempt for different priority where we are particularly interested in the behavior of the relative priorities controlled by the shape parameter α. Figure 6.4 shows that the smaller the α, the lesser the delay. This is because, as demonstrated in Figure 6.3b, for lower value of α (< 1), the scheduled time distribution is squeezed toward the lower values. On the other hand, $\alpha > 1$ stretches the distribution in the opposite direction resulting in higher conditional delays.

The selection failure probability P_{fail} plots in Figure 6.5 show a reverse trend as α changes. This is because, at lower α, since the distribution of scheduled time is shifted toward the lower range, there is a high chance that the two best nodes have nearly the same scheduled time (see equation (6.2)) for the CTS message, which leads to a collision.

With the observation of counter-directional trends of delay in one selection attempt and selection failure probability for different shape parameters, we first compute via analysis the effective delay performance in a successful relay selection

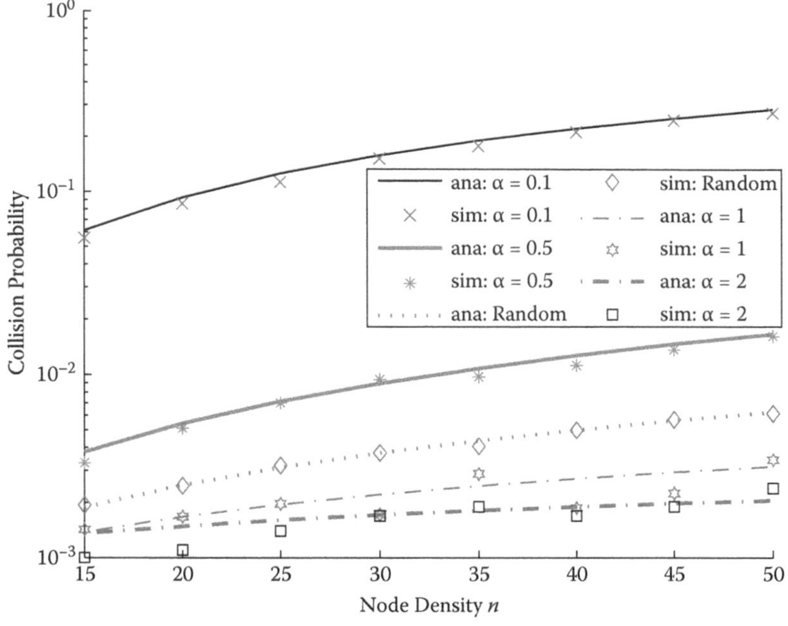

Figure 6.5 Election failure probability P_{fail} versus node density n in different priority approaches.

process. In Figure 6.6, we show the effect of the shape parameter on the effective delay. We observe that an optimum shape parameter can be obtained to achieve the best tradeoff between delay and collision probability, which is also a slowly varying function of the node density. For example, at node density $n = 20$, the optimum shape parameter $\alpha_{opt} = 0.3$ and the effective delay is nearly 37.9 msec, whereas, at $n = 20$, $\alpha_{opt} = 0.5$ and the effective delay is nearly 31.5 msec.

Next, we obtain the effective delay in successfully electing a relay for different prioritization schemes. The intuition from Figure 6.6 is verified in Figure 6.7 via analysis and simulation that there is a critical value of shape parameter for which an optimum selection performance can be achieved. It is observed as well that unless the prioritization scheme is suitably optimized, its performance can be even poorer than the random selection process.

Note that the average one-hop progress in LRD-based, absolute priority forwarding is independent of the shape parameter as the best relay is elected in all cases. We observed that the average progress in random forwarding is nearly less than half of that in the LRD approach (Figure 6.8). Although random forwarding has a reasonably good delay performance, which can be even better than the LRD-based priority unless the shape parameter is chosen judiciously, very poor one-hop progress makes the random forwarding a rather less-promising approach.

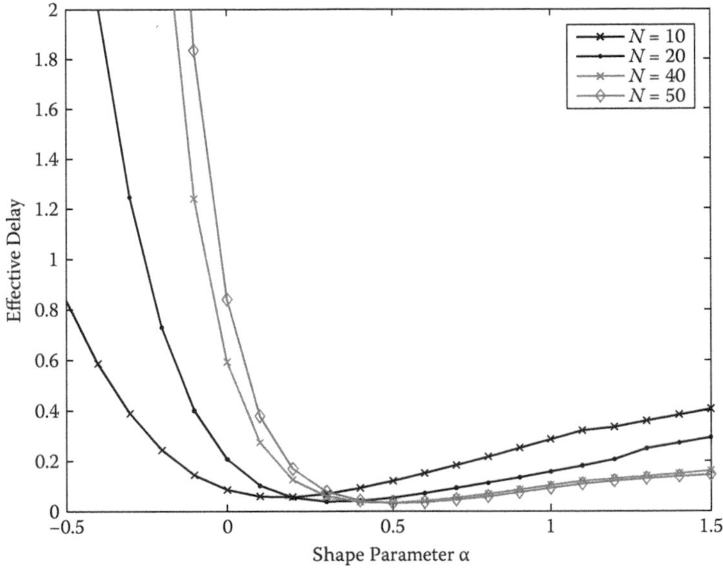

Figure 6.6 Analytically obtained effective delay versus shape parameter and with node density *n* as parameter.

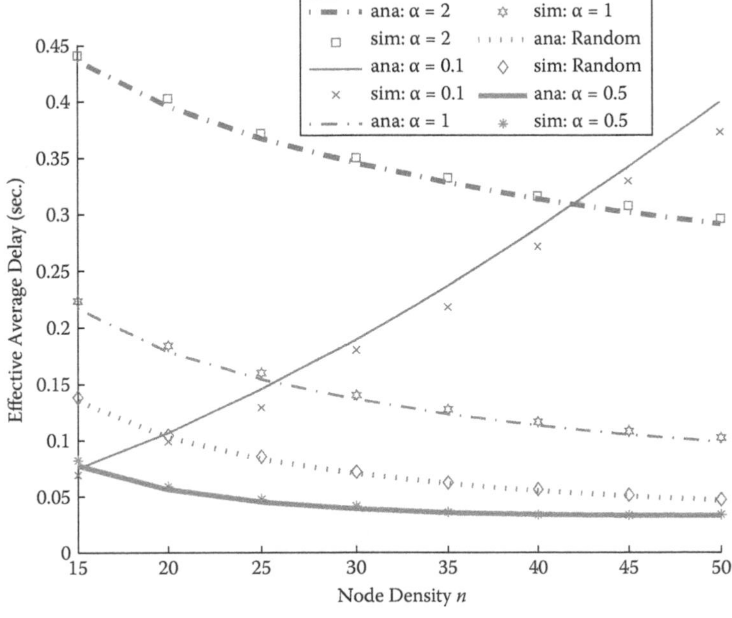

Figure 6.7 Effective delay in successfully electing a relay versus node density in different priority cases.

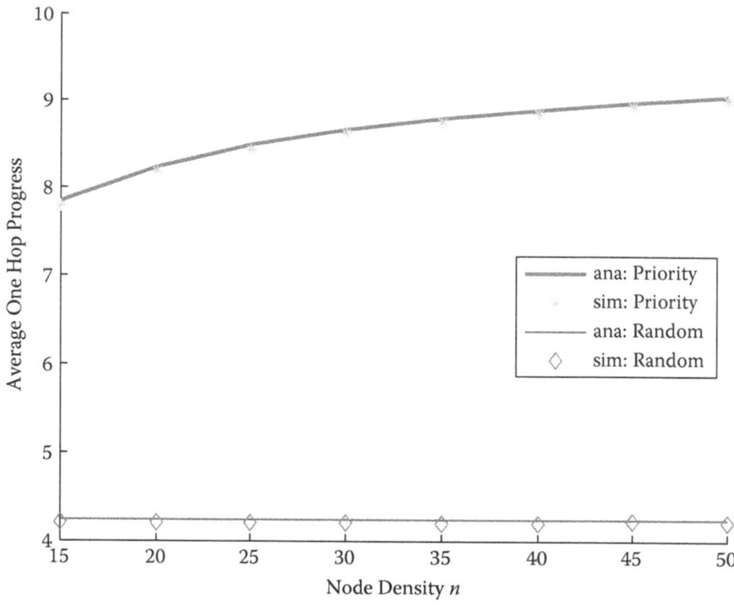

Figure 6.8 One-hop progress in random forwarding and LRD priority-based approaches versus node density, for any shape parameter α.

6.4 Multicriteria, Receive-Side Relay Selection

6.4.1 Optimality Notion in the Multicriteria Case

As noted earlier, multihop forwarding based on the one-hop progress criterion can hardly be optimal because of the unreliable nature of wireless links and other nodal limitations, such as energy, buffer capacity, etc. However, as more than one decision parameters are considered, the ranking of an alternative candidate becomes less obvious than in the single criterion case. Consider, for example, Figure 6.9 where two criteria are used to select the best relay node. With respect to a particular node (node A), the relationship with any other candidate can be classified as follows:

- All nodes in the *dominated zone* are clearly strictly "inferior" compared to node A because they perform strictly poorer on at least one criterion and almost as good on all others.
- All nodes in the *dominating zone* are clearly strictly "superior" compared to node A because they perform strictly better on at least one criterion and at least as good on all others.
- However, nodes in the two *nondominated* zones perform better than node A on a single criterion and poorer on all others. Therefore, nodes in the non-dominated zone cannot be qualified as "inferior" or "superior" to node A.

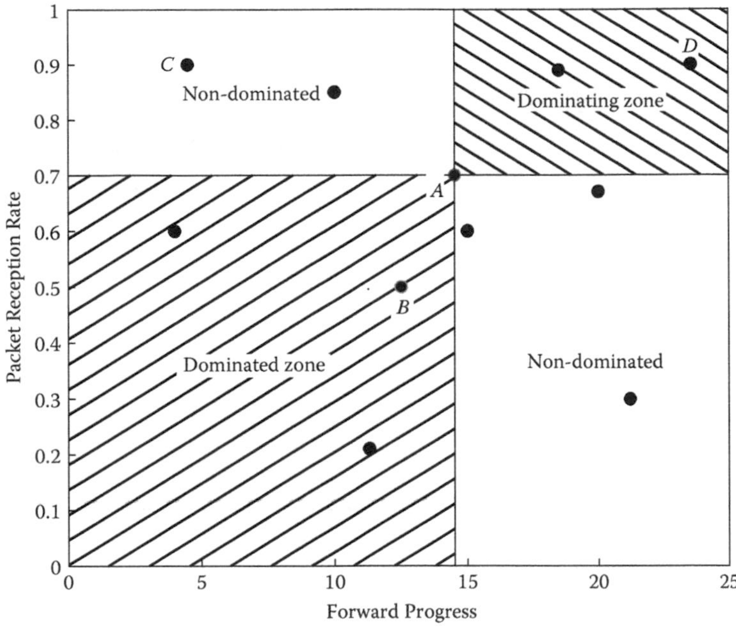

Figure 6.9 Relation between a particular node (A) and other candidates. In general, only candidates in the hatched areas can be strictly compared with node A.

Note that a forwarding decision can be made that maximizes all decision criteria whenever there exists a single candidate that dominates all other candidates (see node D in Figure 6.9). However, in general, a single dominating candidate does not always exist and an additional model is needed to define preference and tradeoffs among multiple criteria.

6.4.2 Multicriteria Mapping Function

Now, we introduce a general preference model in the form of an aggregating function that combines all criteria into a single virtual criterion used to out-rank all weak candidates.

Because the order induced by the dominance relationship on the set of alternative candidates is partial, there may exist, among the set of alternatives, pairs of mutually incomparable candidates. With the mapping function, our objective is to introduce a single ranking scale through the use of an aggregating function that weights all criteria into a single one. Consider a decision based on k numerical criteria for which each candidate i has a performance index represented by the vector $\bar{\Omega}_i = (\Omega_{i1}, \Omega_{i2}, \ldots, \Omega_{ik})$. Without loss of generality, we assume that decision criterion (Ω_i) has a value in the range $[0, \Omega_i^{max}]$ and has to be maximized.

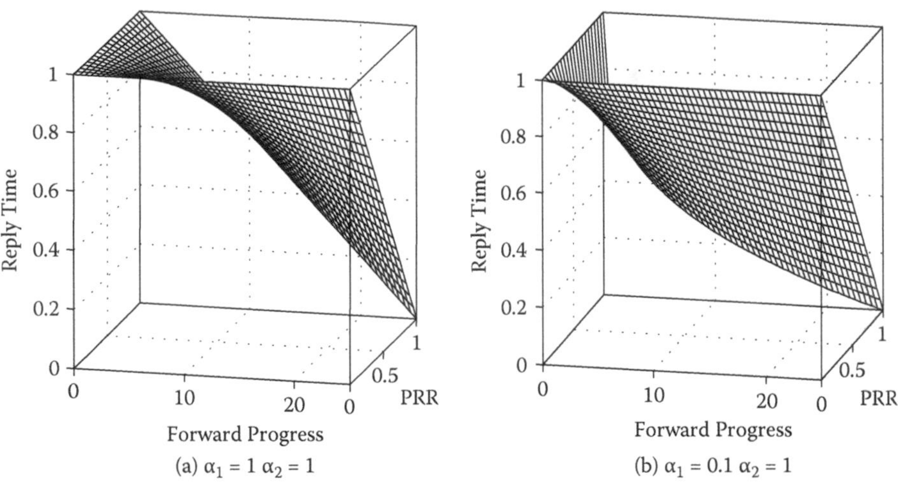

Figure 6.10 Mapping function in a two criteria case.

We then map all decision variables onto the scheduled time by introducing the multidimensional family of functions (see Figure 6.10 for two criteria example).

$$g_{\bar{\alpha}}(\Omega_{i1}, \Omega_{i2}, \cdots, \Omega_{ik}) = a(\bar{\alpha})\Omega_{i1}^{\alpha 1}\Omega_{i2}^{\alpha 2}\cdots\Omega_{ik}^{\alpha k} + b(\bar{\alpha}) \tag{6.11}$$

where $\bar{\alpha} = (\alpha_1, \alpha_2, \ldots, \alpha_k)$ is a k-parameter vector used to weigh the k decision criteria. As in the single criterion case, the scheduled reply time for each candidate is $X_i = g_{\bar{\alpha}}(\bar{\Omega}_i)$.

From the perspective of transmitter-side relay selection, a corresponding cost metric can be derived from $g_{\bar{\alpha}}(.)$ as:

$$C_{\bar{\alpha}}(\bar{\Omega}_i) = \Omega_{i1}^{\alpha 1}, \Omega_{i2}^{\alpha 2}\cdots\Omega_{ik}^{\alpha k} \tag{6.12}$$

Ranking all candidates with respect to $g_{\bar{\alpha}}$ (in descending order) or $C_{\bar{\alpha}}$ (in ascending order) creates a total ordering system on the set of all alternative candidates. That is, for any arbitrary two candidates i and j, $C_{\bar{\alpha}}(\bar{\Omega}_i) \leq C_{\bar{\alpha}}(\bar{\Omega}_j)$ or $C_{\bar{\alpha}}(\bar{\Omega}_i) \geq C_{\bar{\alpha}}(\bar{\Omega}_j)$.

Note that, for any positive real constant $m > 0$, $C_{m\bar{\alpha}}, m\bar{\alpha} = (m\alpha_1, m\alpha_2, \ldots, m\alpha_k)$ produces the same ranking as $C_{\bar{\alpha}}$. Therefore $g_{\bar{\alpha}}$ can be seen as a single *virtual criterion* ($C_{\frac{1}{\alpha_1}\bar{\alpha}}$) which, as in the single criterion case in section (6.3.2), we map onto the time interval $[t_2, t_l]$ for the purpose of receiver-side contention resolution:

$$g_{\bar{\alpha}}(\bar{\Omega}) = a(\bar{\alpha})\left[C_{\frac{1}{\alpha_1}\bar{\alpha}}(\bar{\Omega})\right]^{\alpha_1} + b(\bar{\alpha}) \tag{6.13}$$

Again, we obtain the parameter-dependent coefficients from the limiting conditions for the worst and best candidates:

$$a(\bar{\alpha}) = \frac{t_2 - t_1}{\Pi_1^k [\Omega_i^{\max}]^{\alpha_i}}; \quad b(\bar{\alpha}) = t_1 \tag{6.14}$$

As in the single criterion case, the multidimensional mapping function $g_{\bar{\alpha}}$ is a decreasing function with respect to each dimension considered individually.

6.4.3 Trading Off Greediness for Link Quality

With the general mapping function presented above, we now apply the multicriteria mapping to an example case of a forwarding scheme that finds an optimal trade-off between link quality and greedy forward progress. An investigative approach is required because there is no a priori suggestion on what should be the optimal weights of the two criteria. For example, with $\alpha_1 = \alpha_2 = 1$, we obtain $C_{(1,1)} = d_x{}^*p_x$ (the product of one-hop progress offered by node x and the corresponding packet success probability), which corresponds to the normalized advance (NADV) [5] and maximum expected progress (MEP)[19]. However, as was presented in section 6.3.3, our results show that this is suboptimal, and a substantially better network performance can be obtained by appropriately choosing the weighting parameters.

To see the impact of weight parameters (α_i) on the ranking of alternative relay candidates, consider node A (with $d_A = 14.5$ meter, $p_a = 0.7$) in Figure 6.11. Note how in the case of $\alpha_1 = \alpha_2 = 1$ (Figure 6.11a) a small increase in forward progress can compensate for a large decrease in link quality. On the other hand, with $\alpha_1 = 0.1$ and $\alpha_2 = 1$ (Figure 6.11b), a node at almost a 10 unit distance away from A could offer an almost equally good alternative relay.

Note also that, to find the rules for optimal forwarding decision making, only relative values of the two weight parameters are needed. In other words, we look for the ratio $\frac{\alpha_1}{\alpha_2} \triangleq \lambda$ that optimizes network performance metrics (e.g., energy, packet failure, and delay) both from the perspective of transmitter-side relay selection and receiver-side relay selection. We will investigate the optimum value of λ via network simulations.

6.5 Illustration: Minimum Energy Link-Aware Forwarding

6.5.1 Simulation Model

We have considered randomly deployed nodes, with varied average density p (nodes/m^2). Nodal parameters have been based on Chipcon RFIC CC2420 operating with a BFSK modulation scheme at 900 MHz. All nodes transmit with a

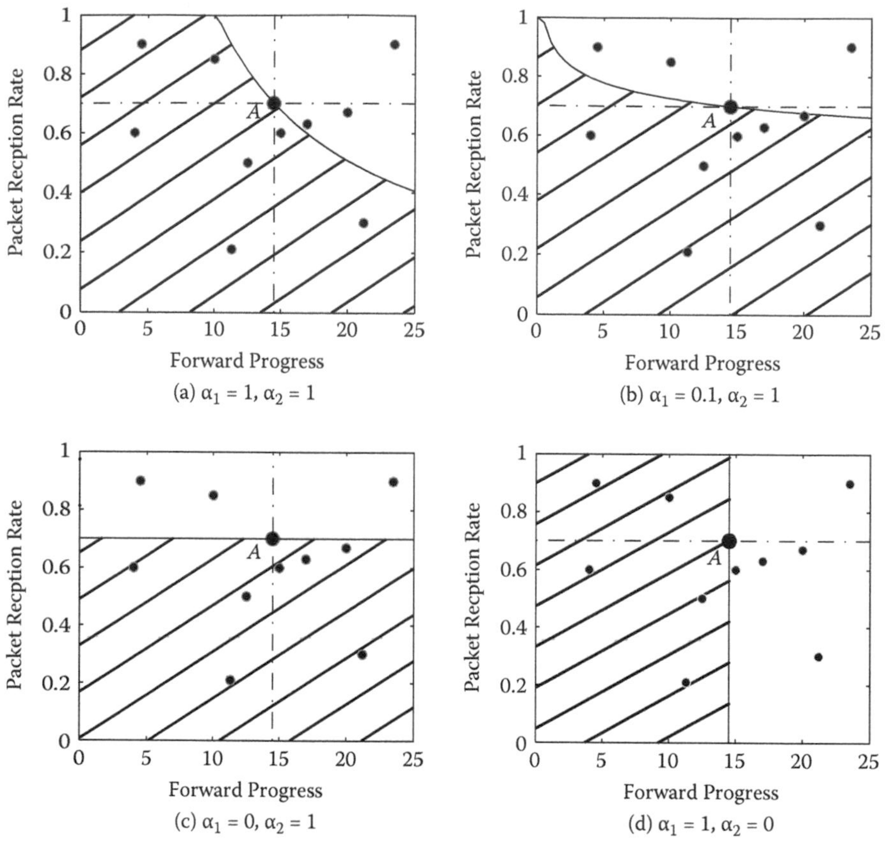

Figure 6.11 Preference relation with respect to a particular node. The set of alternatives is partitioned according to weight given to each criterion.

nominal power (0 dB) and at a rate of 19.2 kbps. A log-normal fading channel with standard deviation of channel disturbance 4 dB and path loss exponent 4.0 has been assumed. Fixed path loss has been calculated considering near field distance of 1 m. Network performance has been studied with approximate end-to-end distance 100 m. The scheduled reply times range from $t_2 = 250$ μsec to $t_1 = 1$ sec. Fixed packet size has been considered for all transmissions (50 bytes for DATA and 4 bytes for RTS). Each message is considered to have 100 data packets. No a priori transmission range has been assumed; all nodes capable of correctly receiving the initial broadcast RTS packet participate in the selection process. Also, it has been assumed that a node is aware of its own geographic or virtual (hop-count based) [18] location information and that of the final destination. Each RTS packet contains position information of both the sender and the final destination.

6.5.2 Performance Metrics

6.5.2.1 End-to-End Packet Failure Rate

To measure the relaying performance with a given tradeoff parameter through an unreliable wireless medium, we consider packet failure rate along the route. As a baseline comparison, we record the number of transmissions required for successful delivery of a message at the final destination. Figure 6.12 shows the packet loss rate with node density, which indicates that beyond certain high node density, irrespective of the tradeoff parameter, the loss performance stabilizes. This is because at very low node density a node tends to find a relay that is associated with a highly error-prone channel. As the density increases, an optimum tradeoff is possible.

Figure 6.13 shows that packet loss along the entire path can be reduced linearly with the tradeoff parameter l. For example, $\lambda = \frac{1}{2}$ reduces the packet failure rate by 50 percent with respect to simple product of hop progress and packet success rate (i.e., with l = 1).

6.5.2.2 End-to-End Forwarding Delay

Now we consider end-to-end delay due to packet transmission/retransmission. In our simulation, once a relay node is elected, up to max_retx retransmissions are allowed. More than max_retx packet failures results in link error and a new relay

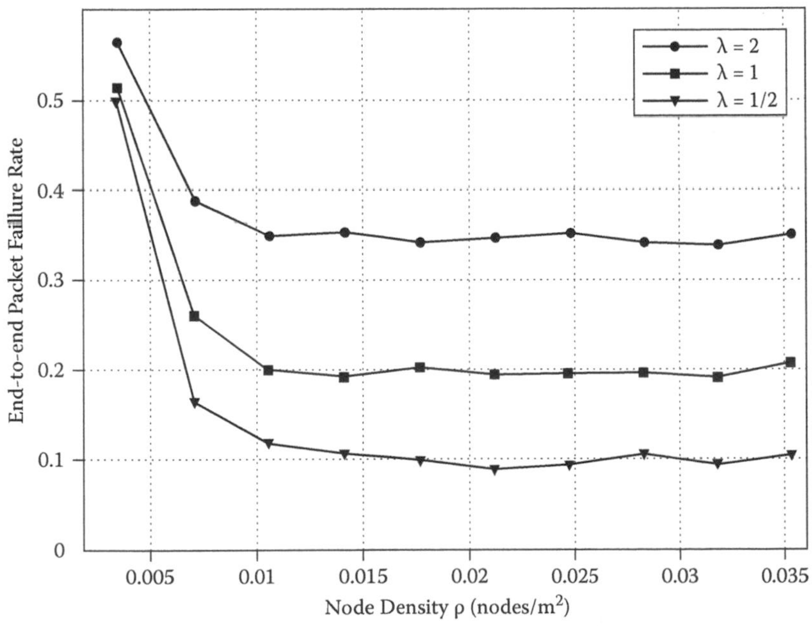

Figure 6.12 End-to-end packet failure rate as function of density.

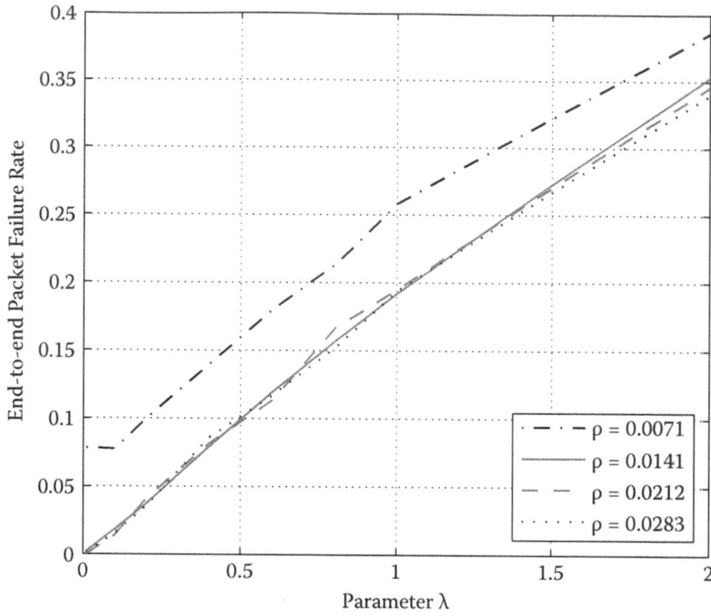

Figure 6.13 End-to-end packet failure rate as function of the tradeoff parameter λ.

selection process is initiated. Also, each successful transmission takes t_{tx} amount of time and each retransmission causes an additional delay t_{out} due to timeout (negative acknowledgment). Figure 6.14 presents end-to-end packet delay as a function of node density, which shows the effect of packet failure on packet delay (compare Figure 6.14 with Figure 6.12). Although Figure 6.13 suggests that packet failure and, as a consequence, end-to-end delay can be made arbitrarily small by selecting smaller tradeoff parameter λ, our next result on energy efficiency shows that there exists a minimum value of λ beyond which adverse energy effect can be seen.

6.5.2.3 End-to-End Energy Consumption

We evaluate the energy efficiency of a given forwarding strategy by the number of transmissions required along the route for a successful end-to-end packet delivery. As expected, the energy requirement due to forwarding decreases with higher node densities, where it is more likely to find a neighbor offering a good combination of hop progress and link quality (Figure 6.15). Figure 6.15 also shows that it is possible to improve energy efficiency by reducing the weight given to hop progress. Clearly, $\lambda = 0.2$ outperforms the simple product form ($\lambda = 1$). It also can be seen that further reduction of the weight given to hop progress results in increasing

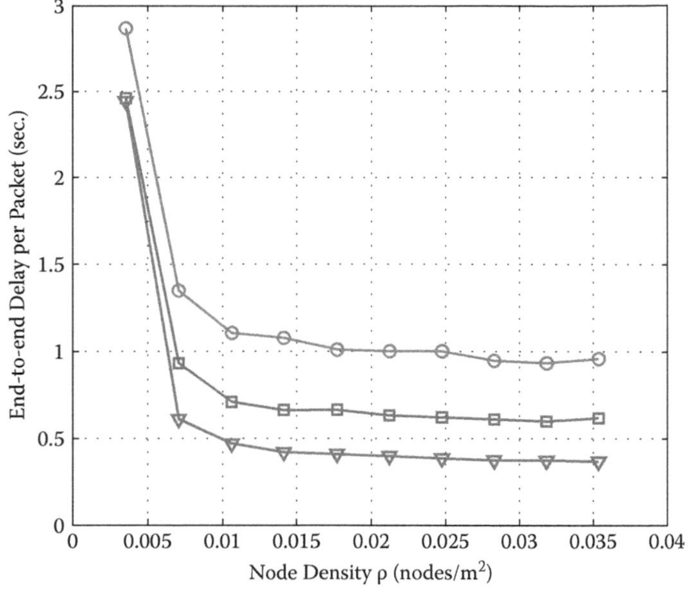

Figure 6.14 End-to-end delay as function of node density. max_retx = 8, t_{tx} = 21.1 msec, t_{out} = 84.4 msec.

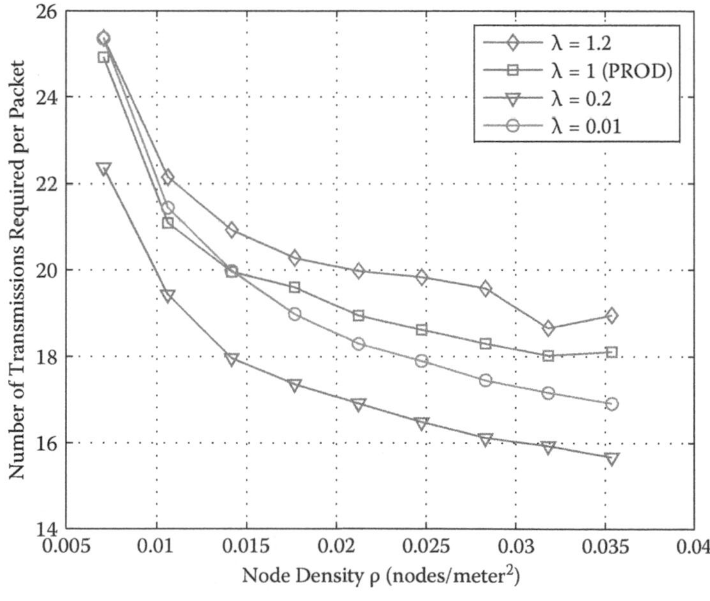

Figure 6.15 Energy consumption (number of required transmission) for end-to-end packet delivery as a function of node density.

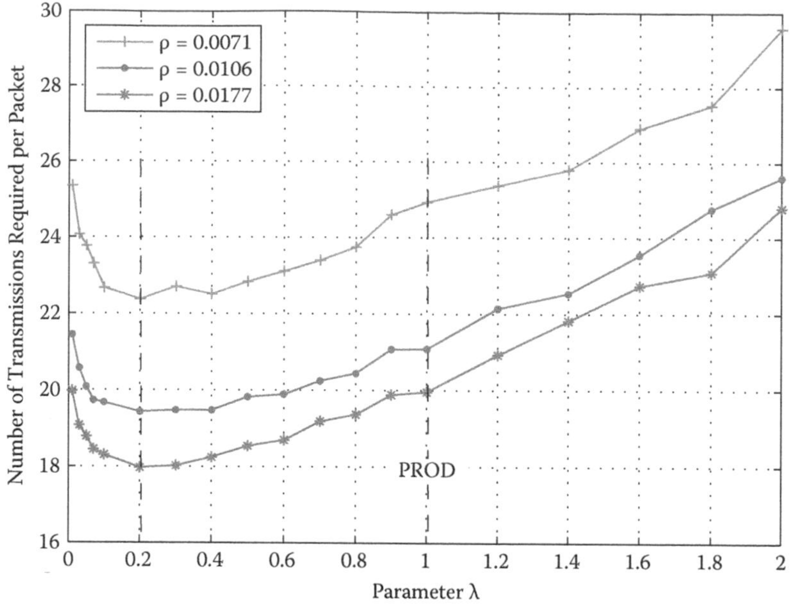

Figure 6.16 Energy consumption (number of required transmissions) as a function of the weight/age parameter α_1 (with $\alpha_2 = 1$).

energy consumption. Figure 6.16 shows that an optimal tradeoff between hop progress and link quality can be found that minimizes the required energy consumption. It shows that the optimal performance is achieved approximately at $\lambda = 0.2$. Notice from Figure 6.13 that this optimal λ can achieve up to *approximately five times reduction* in packet failure rate with respect to the simple product form (i.e., with $\lambda = 1$).

6.6 Conclusion

We have presented a multicriteria, receiver-side relay selection framework for multihop relaying in ad hoc networks. Via intuitive reasoning and examples, we have first shown qualitatively the importance of finding optimum weighted relay selection criteria. A generalized cost metric in the form of multiparameter mapping function has been proposed and used to investigate optimal tradeoff between greedy forwarding and link quality. It has been shown that a much better network performance in terms of total energy consumption for successful end-to-end routing can be achieved through judicious selection of the weighting parameter that optimally trades off between greediness and link quality.

The multicriteria mapping function is quite general and can also be applicable to the transmitter-side relay selection process.

References

[1] B. Karp and H. T. Kung, "GPSR: Greedy perimeter stateless routing for wireless networks," in *Proceedings of the ACM MOBICOM*, Boston, MA, Aug. 2000, pp. 243–254.

[2] R. Zheng, "On routing in lossy wireless networks with realistic channel models," in *Proceedings of the ACM International Workshop on Foundations of Wireless Ad Hoc and Sensor Networking and Computing*, New York, May 2008, pp. 1–6.

[3] K. Seada, M. Zuniga, A. Helmy, and B. Krishnamachari, "Energy-efficient forwarding strategies for geographic routing in lossy wireless sensor networks," in *Proceedings of the ACM SENSYS*, Baltimore, MD, Nov. 2004, pp. 108–121.

[4] M. Zuniga and B. Krishnamachari, "Analyzing the transitional region in low power wireless links," in *IEEE International Conference on Sensor and Ad Hoc Communications and Networks (SECON)*, 2004, pp. 517–526.

[5] S. Lee, B. Bhattacharjee, and S. Banerjee, "Efficient geographic routing in multihop wireless networks," in *Proceedings of the ACM MobiHoc*, Urbana-Champaign, IL, May 2005, pp. 230–241.

[6] M. Zorzi and R. R. Rao, "Geographic random forwarding (GeRaF) for ad hoc and sensor networks: Multihop performance," *IEEE Trans. Mobile Comput.*, vol. 2, no. 4, pp. 337–348, Oct.-Dec. 2003.

[7] H. Fubler, J. Widmer, and M. Kasemann, "Contention-based forwarding for mobile ad hoc networks," *Elsevier Ad Hoc Networks*, vol. 1, no. 4, pp. 351–369, Nov. 2003.

[8] K. Egoh and S. De, "Priority-based receiver-side relay election in wireless ad hoc sensor networks," in *Proceedings of the IEEE IWCMC'06*, Vancouver, British Columbia, Canada, July 2006.

[9] K. Egoh and S. De, "A Multi-Criteria Receiver-Side Relay Election Approach in Wireless Ad Hoc Networks," in *Proceedings of the Military Communications Conference, 2006. MILCOM 2006*, Washington, D.C., Oct. 2006.

[10] T.-C. Hou and V. O. K. Li, "Transmission range control in multihop packet radio networks," *IEEE Trans. Commun.*, vol. 34, no. 1, pp. 38–44, Mar. 1986.

[11] H. Takagi and L. Kleinrock, "Optimal transmission ranges for randomly distributed packet radio terminals," *IEEE Trans. Commun.*, vol. COM-32, no. 3, pp. 246–257, Mar. 1984.

[12] M. Mauve, J. Widmer, and H. Hartenstein, "A survey on position-based routing in mobile ad hoc sensor networks," *IEEE Network Mag.*, vol. 15, pp. 30–39, June 2001.

[13] S. De, "On hop count and Euclidean distance in greedy forwarding in wireless ad hoc networks," *IEEE Commun. Letters*, vol. 9, no. 11, pp. 1000–1002, Nov. 2005.

[14] P. -J. Wan, "A survey on position-based routing in mobile ad hoc networks," in *Network, IEEE*, Nov/Dec 2001, vol. 15, no. 6, pp. 30–39.

[15] R. Tanbourgi, H. Jakel, and F. K. Jondral, "Increasing the One-Hop Progress of Nearest Neighbor Forwarding," in *IEEE Communications Letters*, Jan. 2011, vol. 15, no. 1 pp. 64–66.

[16] C. Yi, P. Wan, X. Li, and O. Frieder, "Fault tolerant sensor networks with Bernoulli nodes," in *Proceedings of the IEEE WCNC*, New Orleans, LA, Mar. 2003.

[17] S. C. Zhang, F. I. Koprulu, R. Koetter, and D. L. Jones, "Feasibility analysis of stochastic sensor networks," in *Proceedings of the Conference on Sensor and Ad Hoc Communications and Networks*, Santa Clara, CA., October 2004.

[18] A. Rao, C. Papadimitrou, S. Ratnasamy, S. Shenker, and I. Stoica, "Geographic routing without location information," in *Proceedings of the ACM MOBICOM*, San Diego, CA, Sept. 2003, pp. 96–108.

[19] M. R. Souryal and N. Moayeri, "Channel-adaptive relaying in mobile ad hoc networks with fading" in *Proceedings of the IEEE SECON* Santa Clara, CA, Sept. 2005.

Chapter 7

Energy Optimization Techniques for Wireless Sensor Networks

Sonali Chouhan

Contents

7.1 Introduction

Advances in very large-scale integration (VLSI) technology have made very small sensor nodes possible. The sensor nodes broadly incorporate sensors, a microcontroller or a microprocessor, limited memory, and a radio. Wireless network of these sensor nodes form a wireless sensor network (WSN). WSN is deployed for a wide variety of applications and includes both indoor and outdoor deployments. Some example applications are landmine detection, monitoring volcano eruption activities, habitat monitoring, vibration and temperature measurement of oil tankers, medical diagnoses, disaster management, structural monitoring, fire monitoring, etc. [1]. WSN can be deployed for various applications where a conventional method for collection of data fails or is difficult to implement. Some examples of such situations are in an unsafe or unreachable environment, such as in landmine detection, monitoring volcano eruption activities, and vibration and temperature measurement of an oil tanker. WSN is also very useful where data collection is crucial, e.g., for medical diagnosis, disaster management, structural monitoring, fire monitoring, etc., and also economical where continuous data collection is required. The easy to deploy nature of WSNs makes them promising in many other areas like home automation, intelligent agriculture, traffic flow management, smart kindergarten, production surveillance, etc. In all such applications, WSN allows deployment of sensing elements close to the phenomenon of interest. It does not create disturbance to the environment, animals, or plants. Once deployed, it can be utilized for multiple purposes. A sensor node may have more than one sensor that allows collection of more than one physical quantity. For example, a sensor network deployed in a jungle for habitat monitoring also can be used simultaneously to collect the temperature of that area.

In most of these applications, the sensor nodes are battery-driven. Battery-driven sensor nodes are easy and fast to deploy as compared to wired nodes because one need not set up an electrical connection network. At the same time, for battery-driven sensor nodes, changing batteries frequently is neither convenient nor economical. Therefore, reducing energy consumption of a sensor node and of an overall wireless sensor network is essential.

To find an energy-efficient WSN, solutions are considered in different directions varying from the system to the component level. At the system level, to design an energy-efficient WSN, the main emphasis of researchers is on network topology, routing, protocols, i.e., systems as a whole. Communication energy can be reduced by proper network infrastructure. Packet-size optimization is one of the ways of reducing energy in data transmission. In data centric networks, energy can be reduced by dynamic power management techniques, such as scheduling protocol. At a lower level of abstraction, each component can be optimized to reduce the power consumption. Depending on the way the energy optimization is done, these methods can be classified as the component-level and the system-level energy optimization techniques. In some of these methods, the energy optimization can be done at the sensor node manufacturer end and some can be done at the user end. Every technique has its own merits. The objective of this chapter is to discuss these techniques and their advantages and limitations. For all of these techniques, it is important to estimate the energy of the sensor nodes and WSN. Therefore, we will discuss the energy models as well in this chapter.

The overall organization of this chapter is as follows. First, we will take a brief overview of the sensor node system architecture and WSN topologies. Next, we will discuss various energy models used to estimate the node energy consumption. Then, the energy optimization techniques, classified as the component level and the system level techniques, will be discussed.

7.2 Wireless Sensor Node and Network

A sensor node is a very small unit that senses a physical quantity and transmits it to another node. It may use its own processing capability to process data. A sensor node typically consists of a power supply unit, a sensor unit, a computation unit, and a radio unit (Figure 7.1). The power supply unit contains a battery to supply the power and a DC-to-DC converter. In a sensor node, different sub parts of the circuit may operate on different supply voltages. The DC-to-DC converter not only regulates the output, but also it allows for meeting the voltage requirement of a circuit operating at a different voltage than the voltage supplied by the battery. The sensor and the analog-to-digital converter typically are parts of the sensor unit. The computation unit typically has one or more microcontrollers/microprocessors and memory. The radio unit consists of a transmitter, a receiver, and an antenna.

The sensor nodes collectively form a WSN. The nodes can be connected in either a hierarchical [2] or a flat multihop topology [3]. In the hierarchical topology, shown in Figure 7.2a, each node is associated with a cluster head (CH). This association of a

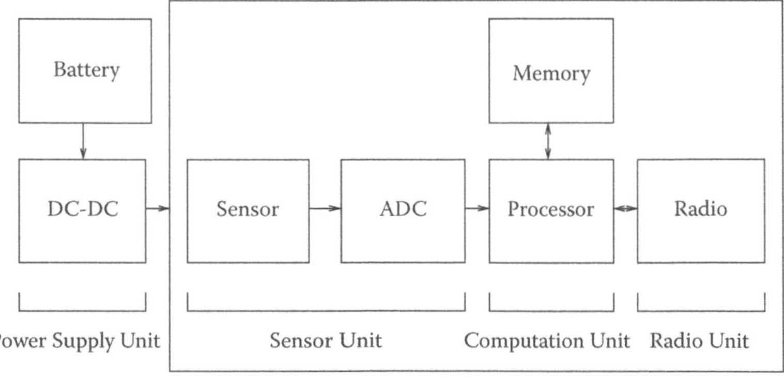

Figure 7.1 Wireless sensor node system architecture.

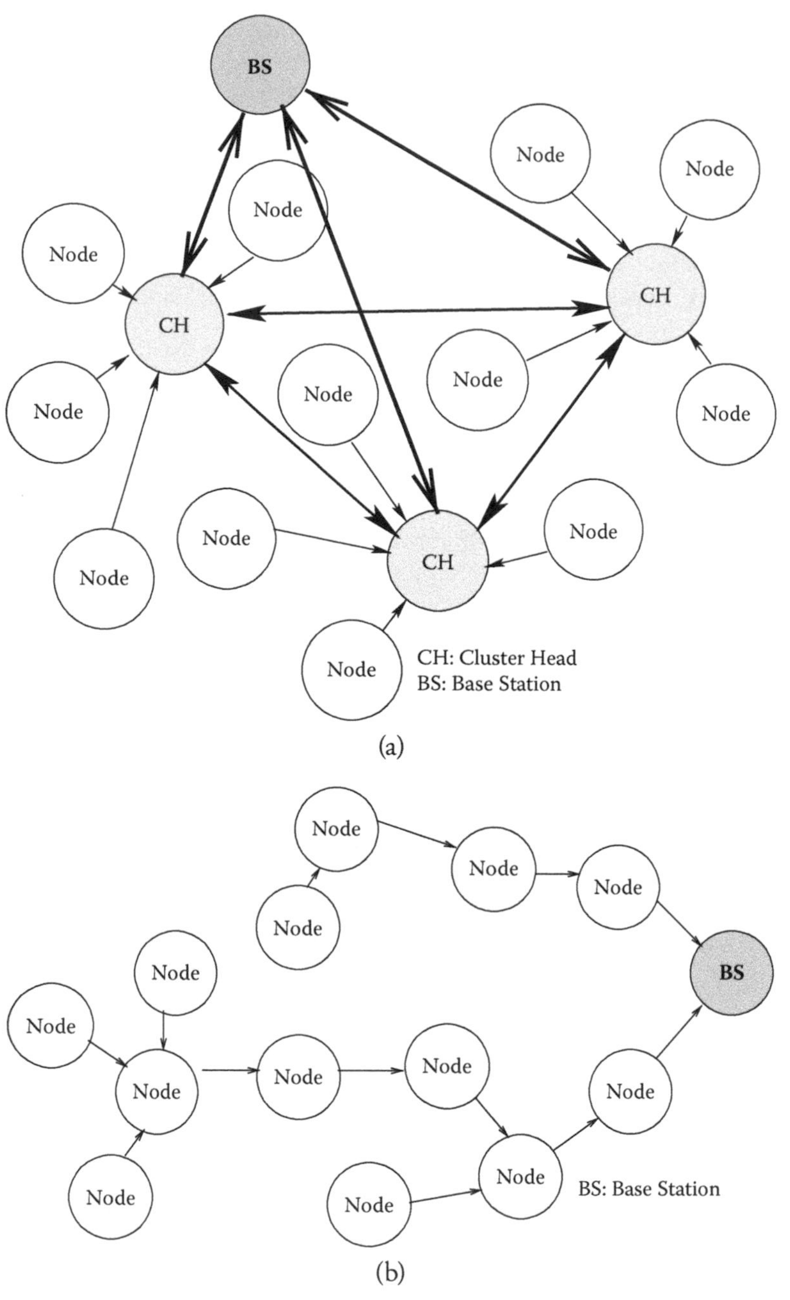

Figure 7.2 Wireless sensor network topologies.

node with a CH may be static or dynamic. The CH is a node with additional functionality. In a particular region, each CH gets data from the nodes associated with it. The CHs are connected to the base station (BS) wirelessly and transmit data to it. The BS generally is a computing machine with high computation and storage capabilities. It collects the data, analyzes it, and performs control operations. According to the need and application, the BS may be a general computing machine. It is an interface between the WSN and rest of the world. In the flat multihop topology (Figure 7.2b) each node collects and sends data to its neighbor. Every node is similar in functionality and computation capabilities. Similar to the hierarchical topology, selection of the neighboring node may be static or dynamic.

To take any measure to reduce the WSN energy consumption, we are required to evaluate the energy consumption of a sensor node for various configuration selections. The individual sensor node energy estimate then can be used to evaluate the energy consumption of a WSN. For this purpose, we need a sensor node energy model. In a sensor node, energy is mainly consumed in the radio unit, the processor unit, the sensor unit, and the battery unit. For a specific application, the sensor used is fixed. Hence, the energy consumed in a sensor unit is nearly constant for the similar usage conditions. Similarly, the energy consumption in the battery unit, due to the presence of the DC-DC converter, also can be assumed to be constant. The major variations in the node energy are due to the computation unit and the radio unit.

In conventional wireless networks (CWNs), the transmission distance is very large, typically on the order of kilometers. At these distances, the computation energy spent in data processing is not significant as compared to the energy spent in signal transmission. On the contrary, in WSNs the nodes are kept in close vicinity. Distance between sensor nodes varies from a few meters to hundreds of meters for different applications. For example, in a WSN deployed for glacial environment monitoring [4], nodes were kept 20 to 25 m apart, whereas for volcano monitoring [5], nodes were deployed 200 to 400 m apart. At these distances, the energy consumption in the radio and computation units is comparable [6]. Therefore, while making any decision for energy optimization, it is important that the computation and the radio energies are taken into account. In the next section, we discuss various approaches for estimating the radio and the computation unit energies.

7.3 Energy Models

For computation of the energy consumption of the wireless sensor networks, different energy models have been proposed. Some of them are for the individual sensor nodes and some are for the WSN. The sensor node energy can be computed as a sum of the individual sensor node unit energies. Some of the WSN simulators compute the energy of the network broadly on the basis of the energy model for individual sensor node and network parameters [7][8].

For the sensor node energy computation, energy models for the major units, e.g., battery, sensor, radio, and computation units, can be developed. The energy in the battery and sensor units remains more or less constant for an application in similar usage conditions. Therefore, the radio and computation unit energies can be taken into account [6][9]. Next, we will discuss the methods used for estimating the radio and computation unit energies.

7.3.1 Radio Energy Model

The radio unit energy is mainly contributed by the radio transceiver circuit energy and transmit signal energy. In a WSN, the transmission distances are typically small. At small distance transmissions, the contribution of the transmit signal energy and the radio circuit energy in the radio energy consumption remains significant. Therefore, taking both the energy components while computing the radio energy in a sensor node is important [6][9].

7.3.1.1 Transceiver Circuit Energy

Typical radio transmitter and receiver circuits are shown in Figure 7.3a,b, respectively. The main components of a typical transmitter circuit are digital-to-analog converter (DAC), low-pass filter (LPF), mixer, frequency synthesizer (FS), power amplifier (PA), and band-pass filter (BPF). The receiver circuit components are mainly BPF, low-noise amplifier (LNA), mixer, FS, intermediate-frequency amplifier (IFA), LPF, and analog-to-digital converter

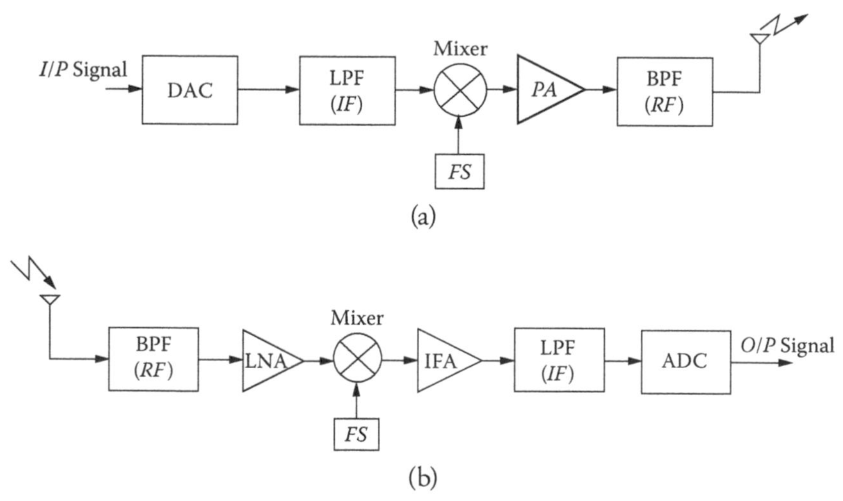

Figure 7.3 Radio unit circuit components.

(ADC). The radio circuit power can be computed by taking these components' individual power into account as:

$$P_{ckt\_compo} = P_{PA} + P_{DAC} + 2(P_{LPF} + P_{FS} + P_{BPF}) + P_{LNA} + P_{IFA} + P_{ADC}. \tag{7.1}$$

The power consumption in a PA is dependent on the signal transmit power as $P_{PA} = \alpha P_{sig}$, where the constant α is related to the drain efficiency η of the RF power amplifier by the relationship $\eta = \frac{1}{(1+\alpha)}$ [10]. The power amplifier efficiency depends on the type of power amplifier used. For example, the class AB PA are typically 30 to 35 percent efficient, whereas class C amplifiers are typically 75 percent efficient. It is to be noted that the PA efficiency varies with the transmitter output power. These typical efficiency values are for the high or saturation output power. In a WSN for small distance transmission, the output power is small and, hence, the PA efficiency is much less. For example, for the CC2420 radio, maximum output power is 1 mW and at this output power the PA efficiency is merely 3.3 percent [11].

The radio circuit energy can be expressed as:

$$E_{ckt\_compo} = P_{ckt\_compo} T_{on} \tag{7.2}$$

where T_{on} is the time for which the radio circuit remains on. In addition to this energy, some energy is consumed in other peripheral components, termed as *base energy*. Therefore, the radio circuit energy can be represented as:

$$E_{ckt} = E_{ckt\_compo} + E_{ckt\_base}. \tag{7.3}$$

The base energy consumption differs for different radio circuits. This can be measured for a specific radio.

7.3.1.2 Transmit Signal Energy

The signal power required in free space can be expressed by Friis transmission equation [12]:

$$P_{sig} = \left(\frac{4\pi}{\lambda}\right)^2 d^n \frac{P_r}{G_r G_t} \tag{7.4}$$

where

d is the distance between transmitter and receiver,
λ is the wavelength of transmitted signal,
P_r is the received power,
G_t and G_r are the transmitter and the receiver antenna gains, respectively,
n is the path loss exponent.

The received power relates to the signal to noise ratio (*SNR*) for data transmission, the number of bits per modulation symbol *b*, bandwidth *B*, and receiver noise figure *NF* differently for different channel models. One may assume additive white Gaussian noise (AWGN), Rayleigh, or any other channel model depending on the environment of the WSN in consideration. For example, the received power for an AWGN channel with $\frac{N_0}{2}$ noise spectral power density is [12]:

$$P_r = SNRbB\frac{N_0}{2}NF \tag{7.5}$$

The corresponding energy can be expressed as:

$$E_{sig} = P_{sig}T_{on}. \tag{7.6}$$

The value of T_{on} and *SNR* can be computed based on the modulation scheme used for the data transmission. For example, for MPSK, T_{on} and SNR_{coded} are expressed as [12]:

$$T_{on} = \frac{L}{bB} \tag{7.7}$$

$$SNR = \begin{cases} (erfc^{-1}(2p_s))^2 & b = 1, \\ \frac{(erfc^{-1}(p_s))^2}{b(sin\frac{\pi}{2^b})^2} & \text{otherwise} \end{cases} \tag{7.8}$$

where *L* is the number of bits transmitted, *erfc* is the complementary error function, and p_s is the channel symbol error probability.

Finally, the radio energy can be computed by summing the circuit and the transmit signal energies.

$$E_R = E_{ckt} + E_{sig} \tag{7.9}$$

The circuit energy and the signal energy with their inputs, contributing to the radio energy per bit, is shown in Figure 7.4.

7.3.2 Processor Energy Estimation

The energy consumption of a processor mainly depends on its architecture and instruction set. For computation energy estimation, no analytical energy model is available because from one processor to another, the architecture and its instruction set may vary significantly. For processor power estimation, different techniques and tools are available. These techniques and tools are discussed next.

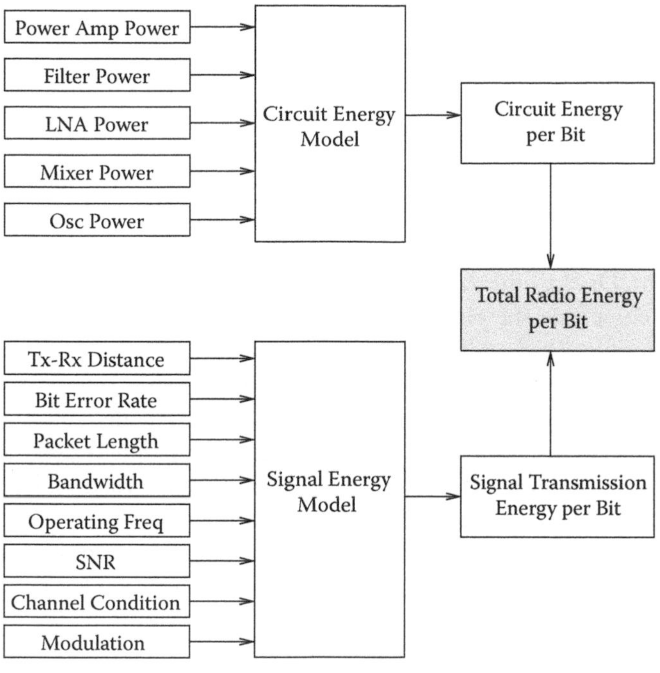

Figure 7.4 Radio energy model.

In battery-operated devices, a key issue is low power consumption. In the battery-operated embedded systems, for computing purposes, application specific integrated circuits (ASICs) are used for meeting the low power requirements. Day by day the embedded systems are becoming more complex as the functionalities implemented are rapidly increasing. In such a scenario, using low power processors instead of ASICs reduces the cost as well as the time to market. This is primarily because of the enormous flexibility that software provides. Along with this, using processors in embedded systems helps in maintaining the versatility of the device. This gives the user the flexibility to implement functions at his end. With the use of processors in battery-operated embedded systems, the need arises to estimate and reduce the power consumption of the processor.

The processor power estimations done at circuit [13]–[18] and gate level [19]–[22] are accurate, but time consuming. This problem becomes more significant in case of complex processors. Simultaneously, circuit- and gate-level power estimation is not very useful for power estimation of software running on the processor. Architectural level power estimation [23]–[25] provides faster results at the cost of a lower accuracy. Also, this estimation requires internal processor details. To achieve a good balance between estimation time and accuracy, the processor software power consumption is mainly estimated at the instruction level [26]–[29] or at the cycle level [30]–[32] of abstraction.

Tiwari et al. [26] were the first to propose a method to estimate the processor power at the instruction level. The core idea of the proposed methodology is to measure the current drawn by the processor in executing different instructions to estimate the power consumed by the processor. Power consumption of each instruction is measured by executing that instruction repeatedly and measuring the average current. Each instruction is assigned a base energy cost through such measurement. On the other hand, while running a program, some inter-instruction effects also come into the picture. This is due to the fact that almost all modern processors are pipelined and, thus, multiple instructions are in different stages of execution simultaneously. In this model, the inter-instruction effects also are taken into account. Total energy cost is the sum of the base energy cost and the inter-instruction energy cost. Instruction-level energy estimation is done for the program by simply summing the energy consumed in each instruction. Energy of each instruction is precomputed for the specific target architecture.

Later, the instruction-level model has been linked to the architecture level [28]. This is beneficial in terms of incorporating architectural level changes easily and achieving speedup in energy estimation without compromising much on the accuracy.

Cycle-level energy estimation is done on the basis of the amount of activities in the microarchitectural unit during a cycle. The amount of activity in each unit varies according to the workload. If the given resource is accessed in a cycle, then its energy can be computed on the basis of the energy model of that resource. Finally, the energy spent can be computed on the unit basis or for the whole processor.

7.4 Component-Level Energy Optimization Techniques

The component-level energy optimization is important as it decides the base energy consumption of the sensor nodes and, hence, of the WSNs. Not only in WSNs, but in many battery-operated devices, keeping the power consumption as low as possible is a must. For satisfying the low power constraint, research is continuously going on for reducing the power consumption of the components used in the radio, computation, sensor, and power supply units. While designing a sensor node, the components are chosen such that they fulfill the functional needs by consuming minimum power. For example, in the commercially available sensor nodes, low power microcontrollers, e.g., the ATmega128L in Mica series sensor nodes [33] [34] and the TI's MSP430 in Telos sensor nodes [35], as well as microprocessors with high processing capability, e.g., the Intel PXA271 in Imote2 [36] and the StrongARM SA-1100 in μAMPS [37] have been used. The processors and microprocessors are being optimized for low power consumption.

The computation unit energy consumption not only depends on the processor energy, but also on the memory configuration. In a sensor node, the memory unit in general consists of a flash memory to store the program and one or more RAM (random access memory) and/or SDRAM (synchronous dynamic RAM) to store data. The increased capacity of the memory unit allows one to program sensor nodes with more complex programs at the cost of a higher power consumption. Similar to the computation unit, the components of the other units of a sensor node also are being optimized for low power consumption. For the component-level power optimization, the optimization techniques can be classified according to the level of abstraction into circuit level, logic level, architecture level, and software and system level. These techniques are out of the scope of this chapter. The interested person may refer to the wide literature available on these techniques [38]–[42]. In addition to the individual component power consumption, the layout of the sensor nodes has been customized to optimize the sensor node power consumption as well.

The main limitation of the component-level energy optimization is that it can be achieved at the sensor node manufacturer level. The user has no or very few options for altering the setting to reduce the energy consumption. The system level energy optimization techniques facilitate a user to customize the sensor node and the WSN parameters according to an application and a deployment environment. In the next section, we will discuss the system level energy optimization techniques.

7.5 System-Level Energy Optimization Techniques

The major advantages of the system level energy optimization techniques are their effectiveness in terms of greater energy savings, and it can be applied at the user end. In this section, we will focus on the system-level energy optimization in WSN.

As discussed in section 7.3, once the application for which the WSN is being deployed has been fixed, the major energy variation occurs in the radio and the computation units. One of the system-level techniques is based on the tradeoff of the radio and computation unit energy consumptions. A dynamic power management (DPM) at the system level allows reduction of energy consumption by keeping unused hardware units in deep sleep. Another system-level energy optimization scheme is dynamic voltage–frequency scaling (DVFS). The other system-level energy reduction techniques can be classified as network-related energy optimization techniques. These techniques include low energy protocols, routing, and network topology. In the following subsections, we will discuss these techniques one by one.

7.5.1 Computation Communication Energy Tradeoff

In this technique the sensor node configuration is chosen based on the tradeoff between the computation energy and the radio energy such that the sensor node energy consumption is less. This is based on the observation that the energy consumed by the radio unit can be reduced by spending some energy in the computation unit. This finally results in the overall energy reduction. A very simple example of this tradeoff can be seen in data compression. Instead of transmitting the data without processing, we can opt for using one of the data compression techniques. In this process, we spend some energy in the computation unit for compressing and decompressing the data and we save energy in the radio unit by transmitting less data. The overall energy saving will be the difference between the savings in the radio energy and the consumption in the computation unit energy. If the energy consumed in the computation unit is more than the energy saved in the radio unit, then there will be no energy savings; instead we will end up with more energy consumption.

For the WSN, different existing source encoding techniques have been explored [43]–[45] to select a suitable encoding for a given condition. Some new encoding schemes that are specific to a sensor network and are intended to reduce power of a radio unit can be used [46][47]. The distributed source coding (DSC) is found to be more suitable for the WSNs because this scheme uses the fact that there is a lot of redundancy in the data due to the high node density in a WSN to limit data exchange between the nodes. In DSC, the nodes send data independently to the base station (BS) and do not communicate with each other. At the BS, the joint decoding is performed. As the decoding is done at the nonenergy constrained BS, in such cases only the computation energy sent in encoding affects the sensor node energy [48]. In the DSC, based on the Slepian–Wolf coding theorem [49] [50], it falls under the category of loss-less compression [51]–[54] and, based on the Wyner–Ziv encoding [55], falls under the category of lossy compression [56].

The computation communication energy tradeoff can be utilized in case of channel encoding [6][9]. The need for using the channel encoding arises from the fact that, for every application, the required bit error rate (BER) could differ significantly. For example, acceptable BER is much lower in medical applications as compared to the environment monitoring applications [57]. For achieving desired BER, normally error correcting codes (ECCs) are used. For achieving a certain BER, the SNR required for transmitting the data with an ECC is less as compared to sending the data uncoded. This saving in the signal power is known as the coding gain. The energy overheads of an ECC relate to energy spent in encoding and decoding of data and energy spent in transmission of "redundant" bits. We can trade off these energy overheads against the energy gain due to the coding gain of an ECC.

For example, the energy consumption of a sensor node with a CC2420 radio and an ARM PXA271 processor is shown in Figure 7.5. In this figure, the energy of a sensor node has been plotted for uncoded data transmission and for data transmission with Reed–Solomon (RS) and Hamming ECCs. This figure shows that even though there is some energy consumed on the computation in encoding the

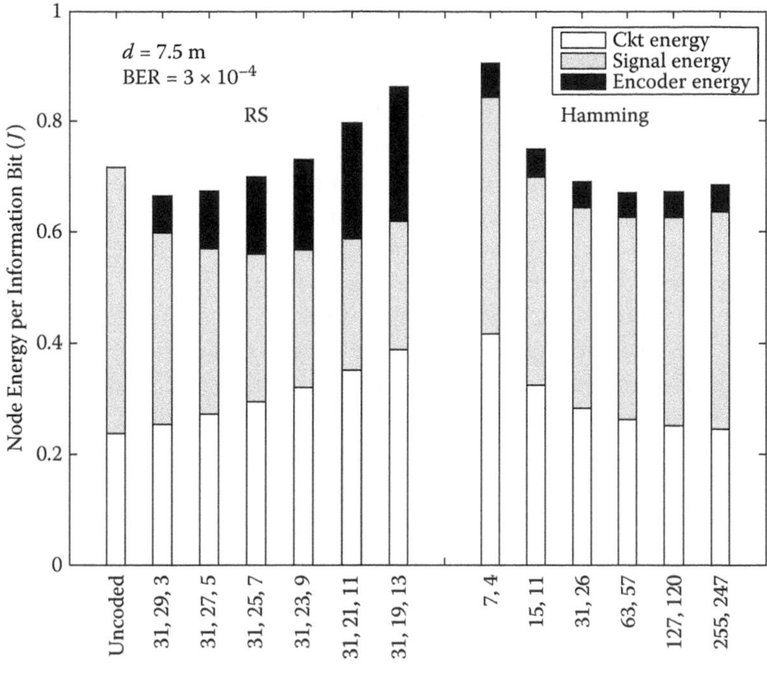

Figure 7.5 Node energy with the Hamming and the RS codes at 7.5 m internode distance.

data with RS and Hamming encoding schemes, ultimately we save energy as the coding gain of the ECC more than compensates for the encoding energy expenditure. The figure shows that, out of the given RS codes, RS (31, 29, 3) is the most energy saving and, in the group of Hamming codes, Hamming (63, 57) is the most energy saving. It is to be noted that this energy graph is for a particular BER, path loss exponent, internode distance, and channel model. For a different network and node configuration, the energy optimal scheme would be different.

7.5.2 Dynamic Power Management

One of the techniques for reducing the power consumption is to shut down the devices when not needed and wake them up when required. This technique is called *dynamic power management* (DPM). To reduce the power consumption, various circuit components, e.g., processor, radio, and memory, support different operating modes. For example, the CC2420 radio has *Voltage Regulator Off, Power Down*, and *Idle* modes in addition to the *Transmit and Receive* modes [58], and the StrongARM SA-1100 processor has three modes of operation: *Run, Idle, and Sleep* [59]. Similarly, the wireless sensor nodes also support different operating modes, e.g., Imote2 supports *Active (radio off)*, *Active (radio on)*, and *Deep Sleep* modes [60].

The energy consumption in transition from one mode to another is significant for the sensor nodes. Therefore, putting a device in the sleep mode is advisable only when the energy saving due to keeping the device in the sleep mode more than compensates for the transition energy consumption. When to change the device operating mode is also important. The decision of putting a device into a particular mode can be taken by the operating system based on the occurrence of an event or some of the network and application parameters [59]–[65].

A task related to an application can be divided into subtasks. In such a situation, one can schedule a group of subtasks such that the power consumption is minimal. This is known as task scheduling. Combining task scheduling with the DPM allows more power savings. For example, let us consider a WSN deployed for sensing the temperature of the surroundings. For this application, we collect temperature sensor data by various sensor nodes, apply compression technique to compress the data on each node, and then transmit the data to another node. To perform these series of tasks, one way is to do the compression and transmission tasks in a pipelined manner. While transmitting previously compressed data, we compress the next sequence of data in the computation unit. By doing this, we transmit the data faster, but computation unit and radio units are simultaneously in active mode. This increases the power consumption of the sensor node. Another way is to do the compression and transmission one after the other. In this situation, we can keep one of the units in the active mode and the other one in idle or sleep mode depending on the data rate. In this case, the rate of data transmission is low, but the power consumption is also low. Typically, the WSN applications are low data rate applications. Also, sometimes these applications are such that the data are available periodically. These characteristics of typical WSN applications allow us to use DPM in conjunction with scheduling to increase the power savings.

7.5.3 Dynamic Voltage–Frequency Scaling

If all components of a device are not active at the same time then the device may not run at the maximum operating frequency and supply voltage. At the run time, by sensing the required voltage and frequency, the supply voltage and/or the operating frequency can be changed. The energy consumption of a device is contributed by the dynamic energy and the static energy consumption. The dynamic energy consumption is due to the charging and discharging of the switching capacitor at the time of a device state change. The static energy consumption occurs mainly due to the leakage current.

The device running at a lower operating frequency and/or supply voltage consumes less dynamic power. This technique is called dynamic voltage–frequency scaling (DVFS). The frequency scaling and voltage scaling can be done together or independently.

The dynamic energy E_d is related to the supply voltage V and operating frequency f as [66]:

$$E_d = CV^2 fT \tag{7.10}$$

where C is the effective switched capacitance. On reducing the operating voltage, the time taken in completing a task T increases linearly, but the dynamic energy reduces quadratically.

The static energy E_s is given by [66]:

$$E_s = I_0 V \tag{7.11}$$

where, I_0 is the leakage current. For the lower order circuit fabrication technologies, the proportion of the leakage current and, hence, the static energy consumption is increasing [67]. The increased time due to the decreased operating frequency finally results in a higher static energy consumption. The energy savings due to the DVFS is the difference between dynamic energy savings and increased static energy consumption. Therefore, the DVFS may not always result in the energy savings. For example, it has been found that on reducing the operating frequency of a SA-1100 processor (used in the μAMPS sensor nodes), the energy per operation increases due to the dominance of the static energy consumption [68].

7.5.4 Network-Level Energy Optimization

At the system level, to design an energy-efficient WSN, the main emphasis is on the WSN design decisions related to the network topologies, routing techniques, and data transmission protocols, i.e., the system as a whole. These design decisions determine the network life. The definition of the network life is application-dependent. It is sometimes measured as the time until the first/last sensor node dies, sometimes as the time until a node/cluster is disconnected from the BS, and sometimes as the failure of a percentage of the deployed sensors [69].

The network topology has a great impact on the performance and the power consumption of a WSN [70]. As discussed in section 7.2, the WSN topology can be configured as a flat or a clustered topology. In case of a low-density WSN, the flat WSN topology may be a good choice as it is less complex. As the node density increases, dividing the network into clusters makes network management easy. Also, the flat or clustered topology depends on the type of application for which a WSN has been deployed. For example, in case of temperature measurement of an area, average of temperature measured by the nodes in close vicinity may be sufficient. In such cases, the cluster head (CH) may compute the average of the temperature data gathered from various nodes and communicate it to the BS. In the clustered WSN topology, the energy consumption depends on the way the clustering is done and on the position of the cluster head and the BS [71]–[74].

In both flat and multihop topologies the energy consumption is a function of number of hops in transmitting data from the source to the destination. The radio energy E_R relates to the transmission distance d as $E_R \alpha d_n$, where n is the path loss exponent [12]. In the case of less number of hops or a single hop, the transmission distance between two nodes will be more, so the radio energy required to transmit the data will be more. On the other hand, on increasing the number of hops, the radio energy consumption will increase due to receiving it more often. Therefore, these energy tradeoffs must be taken into consideration while deciding upon an energy optimal route. The energy optimal route can be determined by the BS or by the node itself. In case of a BS determining the energy optimal route the cost to be pain is in terms of communication bandwidth. This is because the BS has to communicate the route to the concerned node and it has to periodically gather the network statistics on the basis of which BS makes the decision. On determining the energy optimal route by the sensor node, the energy cost of running the algorithm must be taken into account.

Many energy-aware routing algorithms and energy-efficient data transmission protocols have been proposed [75]–[77]. The communication energy can be reduced by proper network infrastructure. The packet-size optimization is one of the ways of reducing energy in data transmission [78][79].

7.6 Summary

Various techniques used for reducing the energy of a WSN have been discussed. These techniques have been classified as the circuit-level and system-level techniques. To reduce the energy consumption of a WSN to increase its life, not only reducing the energy of individual components of a sensor node is essential, but also one can apply one or more system-level techniques, discussed in this chapter. The amount of energy savings differ, depending on the choice of the abstraction level between the circuit and system level. Some of these techniques are used by the sensor node manufacturer and some can be employed at the user end according to the need of the application for which the WSN is being deployed. By carefully studying the application characteristics, one can customize the WSN as well as sensor node configuration to reduce the energy consumption further. Some of the techniques discussed in this chapter are orthogonal to each other, hence, one can use them together to get more energy benefits.

REFERENCES

[1] J. Yick, B. Mukherjee, and D. Ghosal, "Wireless sensor network survey," *Comput. Netw.*, vol. 52, no. 12, pp. 2292–2330, Aug. 2008.

[2] M. Vemula, M. F. Bugallo, and P. M. Djuric, "Target tracking in a two-tiered hierarchical sensor network," in *Proceedings of International Conference on Acoustics, Speech and Signal Processing*, vol. 4, 2006, pp. IV-969–IV-972.

[3] Z. Jin and S. Papavassiliou, "On the energy-efficient organization and the lifetime of multi-hop sensor networks," *IEEE Commun. Lett.*, vol. 7, no. 11, pp. 537–539, Nov. 2003.

[4] K. Martinez, P. Padhy, A. Riddoch, H. Ong, and J. K. Hart, "Glacial environment monitoring using sensor networks," in *Real-World Wireless Sensor Networks Workshop*, 2005, pp. 10–14.

[5] G. Werner-Allen, K. Lorincz, M. Welsh, O. Marcillo, J. Johnson, M. Ruiz, and J. Lees, "Deploying a wireless sensor network on an active volcano," *IEEE Internet Comput.*, vol. 10, no. 2, pp. 18–25, Mar. 2006.

[6] S. Chouhan, R. Bose, and M. Balakrishnan, "A framework for energy consumption based design space exploration for wireless sensor nodes," *IEEE Trans. Computer-Aided Design Integr. Circuits Syst.*, vol. 28, no. 7, pp. 1017–1024, July 2009.

[7] B. L. Titzer, D. K. Lee, and J. Palsberg, "Avrora: Scalable sensor network simulation with precise timing," in *Proceedings of the International Conference on Information Processing in Sensor Networks*, 2005, pp. 477–482.

[8] I. Downard, Simulating Sensor Networks in NS-2. Online at: http://nrlsensorsim.pf.itd.nrl.navy.mil

[9] S. Chouhan, R. Bose, and M. Balakrishnan, "Integrated energy analysis of error correcting codes and modulation for energy efficient wireless sensor nodes," *IEEE Trans. Wireless Commun.*, vol. 8, no. 10, pp. 5348–5355, Oct. 2009.

[10] T. H. Lee, *The Design of CMOS Radio-Frequency Integrated Circuits*. Cambridge, U.K.: Cambridge University Press, 1998.

[11] Q. Wang, M. Hempstead, and W. Yang, "A realistic power consumption model for wireless sensor network devices," in *Proceedings of Conference on Sensor, Mesh and Ad Hoc Communications and Networks*, Reston, VA, 2006, pp. 286–295.

[12] J. G. Proakis, *Digital Communications*, 4th ed. New York: McGraw-Hill, 2001.

[13] F. N. Najm, R. Burch, P. Yang, and I. N. Hajj, "Probabilistic simulation for reliability analysis of CMOS VLSI circuits," *IEEE Trans. Computer-Aided Design Integr. Circuits Syst.*, vol. 9, no. 4, pp. 439–450, Apr. 1990.

[14] F. N. Najm, I. N. Hajj, and P. Yang, "An extension of probabilistic simulation for reliability analysis of CMOS VLSI circuits," *IEEE Trans. Computer-Aided Design Integr. Circuits Syst.*, vol. 10, no. 11, pp. 1372–1381, Nov. 1991.

[15] R. Tjarnstrom, "Power dissipation estimate by switch level simulation," in *International Symposium on Circuits and Systems*, Portland, OR, 1989, pp. 881–884, vol. 2.

[16] A. Salz and M. Horowitz, "IRSIM: An incremental MOS switch-level simulator," in *Proceedings of Design Automation Conference*, 1989, pp. 173–178.

[17] S. M. Kang, "Accurate simulation of power dissipation in VLSI circuits," *IEEE J. Solid-State Circuits*, vol. 21, no. 5, pp. 889–891, Oct. 1986.

[18] L. W. Nagel, "SPICE2: A computer program to simulate semiconductor circuits," Memorandum ERL-M520, Electronics Research Laboratory, College of Engineering, University of California at Berkeley, Berkeley, Tech. Rep., 1975.

[19] F. N. Najm, "Transition density, a stochastic measure of activity in digital circuits," in *Proceedings of Design Automation Conference*, San Francisco, CA, 1991, pp. 644–649.

[20] A. Ghosh, S. Devadas, K. Keutzer, and J. White, "Estimation of average switching activity in combinational and sequential circuits," in *Proceedings of Design Automation Conference*, Anaheim, CA, 1992, pp. 253–259.

[21] C. Y. Tsui, M. Pedram, and A. M. Despain, "Efficient estimation of dynamic power consumption under a real delay model," in *International Conference on Computer-Aided Design*, Santa Clara, CA, 1993, pp. 224–228.

[22] F. N. Najm, "A survey of power estimation techniques in VLSI circuits," *IEEE Trans. Very Large Scale Integr. VLSI Syst.*, vol. 2, no. 4, pp. 446–455, Dec. 1994.
[23] T. Sato, M. Nagamatsu, and H. Tago, "Power and performance simulator: ESP and its application for 100 MIPS/W class RISC design," in *Proceedings of Symposium on Low Power Electronics*, San Diego, CA, 1994, pp. 46–47.
[24] T. Sato, Y. Ootaguro, M. Nagamatsu, and H. Tago, "Evaluation of architecture-level power estimation for CMOS RISC processors," in *Proceedings of Symposium on Low Power Electronics*, San Jose, CA, 1995, pp. 44–45.
[25] P. E. Landman and J. M. Rabaey, "Architectural power analysis: The dual bit type method," *IEEE Trans. Very Large Scale Integr. VLSI Syst.*, vol. 3, no. 2, pp. 173–187, June 1995.
[26] V. Tiwari, S. Malik, and A. Wolfe, "Power analysis of embedded software: A first step towards software power minimization," *IEEE Trans. Very Large Scale Integr. VLSI Syst.*, vol. 2, no. 4, pp. 437–445, Dec. 1994.
[27] V. Tiwari, S. Malik, A. Wolfe, and M. T.-C. Lee, "Instruction level power analysis and optimization of software," *J. VLSI Sig. Proc.*, vol. 13, no. 2-3, pp. 223–238, Aug. 1996.
[28] A. Sama, M. Balakrishnan, and J. F. M. Theeuwen, "Speeding up power estimation of embedded software," in *International Symposium on Low Power Electronics and Design*, Rapallo, Italy, July 2000, pp. 191–196.
[29] A. Sinha and A. P. Chandrakasan, "Jouletrack: A Web-based tool for software energy profiling," in *Proceedings of Design Automation Conference*, Las Vegas, NV, 2001, pp. 340–345.
[30] D. Brooks, P. Bose, S. E. Schuster, H. Jacobson, P. N. Kudva, A. Buyuktosunoglu, J. Wellman, V. Zyuban, M. Gupta, and P. W. Cook, "Power-aware microarchitecture: Design and modeling challenges for next-generation microprocessors," *IEEE Micro*, vol. 20, no. 6, pp. 26–44, 2000.
[31] D. Brooks, V. Tiwari, and M. Martonosi, "Wattch: A framework for architectural-level power analysis and optimizations," in *Proceedings of International Symposium on Computer Architecture*, Vancouver, Canada, 2000, pp. 83–94.
[32] W. Yeand, N. Vijaykrishnan, M. Kandemir, and M. J. Irwin, "The design and use of simplepower: A cycle-accurate energy estimation tool," in *Proceedings of Design Automation Conference*, Los Angeles, CA, 2000, pp. 340–345.
[33] MICA2. Online at: http://www.xbow.com/Products/productdetails.aspx?sid=174
[34] MICAz. Online at: http://www.xbow.com/Products/productdetails.aspx?sid=164
[35] J. Polastre, R. Szewczyk, and D. Culler, "Telos: Enabling ultra-low power wireless research," in *Proceedings of the International Symposium on Information Processing in Sensor Networks*, 2005, pp. 364–369.
[36] imote2. Online at: http://www.xbow.com/Products/productdetails.aspx?sid=253
[37] E. Shih, S. Cho, N. Ickes, R. Min, A. Sinha, A. Wang, and A. Chandrakasan, "Physical layer driven protocol and algorithm design for energy-efficient wireless sensor networks," in *Proceedings of the International Conference on Mobile Computing and Networking*, Rome, Italy, 2001, pp. 272–287.
[38] J. Rabaey and M. Pedram, *Low Power Design Methodologies*. Kluwer Academic Publishers, 1995.
[39] J. Mermet and W. Nebel, *Low Power Design in Deep Submicron Electronics*. Dordrecht, The Netherlands: Kluwer Academic Publishers, 1997.

[40] A. Chandrakasan and R. Brodersen, *Low-Power CMOS Design*. Piscataway, NJ: IEEE Press, 1998.
[41] S. Devadas and S. Malik, "A survey of optimization techniques targeting low power VLSI circuits," in *Proceedings of the ACM/IEEE Design Automation Conference*, 1995, San Francisco, CA, pp. 242–247.
[42] L. Benini, G. D. Micheli, and E. Macii, "Designing low-power circuits: Practical recipes," *IEEE Circuits and Systems Mag.*, vol. 1, no. 1, pp. 6–25, 2001.
[43] G. Hua and C. W. Chen, "Distributed source coding in wireless sensor networks," in *Proceedings of the International Conference on Quality of Service in Heterogeneous Wired/Wireless Networks*, 2005, Lake Buena Vista, FL, p. 6.
[44] M. Sartipi and F. Fekri, "Source and channel coding in wireless sensor networks using LDPC codes," in *Proceedings of Communications Society Conference on Sensor and Ad Hoc Communications and Networks*, October 2004, Santa Clara, CA, pp. 309–316.
[45] D. Marco and D. L. Neuhoff, "Reliability vs. efficiency in distributed source coding for field-gathering sensor networks," in *Proceedings of International Symposium on Information Processing in Sensor Networks*, April 2004, Berkeley, CA, pp. 161–168.
[46] C. H. Liu and H. H. Asada, "A source coding and modulation method for power saving and interference reduction in DS-CDMA sensor network systems," in *Proceedings of American Control Conference*, vol. 4, 2002, pp. 3003–3008.
[47] J. Kim and J. G. Andrews, "An energy efficient source coding and modulation scheme for wireless sensor networks," in *IEEE 6th Workshop on Signal Processing Advances in Wireless Communications*, 2005, New York, pp. 710–714.
[48] J. Chou, D. Petrovic, and K. Ramchandran, "A distributed and adaptive signal processing approach to reducing energy consumption in sensor networks," in *Proceedings of the INFOCOM*, San Francisco, CA, 2003, pp. 1054–1062.
[49] D. Slepian and J. Wolf, "Noiseless coding of correlated information sources," *IEEE Trans. Inform. Theory*, vol. 19, no. 4, pp. 471–480, 1973.
[50] T. Cover, "A proof of the data compression theorem of Slepian and Wolf for ergodic sources," *IEEE Trans. Inform. Theory*, vol. 21, no. 2, pp. 226–228, 1975.
[51] S. S. Pradhan, J. Kusuma, and K. Ramchandran, "Distributed compression in a dense microsensor network," *IEEE Signal Processing Mag.*, vol. 19, no. 2, pp. 51–60, 2002.
[52] Z. Xiong and A. D. Cheng, "Distributed source coding for sensor networks," *IEEE Signal Processing Mag.*, vol. 21, no. 5, pp. 80–94, Sept. 2004.
[53] M. Sartipi and F. Fekri, "Distributed source coding in wireless sensor networks using LDPC coding: The entire Slepian-Wolf rate region," in *Proceedings of Wireless Communications and Networking Conference*, March 2005, New Orleans, LA, pp. 1939–1944.
[54] H. Wang, D. Peng, W. Wang, H. Sharif, and H. Chen, "Cross-layer routing optimization in multirate wireless sensor networks for distributed source coding based applications," *IEEE Trans. Wireless Commun.*, vol. 7, no. 10, pp. 3999–4009, Oct 2008.
[55] A. Wyner and J. Ziv, "The rate-distortion function for source coding with side information at the decoder," *IEEE Trans. Inform.* (Theory), vol. 22, no. 1, pp. 1–10, 1976.
[56] S. Pradhan, J. Chou, and K. Ramchandran, "Duality between source coding and channel coding and its extension to the side information case," *IEEE Trans. Inform. Theory*, vol. 49, no. 5, pp. 1181–1203, IEEE Trans. Inform. Theory, 2003.
[57] R. Bose, *Information Theory, Coding and Cryptography*. New Delhi: Tata McGraw-Hill, 2002.

[58] CC2420 Datasheet. Online at: http://www.ti.com/lit/gpn/cc2420

[59] L. Benini, A. Bogliolo, and G. D. Micheli, "A survey of design techniques for system-level dynamic power management," *IEEE Trans. Very Large Scale Integr. Syst*, vol. 8, no. 3, pp. 299–316, June 2000.

[60] Imote2 Datasheet. Online at: http://www.xbow.com/Products/Product_pdf_files/Wireless_pdf/Imote2_Datasheet.pdf

[61] T. Simunic, L. Benini, and G. D. Micheli, "Dynamic power management for portable systems," in *Proceedings of the International Conference on Mobile Computing and Networking*, Boston, MA, 2000, pp. 49–54.

[62] A. Sinha and A. Chandrakasan, "Dynamic power management in wireless sensor networks," *IEEE Design & Test of Computers*, vol. 18, no. 2, pp. 62–74, Mar/Apr. 2001.

[63] C. F. Chiasserini and R. R. Rao, "Improving energy saving in wireless systems by using dynamic power management," *IEEE Transactions on Wireless Communications*, vol. 2, no. 5, pp. 1090–1100, Sept. 2003.

[64] R. M. Passos, C. J. N. Coelho, Jr, A. A. F. Loureiro, and R. A. F. Mini, "Dynamic power management in wireless sensor networks: An application-driven approach," in *Proceedings of the Second Annual Conference on Wireless On-Demand Network Systems and Services*, 2005, San Moritz, Switzerland, pp. 109–118.

[65] F. Salvadori, M. de Campos, P. S. Sausen, R. F. de Camargo, C. Gehrke, C. Rech, M. A. Spohn, and A. C. Oliveira, "Monitoring in industrial systems using wireless sensor network with dynamic power management," *IEEE Trans. on Instrumentation and Measurement*, vol. 58, no. 9, pp. 3104–3111, Sept. 2009.

[66] J. Rabaey, *Low Power Design Essentials (Integrated Circuits and Systems)*. Berlin: Springer, 2009.

[67] The International Technology Roadmap for Semiconductors. Online at: http://www.itrs.net/

[68] R. Min, M. Bhardwaj, S. Cho, N. Ickes, E. Shih, A. Sinha, A. Wang, and A. Chandrakasan, "Energy-centric enabling technologies for wireless sensor networks," *IEEE Wireless Communications*, vol. 9, no. 4, pp. 28–39, Aug. 2002.

[69] Y. Chen and Q. Zhao, "On the lifetime of wireless sensor networks," *IEEE Communications Letters*, vol. 9, no. 11, pp. 976–978, Nov. 2005.

[70] A. Salhieh, J. Weinmann, M. Kochhal, and L. Schwiebert, "Power efficient topologies for wireless sensor networks," in *Proceedings of the International Conference on Parallel Processing*, Valencia, Spain, September 2001, pp. 156–163.

[71] O. Younis, M. Krunz, and S. Ramasubramanian, "Node clustering in wireless sensor networks: Recent developments and deployment challenges," *IEEE Network*, vol. 20, no. 3, pp. 20–25, May-June 2006.

[72] K. Akkaya, M. Younis, and W. Youssef, "Positioning of base stations in wireless sensor networks," *IEEE Communications Mag.*, vol. 45, no. 4, pp. 96–102, Apr. 2007.

[73] A. A. Abbasi and M. Younis, "A survey on clustering algorithms for wireless sensor networks," *Computer Communications*, vol. 30, no. 14-15, pp. 2826–2841, Oct. 2007.

[74] A. Chamam and S. Pierre, "On the planning of wireless sensor networks: Energy-efficient clustering under the joint routing and coverage constraint," *IEEE Trans. Mobile Computing*, vol. 8, no. 8, pp. 1077–1086, 2009.

[75] J. N. Al-Karaki and A. E. Kamal, "Routing techniques in wireless sensor networks: A survey," *IEEE Wireless Communications*, vol. 11, no. 6, pp. 6–28, 2004.

[76] Y. Yang, R. S. Blum, and B. M. Sadler, "Energy-efficient routing for signal detection in wireless sensor networks," *IEEE Trans. Signal Processing*, vol. 57, no. 6, pp. 2050–2063, 2009.

[77] N. Riaz and M. Ghavami, "An energy-efficient adaptive transmission protocol for ultra-wideband wireless sensor networks," *IEEE Trans. Vehicular Technology*, vol. 58, no. 7, pp. 3647–3660, 2009.

[78] Y. Sankarasubramaniam, I. F. Akyildiz, and S. W. McLaughlin, "Energy efficiency based packet size optimization in wireless sensor networks," in *International Workshop on Sensor Network Protocols and Applications*, May 2003, Anchorage, AK, pp. 1–8.

[79] M. C. Vuran and I. F. Akyildiz, "Cross-layer packet size optimization for wireless terrestrial, underwater, and underground sensor networks," *in Proceedings of the IEEE Conference on Computer Communications*, Phoenix, AZ, 2008, pp. 226–230.

SCAVENGING TECHNIQUES

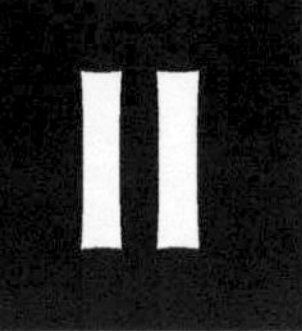

Chapter 8

Design Issues in EM Energy Harvesting Systems

Gianluca Cornetta, David J. Santos, Abdellah Touhafi, and José Manuel Vázquez

Contents

8.1 Energy Harvesting: Techniques and Applications

Energy harvesting (also known as energy scavenging) is the process by which energy is obtained from external sources (e.g., solar power, thermal energy, wind energy, temperature gradients, and kinetic energy), and stored somehow. Energy harvesting techniques have been known for centuries and exploited at the macro-scale level in systems, such as windmills and watermills.

Lately, energy scavenging techniques have been successfully applied to generate energy at the micro-scale level, especially in wearable electronics, active tags, and wireless sensor networks. The extremely exigent power-consumption requirements of these devices put severe design constraints on them that make their implementation really challenging and has forced the designers to develop novel techniques aimed at extending battery lifetime. As a matter of fact, nowadays, we are experiencing a transition to a completely new approach to low power design, in which the designer is shifting his attention from the circuit to the power source.

Ultra low power circuit techniques are now well understood; however, efficient power delivery is still a challenging task. Low power techniques alone are not sufficient to improve battery lifetime beyond the actual limits, but the possibility of applying energy harvesting techniques at the micro-scale level has disclosed the path toward the implementation of ubiquitous sensor nodes and many other applications in which it is crucial to reduce battery depletion.

Energy scavenging still presents many challenges and it is still not possible to implement fully autonomous and batteryless circuits. The main reason is that harvested power is obtained from ambient sources and so it tends to be unregulated, intermittent, and small. Nonetheless, energy harvesting techniques can be used in battery-operated circuits and have been proved to be effective in helping to extend battery lifetime.

As an example of battery autonomy requirements we may consider medical applications in which electronic devices are implanted or attached to the body. Implanted medical devices, in-ear devices, and surface of skin devices have different requirements in terms of power consumption and battery lifetime. Design constraints depend on the kind of device and the application, e.g., 10 μW power consumption and battery lifetime of 15,000 hours are typical for implanted devices. In the case of in-ear devices, size is more important than power consumption. Typical power consumption and battery requirements for these devices would be 1 mW and 1,500 hours, respectively. Finally, typical power dissipation and battery lifetime for surface-of-skin devices are 10 μW and 150 hour, respectively. The lower bound for the energy production of harvesting devices is set by those medical applications that consume power in the range of a few milliwatts. Some nonmedical products, including calculators, wristwatches, radios, and Bluetooth headsets, already use micro-harvesting sources. There are also many applications, such as remote sensors nodes, that consume on the order of a few milliwatts as well, but have not yet been adapted to micro-harvesting.

The most promising micro-harvesting technologies extract energy from vibration, temperature gradients, and light. Unfortunately, all of them rely on expensive and bulky MEMS (microelectromechanical systems) processes and solar cells. A fourth possibility—scavenging energy from RF (radio frequency) emissions—is interesting, but the energy availability is about in the order of magnitude less than that of the first three. Nonetheless, scavenging energy from the electromagnetic (EM) waves is a very cheap alternative to the previous techniques because it can be implemented with simple and inexpensive CMOS (complementary metal oxide semiconductor) circuits.

Implementing an energy recovering system to scavenge energy from radio frequency signals is a challenging task because it not only involves the design of a high-efficiency power harvester, but it also implies a careful system design to optimize power delivery from the antenna to the power harvester. To minimize the losses of the antenna, it is of paramount importance to carefully choose the antenna geometry and the polarization more suitable for the application. In addition, to overcome signal degradation due to path losses and channel fading, and assuring correct operation within the distance range set by the specifications, a boosting network is necessary to generate the power-on voltage for the energy-harvesting circuit. The boosting network must be matched to the energy harvester to guarantee maximum power transfer.

The core building block of an EM energy harvester is the RF-DC (radio frequency to direct current) rectifier. This circuit senses an incoming RF signal and amplifies and rectifies it to produce an unregulated DC voltage. Further processing is needed to produce a regulated DC output (e.g., using a band-gap circuit) that supplies stable power to the rest of the circuit. The major challenge at this stage is designing a power-conditioning circuit that draws minimum current from the supply capacitor. Finally, particular care must be put in the design of the communication circuits, especially on the transmit side where backscatter modulation techniques are used to transmit a signal exploiting the energy of the incoming carrier without drawing power from the supply.

The design of the energy recovering system is very challenging and many factors must be taken into account during the design phase, including target process (e.g., the availability of Schottky diodes, multiple-threshold transistors, or lossless-substrate processes, such as Silicon on Insulator or Silicon on Saffire, heavily affects the design choices), input sensitivity, and antenna and matching network design. In this chapter, we will review the basic techniques used in practical designs as well as the major design issues, outlining both the practical and theoretical aspects, and performance bounds.

The rest of the chapter is structured as follows. In section 8.2, we review the major issues in voltage rectifier design and determine the lower bound for correct operation under both ideal matching and simple L-match between antenna and rectifier. In section 8.3, we review the techniques to maximize the energy captured by the antenna and to maximize the power transfer to the RF-to-DC rectifier. We also show a simple *RLC*-lumped model to estimate antenna performance. In section 8.4, several practical implementations of the RF-to-DC rectifier are evaluated using 90 nm and 130 nm technology nodes, and design equations for the Dickson's charge pump are given. Section 8.5 deals with voltage regulators and power-conditioning system design, whereas section 8.6 discusses backscatter modulation schemes. Finally, in section 8.7, we draw our conclusions.

8.2 Design Issues of RF Energy Harvesting Schemes

As stated before, RF energy harvesting is an alternative to traditional power sources to power up small sensors. One way to implement a chip-sized wireless power source is with a telemetry system that relies on an antenna that inductively couples power onto the chip [1]–[3].

Recently, Intel has demonstrated the possibility of scavenging energy from VHF and UHF signals using standard off-the-shelf components [4], whereas Yan et al. [5] have a scheme to harvest power from GSM (Global System for Mobile Communications) signals. However, despite the differences, all the schemes rely basically on passive voltage rectifiers or voltage multipliers originally developed for radio frequency identification (RFID)-tag applications, such as presented by Karthus and Fischer [6]. Figure 8.1 depicts the typical architecture of an RF-energy scavenging circuit.

The coupling element is an antenna, typically dipole or patch. A voltage multiplier converts the input AC voltage into an output unregulated DC voltage. Such voltage is used to drive a series of voltage regulators to provide the transceiver with a regulated supply voltage V_{reg}. A matching network ensures maximum power transfer between the antenna and the transceiver as well as voltage boosting. Most passive tags transmit by backscattering the carrier signal from the reader [7]. A backscatter modulator is used to modulate the impedance seen by the transceiver's antenna during transmission.

8.2.1 Voltage Multiplier

A voltage multiplier is a diode-based circuit that performs half-wave or full-wave rectification as well as voltage multiplication. Figure 8.2a depicts a basic voltage multiplication scheme: the Dickson charge pump.

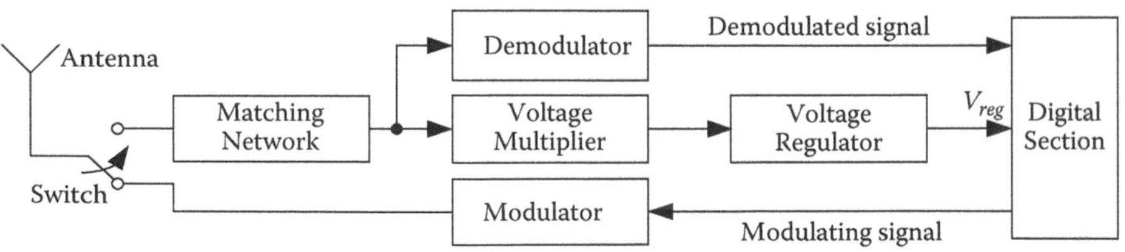

Figure 8.1 Architecture of a typical RF-scavenging circuit.

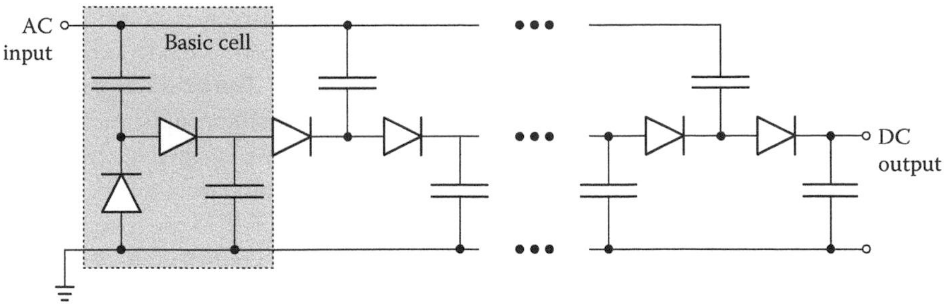

Figure 8.2 Schematic of a Dickson's charge pump.

The coupling capacitors and the diode junction capacitance act as a voltage divider for the AC signal, while diode leakage currents and series resistance limit the maximum achievable DC output.

For a typical 50 Ω antenna, and, say, a –20 dBm-received RF signal power, the input voltage amplitude is 32 mV. The peak voltage of the AC signal is much smaller than the diode threshold. In order to sufficiently drive the rectifier, a voltage-boosting network based on resonant a LC tank has to be employed to match the circuit to the antenna and to produce a larger voltage swing. In addition, in order to improve the rectifier efficiency and to reduce the number of stages, Schottky diodes with very low threshold voltage must be used. Also, the small reverse recovery time, dictated by their junction capacitance rather than by the minority carrier recombination, make Schottky rectifiers very suitable for high frequency applications and very appealing for their little reverse current overshoot as they switch from forward to reverse bias. To obtain maximum output voltage, low series resistance Schottky diodes must be used and they must be carefully laid out so as to minimize junction capacitance, while maximizing coupling capacitance.

However, despite all these advantages, the number of applications of Schottky multipliers is limited by their high reverse currents and higher temperature sensitivity compared to traditional *p-n* junction rectifiers.

Schottky diodes are used for their low small-signal conduction resistance and low junction capacitance. Unfortunately, this device is not available in conventional manufacturing processes; as a consequence, Schottky multipliers are not compatible with standard monolithic CMOS circuits.

8.2.2 Impedance Matching

The rectifier must extract enough DC power from the incoming electromagnetic wave to power up the rest of the circuit [6]–[9]. Unfortunately, RF-to-DC rectification is hard to perform when the input power level is low, since the incoming voltage may fall below the threshold voltage of the transistors that form the rectifier. The rectifier dead zone severely decreases the power conversion efficiency when the input voltage is very low; for this reason, a boosting matching network is necessary. The antenna that provides the source impedance Z_s must be matched to the rectifier input impedance Z_{in} to maximize the power transfer to the rectifier. The matching network performs an impedance transformation that increases the rectifier input impedance and, hence, the RF voltage amplitude. Nonetheless, a network of passive reactances has a narrowband behavior and can perform optimum matching only over a limited frequency band. Unfortunately, the resonant frequency of an RFID tag is very sensitive to environmental conditions; this, in turn, may lead to considerable frequency shifts with respect to nominal values. In order to ensure a safe operation under all the environment conditions and to achieve high communication speeds between tags and reader, it is desirable to design wideband devices; however, as stated by the Bode–Fano limit [10][11], a tradeoff exists between impedance-transformation ratio and bandwidth. The bandwidth over which an arbitrarily good matching can be achieved in the case of a parallel RC (i.e., a complex) load impedance (Figure 8.3a) and assuming a lossless matching network and a purely resistive source impedance is related to the ratio of reactance to resistance (i.e., the load capacitor quality factor):

$$\left| -\int_0^{\infty} \ln(|\Gamma(\omega)|)\, d\omega \right| \leq \frac{\pi\omega_0}{Q_{L0}} \tag{8.1}$$

where Γ is the reflection coefficient of the load and $Q_{L0} = \omega_0\, RC$ is the load quality factor calculated at the center frequency ω_0 of the band of interest.

Referring to equation (8.1), it can be observed that the maximum value of the integral is limited by $\pi\omega_0/Q_{L0}$. To fully utilize the given limit for a desired angular bandwidth $B = \omega_2 - \omega_1$, $|\Gamma|$ should always be 1, except for the band of interest B. This means that, as depicted in Figure 8.3b, a maximum mismatch is desirable outside the band of interest, whereas in the target bandwidth $|\Gamma|$ must be kept at a constant value $|\Gamma_B|$ as small as possible. According to these considerations, $\ln(|\Gamma|) = 0$

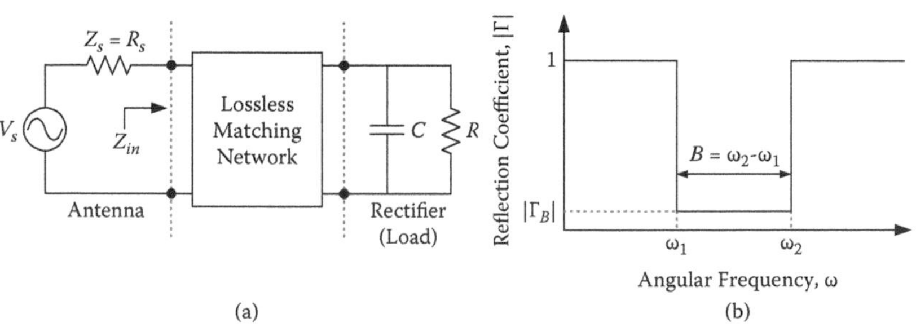

Figure 8.3 (a) Circuit for Bode-Fano limit calculation, (b) reflection coefficient for best utilization of $\pi\omega_0/Q_{L0}$.

outside the band of interest, and contributes nothing to the integral in equation (8.1). Consequently,

$$\int_0^{\infty} \ln(|\Gamma(\omega)|)d\omega = B\ln(|\Gamma_B|) \tag{8.2}$$

leading to a theoretical upper bound for the achievable impedance matching bandwidth:

$$B \le \frac{\pi\omega_0}{Q_{L0}} \frac{1}{\ln\left(\frac{1}{|\Gamma_B|}\right)} = \frac{\pi}{RC} \frac{1}{\ln\left(\frac{1}{|\Gamma_B|}\right)} \tag{8.3}$$

From equation (8.3), two important conclusions can be drawn. First, since $\ln(0) = -\infty$, it is impossible to achieve a perfect match unless $B = 0$. Secondly, the higher the Q of the load, the harder is the matching over a wide bandwidth.

Solving equation (8.3) for $|\Gamma_B|$ and recalling that $\ln(x) = -\ln(1/x)$, and that $B = \omega_2 - \omega_1 = \Delta\omega = 2\pi\Delta f$, we obtain a lower bound for the in-band reflection coefficient:

$$\Gamma_B \ge e^{-\frac{1}{2RC\Delta f}} \tag{8.4}$$

Figure 8.4 is a plot of the in-band reflection coefficient $|\Gamma_B|$ versus the target bandwidth and the RC product. Acceptable matching with high bandwidths is only achievable for small RC products. This means that for a given load capacitance, the parasitic load resistance R must be as small as possible in order to maximize the quality factor of the load.

In general, an energy harvesting circuit relies on a simple first-order L matching network [5][12] to boost the input voltage to the rectifier. It has been demonstrated in [13][14] that, when the antenna is matched to the load with a first-order L matching network, the achievable bandwidth becomes:

$$B = \frac{2\omega_0}{Q_{L0}} \frac{1}{\sqrt{\frac{1}{|\Gamma_B|^2} - 1}} \tag{8.5}$$

In Mandal and Sarpeshkar [14], the bandwidth reduction due to the matching network was represented by a function $f_b(|\Gamma_B|)$, defined as the ratio between the Bode–Fano limit expressed by equation (8.3), and the L-match band expressed by equation (8.6):

$$f_b(|\Gamma_B|) = \frac{\pi}{2} \frac{\sqrt{\frac{1}{|\Gamma_B|^2} - 1}}{\ln\left(\frac{1}{|\Gamma_B|}\right)} \tag{8.6}$$

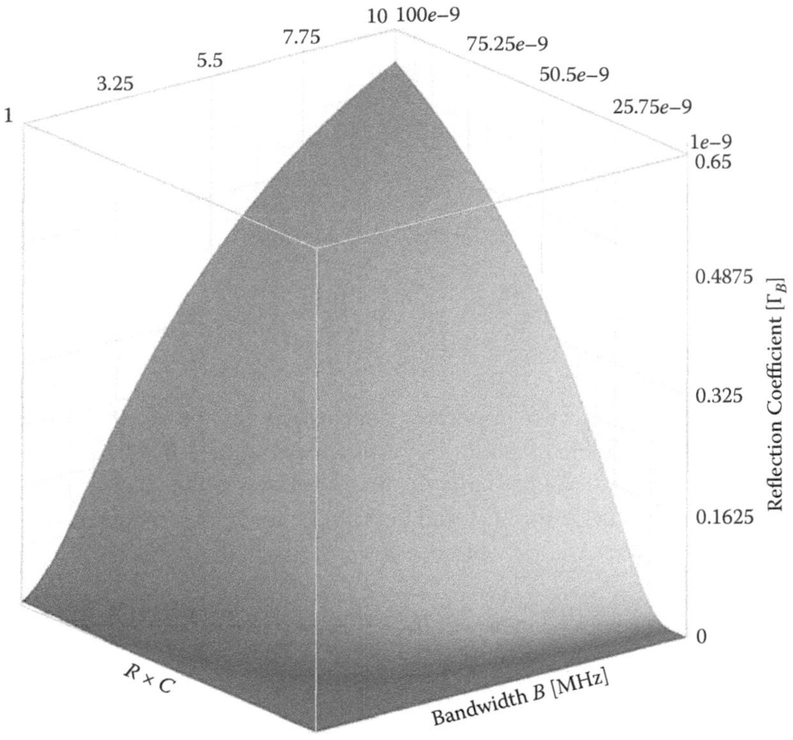

Figure 8.4 In-band reflection coefficient versus bandwidth and RC product.

The minimum value of $f_b(|\Gamma_B|)$ is 3.9, which occurs at $|\Gamma_B| \approx 0.451$. This means that the L-match introduces a bandwidth loss of a factor of 4 with respect to the maximum theoretical limit expressed by equation (8.3).

8.2.3 Power-Up Threshold

The power-up threshold P_{th} is the minimum power available at the antenna terminals, which is needed by the rectifier to generate the supply voltage required by the load to operate correctly. The available power P_r at the antenna terminals is given by the Friis transmission equation:

$$P_r = G_r P_{rad} \frac{\lambda^2}{4\pi} \tag{8.7}$$

where $P_{rad} = P_t G_t / 4\pi r^2$ is the radiated power density around the receiving antenna, P_t the transmit power, G_t the transmit antenna's gain, G_r the receive antenna's gain, and λ the received signal's wavelength. Part of the incoming power captured at the antenna's terminals is dissipated on the real part R of the rectifier input impedance. The amount of the dissipated power P_{diss} is:

$$P_{diss} = \frac{V_{in}^2}{2R} \tag{8.8}$$

where V_{in} is the amplitude of the RF voltage across the load. When the antenna is matched to the load, all the available power P_r is dissipated on the load; however, this is not the case because a mismatch, quantified by the reflection coefficient, $|\Gamma|$, always exists, so that $P_{diss} = (1 - |\Gamma|^2)P_r$ is the power effectively delivered to the load.

The input admittance of the voltage rectifier can be expressed as: $Y_{in} = 1/R + jY$; consequently, from equation (8.7) and equation (8.8), and remembering that the load quality factor is $Q_L = YR$, the input voltage is given by:

$$V_{in} = \sqrt{\frac{(1-|\Gamma|^2)G_r P_{rad} Q_L \lambda^2}{2\pi Y}} \tag{8.9}$$

When designing the receiver, the antenna must be nearly isotropic because the receiver has no a priori location information about the transmitter. Consequently, antenna gain G_r cannot be increased. Wavelength λ cannot be increased because it is desirable to keep the antenna's size small and transmission bandwidth high. The only degree of freedom in equation (8.9) is offered by the rectifier resistance R. Therefore, in order to increase the rectifier input voltage V_{in}, it is desirable to increase R (i.e., to increase the quality factor Q_L of the rectifier input admittance Y_{in}).

When the antenna is matched to the rectifier (i.e., when $|\Gamma| = 0$), all the available input power is dissipated on the load, hence, from equation (8.8), the threshold P_{th0} required to power up the circuit at a given frequency ω_0 is given by:

$$P_{th0} = V_{t0}^2 C \frac{\omega_0}{2Q_{L0}} \tag{8.10}$$

where V_{t0} is the input voltage when P_r is the rectifier input power, and $Q_{L0} = \omega_0 RC$. The power-up threshold for nonzero bandwidths is given by:

$$P_{th} = \frac{V_{t0}^2}{2R(1-|\Gamma_B|^2)} = \frac{P_{th0}}{(1-|\Gamma_B|^2)} \tag{8.11}$$

The minimum value $P_{th,\min}$ of the power-up threshold voltage is determined by the Bode–Fano limit expressed by equation (8.3). In Mandal and Sarpeshkar [14],

$P_{th,\min}$ has been related to a function F_1 of the fractional bandwidth B/ω_0 and of the quality factor of the load Q_{L0}:

$$P_{th,\min} = \frac{P_{th0}}{1-e^{\left(-\frac{2\pi\omega_0}{BQ_{L0}}\right)}} = \omega_0 C V_{t0}^2 F_1\left(\frac{B}{\omega_0}, Q_{L0}\right) \tag{8.12}$$

Analogously, the power threshold in the case of a simple L-match is obtained from equation (8.6), and can be related to a function F_1 of the fractional bandwidth B/ω_0 and of the quality factor of the load Q_{L0}:

$$P_{th} = P_{th0}\left[1+\left(\frac{BQ_{Lo}}{2\pi\omega_0}\right)^2\right] = \omega_0 C V_{t0}^2 F_2\left(\frac{B}{\omega_0}, Q_{L0}\right) \tag{8.13}$$

where:

$$F_1\left(\frac{B}{\omega_0}, Q_{L0}\right) = \frac{1}{2Q_{L0}\left(1-e^{\left(-\frac{2\pi\omega_0}{BQ_{L0}}\right)}\right)}$$

and

$$F_2\left(\frac{B}{\omega_0}, Q_{L0}\right) = \frac{1+\left(\frac{BQ_{L0}}{2\pi\omega_0}\right)^2}{2Q_{L0}}$$

If V_{t0} and C are set, in the case of ideal Bode–Fano matching the power-on threshold decreases monotonically as Q_{L0} increases for a given fractional bandwidth. Conversely, in the case of simple L-match, there is an optimal value of Q_{L0} that minimizes the power-on threshold. This is given by:

$$Q_{L0,opt} = \frac{2\omega_0}{B} = \frac{2}{\left(\frac{B}{\omega_0}\right)} \tag{8.14}$$

Therefore, the fractional bandwidth B/ω_0 is important because it determines the performance of a power extraction system. In general, impedance-matching becomes a major issue when Q_{L0} becomes comparable to ω_0/B.

8.3 Antenna and Matching Considerations

As stated before in section 8.2.3, the radiated power around the receiver's antenna determines the power-up threshold of the receiver's circuit. Consequently, power losses at the antenna and in the matching network between the antenna and the RF-to-DC rectifier put severe constraints on the operating specifications of the receiver circuit. The power available at the receiver depends on its distance r from the transmitter; however, it must be pointed out that the behavior of the magnetic fields changes significantly according to r. More precisely, certain properties of the electromagnetic fields dominate at one distance from the radiating antenna, whereas, as the distance increases, other properties start to dominate. These regions are known as *near field* and *far field*, respectively [15]–[17]. The width of these regions depends on the operating frequency: the near field region spans from the transmitter's antenna until one wavelength, whereas the far field region spans from two wavelengths to infinity. The region comprised of between one and two wavelengths from the radiating antenna is known as the *transition zone.* In the far field region, electric and magnetic fields act normally, i.e., they radiate energy to infinite distances, electric field $\boldsymbol{E}$ and magnetic field $\boldsymbol{H}$ are equal in any point in space, and the electromagnetic radiation falls off in amplitude by $1/r$. This means that the total energy per unit area at a distance r from the antenna is proportional to $1/r^2$. By contrast, the behavior in the near field region is completely different, and either reactive or radiative effects may dominate depending on the distance from the source antenna. The bound between the reactive and the radiative near-field region is $r = \lambda/2\pi$. When $r < \lambda/2\pi$, the average power flow over time is $P_{avg} = \frac{1}{2}\Re\mathrm{e}(\boldsymbol{E} \times \boldsymbol{H}^*)$. Since $\boldsymbol{E}$ and $\boldsymbol{H}$ fields are in quadrature, there is no real power flux in this region; consequently, the reactive near-field region is also known as the *energy storing zone* because, in this region of the space, the energy is imaginary (i.e., due to purely capacitive or inductive effects). Computing the power density in the near-field region is a challenging task because it is hard to predict which is the dominant component of the electromagnetic field in a given region of the space; in addition, the phase relationship between $\boldsymbol{E}$ and $\boldsymbol{H}$ must be measured in order to compute the power.

Due to the reactive behavior in the reactive near-field region, the energy not absorbed by a nearby receiver is delivered from the near field back to the transmitting antenna. This in turn originates regenerative self-inductive and self-capacitive effects that may produce the reversal of the antenna current and alter the electric and magnetic fields distribution in the proximity of the source antenna. Conversely, the radiative near field does not contain reactive field components from the source antenna, since it is so far from the antenna that back-coupling of the fields is out of phase with the antenna signal, and, thus, the regenerative self-inductive and self-capacitive effects on the source antenna cannot take place. Therefore, the energy in the radiative near field is all radiant energy, although the relationship between the magnetic and the electric fields is still hard to measure and different from the far field. Below, we briefly describe the major issues related to antenna design and wave propagation.

8.3.1 Radiation and Coupling

As stated before, radiation and coupling are two different means of power transfer in a wireless medium. Coupling normally takes place in the near-field region and can be either inductive or capacitive. If there is a load in the near field, the energy will be transferred from source to load through coupling (e.g., a transformer is a circuit in which primary and secondary are not physically in contact and the energy is transferred through magnetic coupling). If there is no load, the energy flows back to the source.

Inductively coupled systems are limited to short transmission ranges compared with the size of the antenna. Practical systems usually use antennas whose sizes range from a few centimeters to one meter and operating frequency in the LF (low frequency) or HF (high frequency) region of the spectrum (i.e., 125/134 kHz, or 13.56 MHz). Therefore, the wavelength, about 2000 m in the LF region or 20 m in the HF region, is much longer than the antenna size.

Radiation normally takes place in the far-field region. In this case, energy propagates or radiates away from the source and never goes back to the antenna regardless of whether a load is present or not; this means that when a load is present it absorbs the radiated energy without affecting the source.

Electromagnetic-coupled systems use antennas comparable in size with the wavelength. Practical systems operate in the UHF (ultra high frequency) region of the spectrum and operating frequency of either 900 MHz or 2.4 GHz. Thus, the antenna size ranges from 10 to about 30 cm. As we will see in the next section, far-field practical systems, performance is severely limited by signal propagation issues.

8.3.2 Power Transfer

In this chapter, we are mainly concerned with far-field operation; therefore, we will consider only the power transfer mechanisms that take place sufficiently far from the source antenna. As mentioned before in this chapter, the power transfer from the transmitting to the receiving antenna can be computed assuming a perfect matching and using the Friis transmission equation [18]:

$$\frac{P_r}{P_t} = G_t G_r \left(\frac{\lambda}{4\pi r} \right)^2 \tag{8.15}$$

The Friis transmission equation has two major limitations that must be taken into account when performing the link budget. First, equation (8.15) is valid only in the case of *far field*, when the distance r between transmitting and receiving antennas is

$$r \geq \frac{2a^2}{\lambda} = 2\lambda$$

assuming the case of a simple dipole antenna with length $a = \lambda$. If the operating frequency is 2.4 GHz, the wavelength λ is 0.125 m, so the minimum distance between the antennas should be at least 0.25 cm. Second, equation (8.15) is valid only for free space propagation, which means that multipath fading, interference, atmospheric losses, etc., are not considered and should be included in the link budget. Furthermore, the matching between the antenna and the transmitter or receiver is not perfect, and mismatch losses must be considered as well when performing the link budget.

The DC power consumption of the receiver P_{DC} and the efficiency η of the RF-to-DC converter are the key factors that limit the operation range of a far-field practical system. According to the Friis transmission equation, the power P_r available at the input terminals of the receiver antenna decreases according to the square of the distance r. Furthermore, in order to guarantee the receiver's correct operation, it is indispensable that $\eta P_r \geq P_{DC}$. It is not possible to increase the transmission power to overcome this limitation because it is constrained by governmental regulations. For example, in Europe, the effective radiated power (ERP), i.e., the P_tG_t product, is limited to 2W [19]. Assuming a 2.4 GHz operating frequency and a linear-polarized antenna with a typical gain G_r of 2.5 dBi and typical polarization losses of 3 dB, the plot of the available power P_r as a function of the distance becomes that of Figure 8.5.

Figure 8.5 suggests that, in the case of free-space propagation, i.e., without considering the effects of multipath fading due to the presence of obstacles and

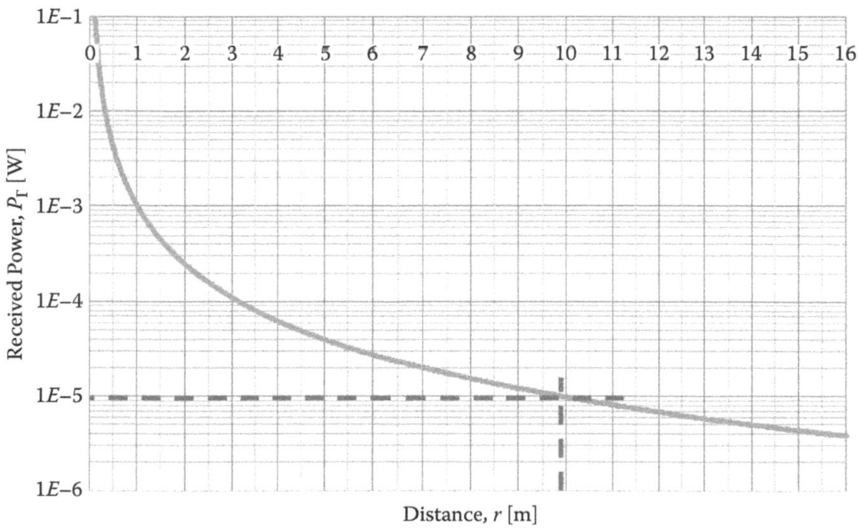

Figure 8.5 Available received power at the antenna terminals as a function of the distance.

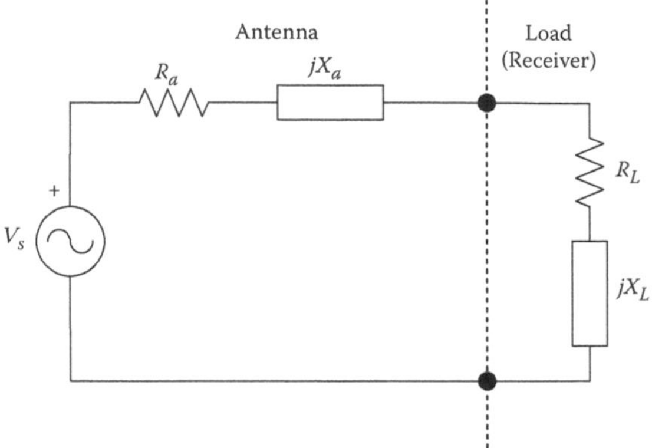

Figure 8.6 Simplified model of the receiving circuit.

reflections, the power that reaches the receiver is a quadratic function of the distance and can be as small as 9.98 μW at a distance of 10 m from the transmitter. Assuming an RF-to-DC conversion efficiency $\eta = 20\%$, the DC power consumed by the receiver for correct operation at a distance of 10 m from the transmitter must be approximately 2 μW, which puts severe design constraints on the receiver if the goal is designing a batteryless circuit.

The key to achieving long transmission ranges is combining several techniques at different stages of the receiver circuit in order to maximize the energy captured by the antenna as well as the RF-to-DC conversion efficiency, and to minimize the power consumption. Many of these techniques are used in RFID passive tags. The use of a Dickson voltage multiplier to improve the conversion efficiency and of Schottky diodes to relax the requirement of the matching network between antenna and multiplier was first proposed by Karthus and Fischer [6] to achieve a transmission distance of 9.25 m with an EIRP* (effective isotropic radiated power) of 4 W at an operating frequency of 869 MHz. However, a real breakthrough in performance was achieved by Curt et al. [8] using Silicon-on-Saffire technology with very low parasitic capacitances and transistor threshold voltages.

To enable maximum transmission range, the power transfer between transmitter, receiver, and antennas must be maximized. This can be easily achieved by means of *conjugate matching*. Figure 8.6 depicts a simplified lumped-element model of a wireless receiver (for the transmitter analogous considerations are valid), where V_s is the peak value of the phasor of the incoming signal, $v(t) = \Re e\{V_s e^{j\omega t}\}, Z_a = R_a + jX_a$, is the impedance of the receiving antenna (we assume a fixed source impedance Z_a), and $Z_L = R_L + jX_L$ is the impedance of the load, i.e., the receiving circuit.

* The equivalent isotropic radiated power (EIRP) is the power that an ideal isotropic antenna would radiate to produce the peak power density observed in the direction of maximum antenna gain.

The *maximum power transfer theorem* states that, for a linear network with a fixed source impedance like the one of Figure 8.6, the maximum power is delivered from the source to the load when *the load impedance is the complex conjugate of the source impedance*, namely $Z_L = Z_a^*$, which means $R_L = R_a$, and $jX_L = -jX_a$.

Under these assumptions, the available source power, that is the average power dissipated in the load, is given by:

$$P_r = \frac{|V_s|^2}{8R_a} \tag{8.16}$$

If there is no conjugate match between the load and the antenna, the power P_{diss} effectively dissipated on the load is

$$\frac{P_{diss}}{P_r} = \left|\frac{Z_s - Z_L^*}{Z_s + Z_L}\right|^2 \tag{8.17}$$

where Z_s is the impedance of the antenna and Z_L is the input impedance of the load circuit.

8.3.3 Polarization

The antenna polarization is the shape of the curve traced by the vector representing the electric field $\boldsymbol{E}$ [15]. Three types of polarization exist: linear, circular, and elliptical. Normally, antennas are linearly polarized; however, when transmitter and receiver are moving, the movement may cause misalignments between transmitter and receiver antennas. If the polarization of the receiver's antenna is not the same as the polarization of the incoming wave, less power than the maximum available power is received. If the incident field is

$$E^i = p_i E_0 e^{-jk\cdot r} \tag{8.18}$$

where $\boldsymbol{r} = x\hat{\boldsymbol{x}} + y\hat{\boldsymbol{y}} + z\hat{\boldsymbol{z}}$ is the propagation direction, $\boldsymbol{k} = k_x\hat{\boldsymbol{x}} + k_y\hat{\boldsymbol{y}} + k_z\hat{\boldsymbol{z}}$ is the propagation constant in a given region of the space (so that $\boldsymbol{k}\cdot\boldsymbol{r} = k_x x + k_y y + k_z z$ is the value of the propagation constant in the direction of propagation), and the polarization of the receiver's antenna is $\hat{\boldsymbol{p}}_a$, then the polarization efficiency η_{pol} is given by:

$$\eta_{pol} = |\hat{\boldsymbol{p}}_i \cdot \hat{\boldsymbol{p}}_a| = |\cos(\psi_{pol})|^2 \tag{8.19}$$

where ψ_{pol} is the angle between the polarization of the incident field.

Consequently, from equation (8.19), in the case that the antennas' polarizations are orthogonally aligned, the receiver will not be able to detect the incoming signal.

For this reason, circularly polarized antennas on the transmitter side offer better performance.

With a circularly polarized antenna, the field rotates carrying out one complete revolution during each wavelength; for this reason, such antennas radiate energy in every plane and not only in one plane like a linear antenna. This characteristic makes circularly polarized antennas much less sensitive to phasing issues and multipath fading than their linearly polarized counterpart.

For example, let us assume we use two right-hand circularly polarized (RHC) antennas both at the transmit and the receive side. The polarization vector of the incident electric field that propagates along the z axis is $p_i = \frac{1}{\sqrt{2}}(x - jy)$, consequently (taking into account that on the receiver's side the incoming field propagates through $-z$), the polarization vector of the receiver's RHC antenna is $p_a = \frac{1}{\sqrt{2}}(x + jy)$ Therefore, according to equation (8.18), the polarization efficiency is $\eta_{pol} = |\hat{p}_i \cdot \hat{p}_a| = 1$.

8.3.4 Performance Estimation

As stated before, maximum power transfer from the antenna to the chip occurs only in the case of conjugate matching. Normally, the antenna impedance is inductive and, as depicted in Figure 8.6, also presents a resistive component R_a that models the losses due to parasitic resistances. The chip impedance is capacitive in nature due to the supply capacitor of the voltage rectifier. Typical values of the supply capacitors range from 1 to a few pF (picofarad). For example, a 1 pF capacitor has an impedance of $-j200\ \Omega$ at a frequency of 915 MHz. The chip load resistance is about 10 Ω. In the case of commercial implementations [20], the input impedance specified by the manufacturer is 380 D in parallel with a capacitance of 2.8 pF, or equivalently a series impedance 9.786 to j60.192 Ω.

It is easier to model the performance of an antenna using a resonant *RLC* circuit. The lumped-element model allows a first estimation of the quality factor of the antenna.

Figure 8.7a depicts the equivalent resonant *RLC* circuit of the antenna and of the chip. The system input impedance Z_{in} is

$$Z_{in} = R_a + R_c + X_a + X_c = R_a + R_c + j\left(\omega L - \frac{1}{\omega C}\right)$$

At the resonant frequency ω_0, the imaginary part of the input impedance is cancelled, so $\omega_0 = 1/\sqrt{LC}$. Observe that this is also one of the conditions for maximum power transfer between the antenna and the load. However, in a suitably restricted range of frequencies near resonance, the series *RC* section of the resonator can be replaced with a parallel one (Figure 8.7b). The equivalence between the series and the parallel section can be computed, as reported by Lee [21]:

$$R_p = R_c(Q^2 + 1) \tag{8.20}$$

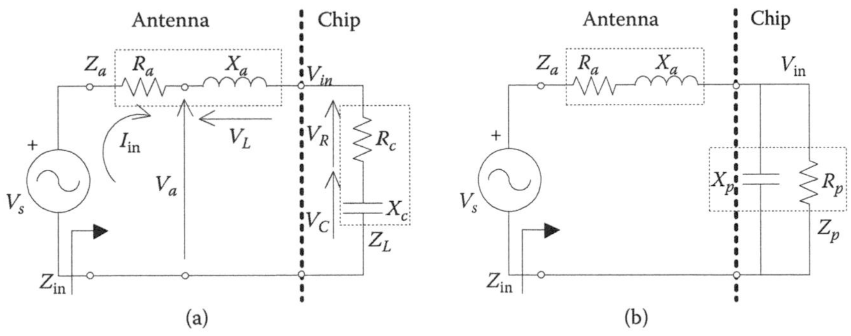

Figure 8.7 Antenna and chip equivalent *RLC* circuit: (a) series arrangement and (b) parallel transformation of the load.

Analogously, we obtain:

$$X_p = X_c\left(\frac{Q^2+1}{Q^2}\right) \tag{8.21}$$

where $Q = 1/\omega_0 R_c C = X_c/R_c$.

Remembering that $P_{diss} = (1-|\Gamma|^2)P_r$, and that the *equivalent isotropic radiated power* is $EIRP = G_t P_t$, from equation (8.8) and equation (8.15), the input voltage to the rectifier is

$$V_{in} = \sqrt{2R_p(1-|\Gamma|^2)\cdot EIRP \cdot G_r\left(\frac{\lambda}{4\pi r}\right)^2} \tag{8.22}$$

Introducing in equation (8.21) the condition of resonance and substituting into Γ, we obtain:

$$V_{in} = \sqrt{2R_p\cdot\left[\frac{4R_aR_c}{(R_a+R_c)^2}\right]\cdot EIRP \cdot G_r\left(\frac{\lambda}{4\pi r}\right)^2} \tag{8.23}$$

Figure 8.8 depicts the plots of equation (8.23) for different values of the chip resistance R_c and of the chip reactance X_c as a function of the antenna resistance. Figure 8.8 shows that a maximum in V_{in} at resonance condition ($X_a = -X_c$) occurs when R_a and R_c are equal. Interestingly, this is also the condition for maximum power transfer between the antenna and the load. Furthermore, Figure 8.8a also shows that higher V_{in} peaks are obtained for lower chip resistances R_c. From equation (8.20), it is clear that a low R_c increases the quality factor and, hence, R_p. The increase

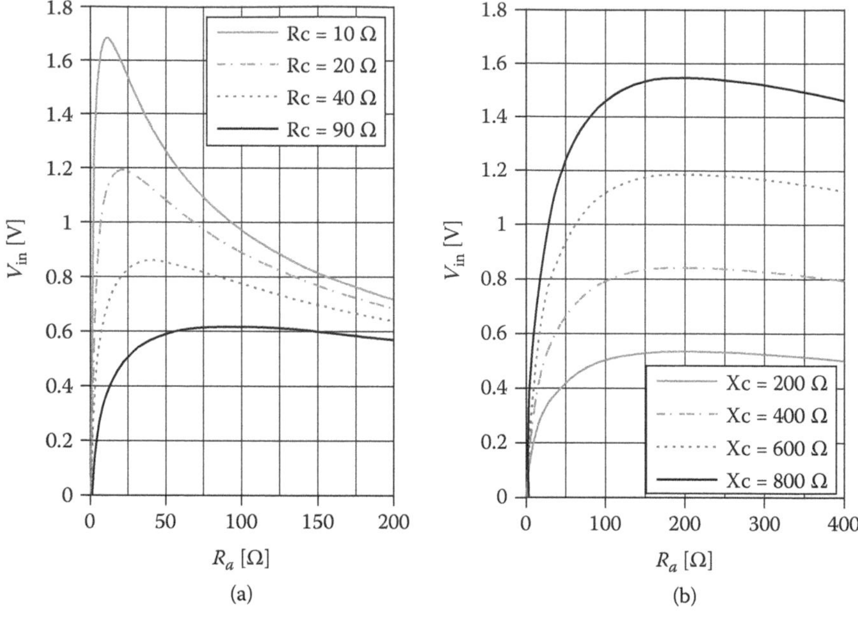

Figure 8.8 Input voltage as a function of the antenna resistance (G_r = −0.5 dB, r = 1m, EIRP = 4W, f = 2.4 GHz): (a) for different R_c values and fixed X_c = 200 Ω, and (b) for different X_c values and fixed R_c = 200Ω.

of R_p is then translated in a higher V_{in}. In addition, from Figure 8b, the higher the chip reactance X_c, the higher R_p and, therefore, the higher V_{in}. However, the sharpness of V_{in} increases more softly than in the case of R_c.

From Figure 8.7a, the voltage across the capacitor at resonance is

$$|V_C| = |V_L| = \frac{|I_{in}|}{Y} = \frac{|V_a|\,Z_c}{R_c} = \frac{|V_a|}{\omega_0 C R_C} = Q\,|V_a|$$

because at resonance the energy is stored by the capacitor and the power is dissipated by the resistor [21]. Consequently, the voltage across the chip input impedance is given by:

$$V_{in} = V_C + V_R = (1+Q)\,|V_a|$$

This equation shows that the chip input voltage can be made arbitrarily high by increasing the quality factor Q; however, as depicted in Figure 8.9, increasing Q affects the frequency response of the input current I_{in}.

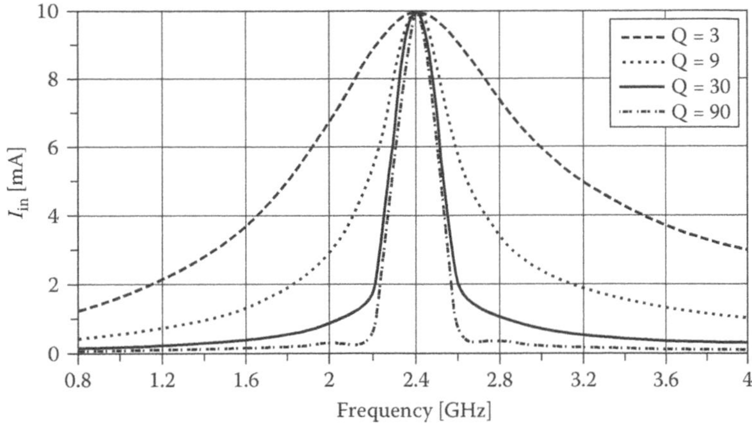

Figure 8.9 Frequency response of input current for different values (V_a= 0.1 V, R_c= 10 Ω).

The frequency response of the series *RLC* circuit also can be described in terms of the quality factor *Q*:

$$Q = \frac{f_0}{BW}$$

where f_0 is the resonant frequency and $BW = f_2 - f_1$ is the 3-dB band, i.e., the range of frequencies in which the input current I_{in} falls 3 dBs from its value at the resonant frequency.

Finally, besides the loss resistance of R_a, an antenna also exhibits a *radiation resistance* R_r. The radiation resistance is caused by the radiation of electromagnetic waves from the antenna and depends exclusively on the antenna geometry and not on the materials of which it is made. Figure 8.10 depicts several antennas' topologies commonly used in many applications.

The radiation resistance of an electric dipole of length *l*, assuming that the current distribution of the antenna is uniform and equal to the feed point current, is $R_r = 20(kl)^2\Omega$, where $k = \omega\sqrt{\varepsilon\mu}$ is the propagation constant. In the case of a loop antenna of radius *a* and under the same assumptions, the radiation resistance is given by $R_r = 20\pi^2(ka)^4\Omega$. This is also the radiation resistance of a square loop antenna whose side length is equal to *l*, provided that $\alpha = l/\sqrt{\pi}$.

8.3.5 Matching

The load of the antenna always has a capacitive nature because it is mainly formed by the voltage rectifier that performs RF-to-DC conversion. For this reason, simple

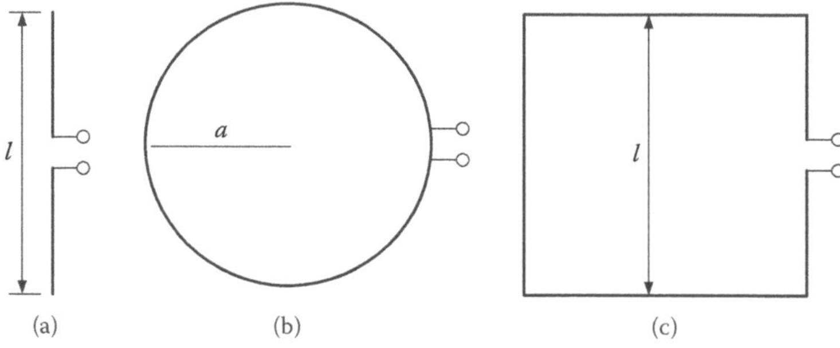

Figure 8.10 Antennas' topologies: (a) dipole, (b) loop, and (c) square loop.

matching can be achieved using an antenna with inductive behavior (such as, e.g., a loop antenna), tuning the antenna to resonate with the load capacitance at the operating frequency. However, other kinds of antennas, such as dipoles, normally have a capacitive input impedance [15]. In that case, a matching network is necessary to implement conjugate matching and assure maximum power transfer to the load.

8.4 RF-to-DC Rectifier Topologies

In this section, we will review the main characteristics of several rectifier topologies. We will perform a comprehensive Spice simulation using 130 and 90 nm technology nodes. As stated before, Schottky diodes are the best option to implement the rectifier circuit due to their low threshold voltage and their superior switching characteristics, which make them very appealing for high-frequency applications [22]. The improved performance of the Schottky barrier diodes with respect to conventional *p-n* diodes is due to the fact that in the former the dominant transport mechanism relies on the majority carriers, whereas in the latter, minority carrier transport is dominant. For this reason, we have no stored charge effects and a faster switching time. A Schottky barrier diode is formed by the union of a metal layer (the anode) and a lightly doped n-type or p-type semiconductor region (the cathode). At the junction between the two materials, a difference of potential exists that originates a barrier that impedes the flow of electrons. The potential barrier height at the interface is controlled by the signal applied to the anode making, the barrier either lower (to allow current flow) or rise (to block current flow between anode and cathode.

Unfortunately, Schottky diodes do not come with standard CMOS processes; thus these components are not suitable for low-cost designs unless a methodology is developed to implement them with a standard fabrication process. Several implementations relying on standard CMOS process have been reported [23]–[25]; however, the major challenge is finding the best layout techniques in order to minimize

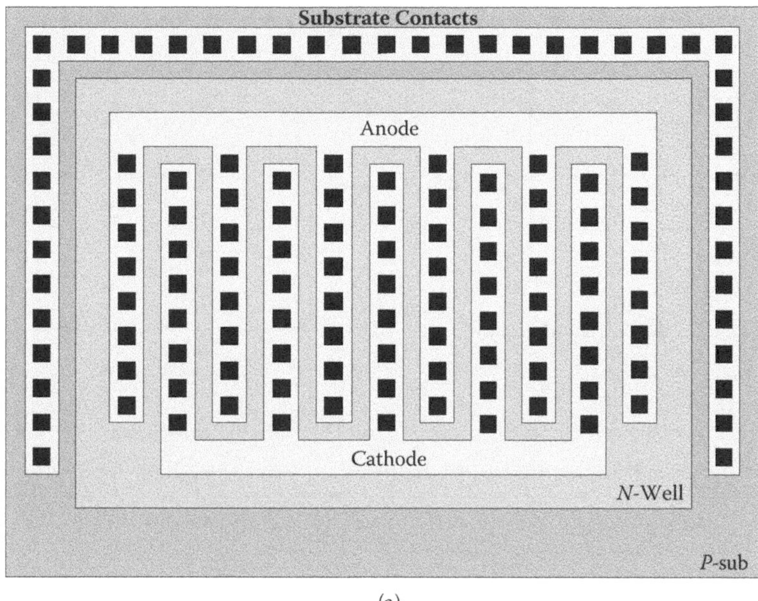

(a)

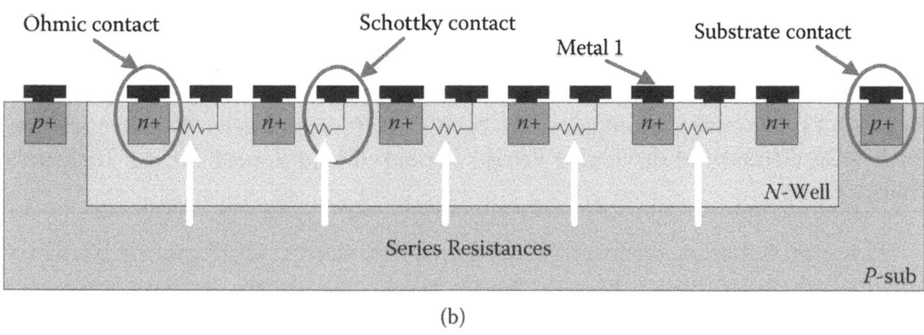

(b)

Figure 8.11 Schottky diode: (a) top view of the layout, and (b) cross section of the layout.

the effects of the series, parasitic resistance and improve the frequency response of the device.

Figure 8.11 depicts the top view and the cross section of the Schottky diode layout. The series resistance of the Schottky diode is reduced by interdigitating the fingers of the ohmic and Schottky or rectifying contacts. The spacing between the Schottky and ohmic contacts must be reduced to the minimum distance allowed by process lithography. Furthermore, interdigitating the fingers greatly reduces the distance from the anode to cathode, eliminating the likelihood of electrons being swept down to the substrate creating charge accumulation.

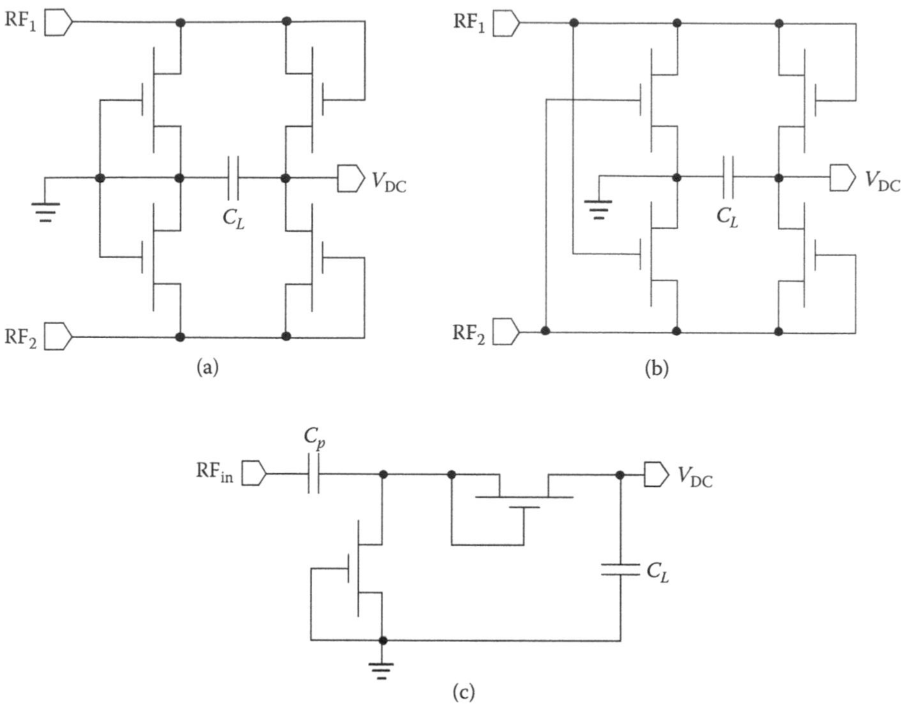

Figure 8.12 Rectifier topologies: (a) NMOS differential-drive bridge rectifier, (b) NMOS differential-drive gate cross-connected bridge rectifier, (c) Dickson's charge pump.

However, Schottky devices fabricated with standard CMOS processes still are not the best design option because their performance greatly depends on process parasitic effects and fabrication tolerances, and it is very difficult to replicate the device behavior even in the same fabrication lot. For this reason, transistor-based rectifiers are still the best option for low-cost implementation in CMOS technology.

In this section, we will review the main characteristics of three possible implementations of CMOS RF-to-DC rectifiers. The topologies reviewed are depicted in Figure 8.12. We will compare only fully NMOS (n-type metal oxide semiconductor) rectifiers since they perform better than NMOS-PMOS (p-type metal oxide semiconductor) implementations [26]. In fact, in an NMOS-PMOS differential-drive gate cross-connected bridge, the charge held into the storage capacitor C_L will flow back to the antenna through the PMOS transistor when the antenna voltage is smaller than the voltage on the storage capacitor, and this, in turn, will lead to lower efficiency and larger voltage variations on the load.

The simulations have been performed with zero-V_t transistors (to mitigate the voltage drop across the transistors and improve the conversion efficiency) with an aspect ratio $W/L = 500$ (we choose $W = 250\mu m$ and $L = 0.5\mu m$). The load capacitor

Table 8.1 Simulation Results for the Rectifier Topologies Under Test

Technology Node	*Rectifier Topology*	*Settling Time [ns]*	*Unregulated DC Output voltage [V]*	*Pump Capacitor [pF]*	*Load Capacitor [pF]*	*Number of Stages*
130 nm	Differential drive bridge	50	1.26	1	5	4
	Differential drive gate cross-connected bridge	50	1.1	1	5	7
	Dickson	100	1.2	0.8	10	3
90 nm	Differential drive bridge	250	1.32	1	5	4
	Differential drive gate cross-connected bridge	200	1.23	1	5	7
	Dickson	250	1.34	0.7	10	3

is 5 pF for the bridge converter and 10 pF for the Dickson's charge pump, and we assume an RF peak input voltage $V_p = 0.4V$ at a frequency of 2.4 GHz. The simulation results are reported in Table 8.1.

8.4.1 Voltage Multiplier Principle of Operation and Design

From the simulation results reported in Table 8.1, it can be deduced that the implementation that offers the best performances in terms of number of stages, hardware complexity, and DC output voltage is the Dickson's voltage multiplier (or charge pump). For this reason, it is worth showing the design equations that lead to optimum circuit implementation.

The principle of operation of the Dickson's charge pump resembles that of a "bucket brigade" delay line by pumping charges along the diode line as the pump capacitors C_p are successively charged and discharged during half the AC signal's period. However, unlike the "bucket brigade" line, the DC voltages along the diodes chain are not reset after each pumping cycle and average nodes potentially increases as the signal moves toward the output.

Referring to Figure 8.2, the difference between the output potentials of nodes n and n+1 is

$$V_{n+1} - V_n = \Delta V_{ac} - V_{th} - V_{CD} \tag{8.24}$$

where ΔV_{ac} is the voltage swing at each node due to capacitive coupling with the input AC signal, V_{th} is the voltage drop on each diode (due to the threshold voltage of the NMOS transistor with which a diode is implemented), and V_{Cp} is the voltage by which the pump capacitors C_p are charged and discharged in each cycle when the multiplier is supplying to the load a DC output current I_{out}. If C_S is the transistor stray capacitance, the resulting capacitive divider at each node leads to:

$$\Delta V_{ac} = \left(\frac{C_p}{C_s + C_p} \right) \cdot V_{ac}$$

Furthermore, since the total charge pumped by each diode during each cycle is $(C_s + C_p)\ V_{Cp}$, the current supplied by the multiplier to the load with an input AC voltage V_{ac} at a frequency f is $I_{out} = f(C_s + C_p)\ V_{Cp}$. By substitution in equation (8.24), we obtain:

$$V_{n+1} - V_n = \left(\frac{C_p}{C_s + C_p} \right) V_{ac} - V_{th} - \frac{I_{OUT}}{(C_s + C_p) f}$$

So, for the N-stage multiplier and taking into account the input DC voltage is zero, we obtain:

$$V_N = V_{OUT} = N \cdot \left[\left(\frac{C_p}{C_s + C_p} \right) V_{ac} - V_{th} - \frac{I_{OUT}}{(C_s + C_p) f} \right] \tag{8.25}$$

From equation N (8.25), it can be deduced that the voltage multiplication takes place only if:

$$\left(\frac{C_p}{C_s + C_p} \right) V_{ac} - V_{th} - \frac{I_{OUT}}{(C_s + C_p) f} > 0$$

It is important to observe that this equation does not depend on the number of stages of N, so in practice, there is no limit to the number of stages that can be cascaded. In addition, provided the above voltage multiplication condition is satisfied, the current drive capability of the voltage multiplier is also independent of the number of stages.

The output of the multiplier also exhibits a ripple voltage V_R due to the fact that the multiplier load is not purely capacitive, but also has a resistive component R_L. For this reason, the output capacitance C_L partially discharges through R_L.

Nonetheless, the load capacitance is sufficiently high to guarantee that in a practical implementation the ripple voltage V_R is negligible with respect to the unregulated output voltage V_{OUT}, namely:

$$V_R = \frac{I_{OUT}}{fC_{OUT}} = \frac{V_{OUT}}{fR_L C_{OUT}}$$

Practical voltage multipliers also will exhibit an additional ripple component due to the capacitive coupling among the diodes and the AC signal.

8.5 Voltage Regulators

An energy recovery system needs a voltage regulator to stabilize the DC voltage generated by the RF-to-DC converter and provide a "clean" power supply for the system [1] [27]. The voltage regulator must be designed to deliver the current and voltage required by the system under the constraints fixed by the target technology while drawing minimum current from the supply.

The RF-to-DC converter must provide the back-end circuit with a power high enough to guarantee correct operation under nominal conditions. However, the energy harvested by the voltage multiplier does not depend exclusively on the circuit topology and on the numbers of stages. In fact, the input power depends on the distance between the transmitter and the receiver. Nonetheless, the design specifications impose a constraint on receiver sensitivity, which means that if the input power exceeds the sensitivity, the voltage multiplier will produce an output unregulated DC voltage that exceeds the nominal supply voltage of the back-end circuits.

In order to prevent the circuits from being damaged, the supply voltage should in no case be higher than the nominal voltage specified by the target technology (i.e., 1.3 V for the 130 nm technology node, and 1.1 V for the 90 nm technology node).

The easiest way to implement a simple regulated supply voltage for the back-end circuits is by using a crude clamp circuit [28]–[30] like that depicted in Figure 8.13a. The operation principle is very simple: The clamp circuit is powered by the voltage generated by the multiplier, and the NMOS diode stack (transistors M1 to M4) is sized to produce a regulated output voltage of approximately $4V_{th}$. If the voltage from the multiplier exceeds $4V_{th}$, then the output bypass device (transistor M7) is activated draining current from the capacitive load and clipping the output voltage. The diode-connected transistors are sized so that they subtract minimum current to the supply (bias currents below 1 μA are desirable), and for this reason they are all long-channel devices with their widths taken as small as possible. On the contrary, the output bypass device is designed to operate as close as possible to weak inversion and has minimum channel length and a large width. Also, observe that the voltage regulator is not directly connected to the back-end circuit (modeled by the load

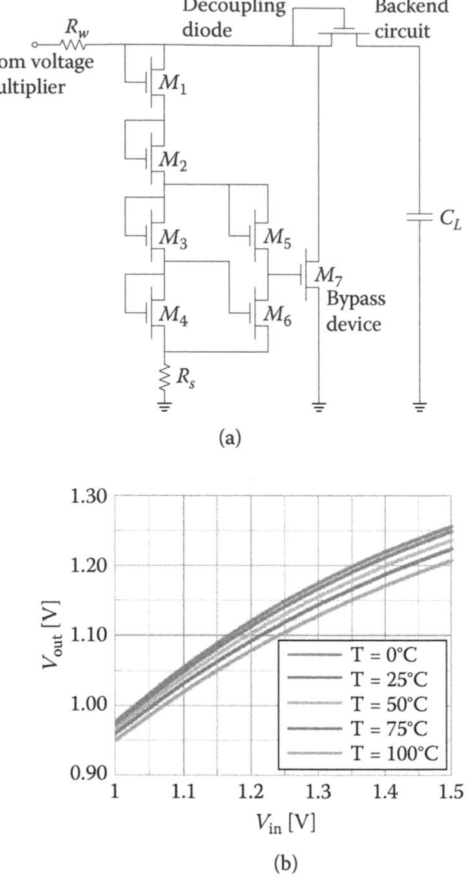

Figure 8.13 Crude clamp regulator: (a) schematic and (b) DC behavior for several temperatures.

capacitance C_L), but is decoupled from the load by a diode to prevent back-feeding from circuit to antenna when the input RF power is zero.

Finally, Figure 8.13b depicts the simulation results of the circuit of Figure 8.13a implemented with a 130 nm technology node and assuming a load capacitance of 200 fF. Resistance R_W models the losses of the power supply line. The graphics represent the output voltage when the input supply sweeps from 1 to 1.5 V and the temperature from 0 to 100° C. In these operating conditions, the circuit of Figure 8.13a draws from the supply only 269 nA.

The clamp regulator of Figure 8.13 is not an optimal solution. In practical implementations, it is used as a preconditioning stage that performs voltage clipping on the output of the voltage rectifier to make it fit into the DC input range of the voltage regulator. Several implementations have been reported in the literature

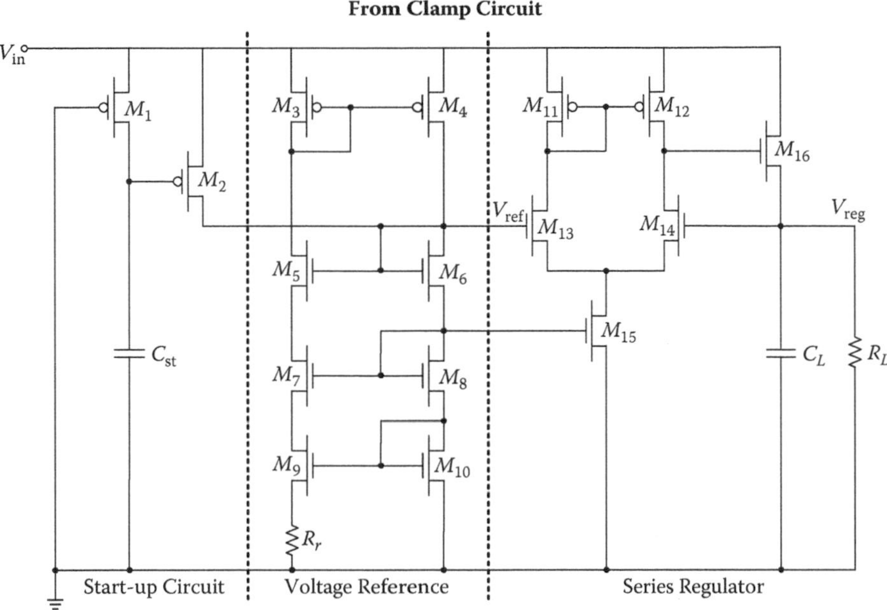

Figure 8.14 Complete schematic of a typical low-power voltage regulator.

[31]–[33]; however, regardless of some slight differences, all the proposed circuit topologies rely on the architecture depicted in Figure 8.14. The voltage regulator of Figure 8.14 is composed of three stages: a start-up circuit, a simple double-cascode voltage reference, and a series regulator. The circuit has been designed using the 130 nm technology node and the transistors have been sized to operate in weak inversion and to provide an output regulated voltage of approximately 0.6 V.

The start-up circuit is composed of transistors M1 and M2 as well as the storage capacitor C_{st}. The gate of M1 is grounded so it is always on, whereas M2 acts like a switch. Initially, M2 is on and leaks current in the gates of M5 and M6 assuring that gates of M5 and M6 are not at ground. Transistor M2 switches off as soon as the storage capacitor is charged at a given voltage.

Transistors M5 to M10 build up a triple cascode connection to increase the output resistance and generate a reference voltage V_{ref} of approximately 0.6 V. Higher reference voltages can be achieved by suitably sizing resistance R_r. The transistors of the current mirror are designed to operate in weak inversion (or subthreshold region) and draw very little current from the supply (about 630 nA). In the subthreshold region, the drain current [34] is

$$I_D = I_{D0} \cdot \frac{W}{L} \cdot e^{q(V_{gs}-V_{th})/(n \cdot kT)} \tag{8.26}$$

where $kT/q = 0.026\text{V}$ is the thermal voltage and n the slope parameter (in bulk CMOS, n is around 1.6).

If the aspect ratio W/L of transistor M9 is made Q times larger than that of M10 and both have the same length, L the gate-to-source voltage V_{gs} of M9 and M10 can be expressed in terms of subthreshold current I_D as:

$$V_{gs9} = n \cdot \frac{kT}{q} \cdot \ln\left(\frac{I_D \cdot L}{I_{D0} \cdot Q \cdot W}\right) + V_{th} \tag{8.27}$$

and

$$V_{gs10} = n \cdot \frac{kT}{q} \cdot \ln\left(\frac{I_D \cdot L}{I_{D0} \cdot W}\right) + V_{th} \tag{8.28}$$

Observing that $V_{gs10} = V_{gs9} + I_D R_r$ and solving for subthreshold current I_D using equation (8.27) and equation (8.28), we obtain:

$$I_D = \frac{n \cdot kT}{qR_r} \ln Q \tag{8.29}$$

which is independent of the DC supply voltage.

The series regulator is based on a differential amplifier (transistors M11 to M15) and a negative feedback NMOS transistor M16 (biased in saturation) to stabilize the regulated output voltage V_{reg} around the reference voltage V_{ref}.

Finally, Figure 8.15 represents the DC response of the circuit of Figure 8.14 for several operating temperatures when the input voltage is swept from 0 to 1.5 V and when the load resistance is 1 MΩ. Under these operating conditions, the output current is approximately 630 nA.

8.6 Backscatter Modulation Schemes

Autonomous wireless passive systems that harvest energy from an external source must be designed to consume very little power; for this reason, the design of a power-efficient transmission subsystem is a major issue. The common approach for data transmission in ultralow power systems, such as RFID passive tags, is using backscatter modulation techniques. In this transmission scheme, the tag transmits data by changing (modulating) its reflection coefficient to incoming radiation. The reflection coefficient is either varied in magnitude, producing amplitude shift keying (ASK), or in phase, producing phase shift keying (PSK), or both. Varying

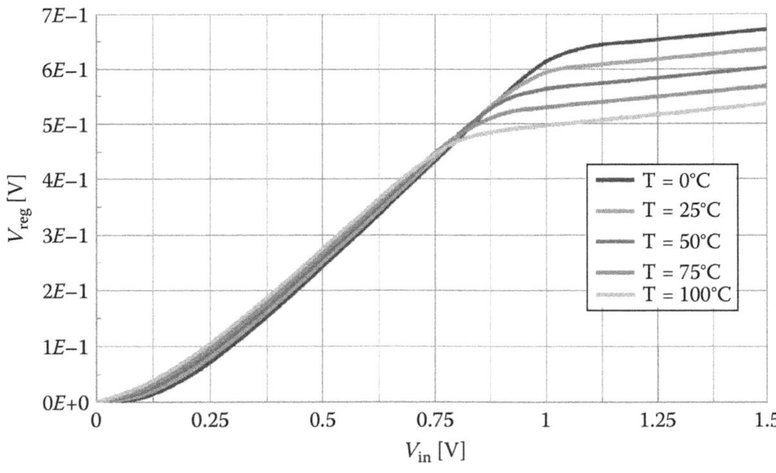

Figure 8.15 Voltage regulator DC response for several operating temperatures.

the reflection coefficient modulates the amount of power reflected by the receiver back to the transmitter, thereby transmitting information. This method is the best option when a tag does not have to broadcast information to the surrounding tags.

8.6.1 ASK Modulation

A simple ASK modulator can be implemented simply by using a CMOS switch and a capacitor to change the reflection coefficient seen by the antenna. In the ideal backscattering ASK modulator, the input impedance can be switched between the matched state (with maximum power transfer between the antenna and the load), and a complete reflection state in which the impedance seen by the antenna is either a short circuit or an open circuit. The switch is controlled by the binary bit stream that, thus, modulates the amplitude of the backscattered RF signal. In practical implementations, a short circuit is easier to implement and more reliable at high frequencies.

8.6.2 PSK Modulation

For PSK, the information is encoded into the different phase states of the backscattered signal. To change the phase, the modulator simply switches the input impedance seen by the antenna between two complex and conjugate impedances. This can be achieved by changing the input capacitance of the modulator using, for example, a varactor. Ideally, to ensure a constant available power to the transmitter, the impedances seen by the radiation resistance of the antenna during the two modulation states must have a constant module. In practice, even if capacitance values that ensure a constant module can be found, the real parts are not constant.

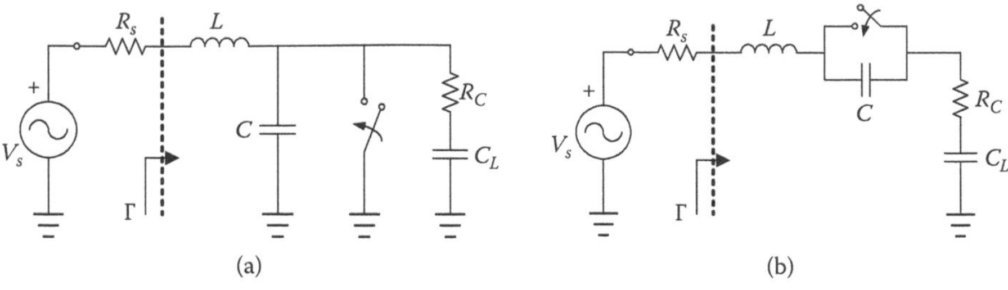

Figure 8.16 Simplified schematic of (a) the ASK modulation scheme and (b) the PSK modulation scheme.

This results in a variation of the reflection coefficient and, therefore, amplitude modulation occurs.

8.6.3 PSK versus ASK

Figure 8.16 shows two possible lumped representation models for an ASK and a PSK backscatter modulator. The PSK modulator has superior performance in terms of low power consumption when compared to its ASK counterpart because the ASK modulation scheme requires an increase of 0.74 dB in the transmission power to achieve the same BER (bit error ratio) [35]. Nonetheless, a difference of 0.74 dB is rather small to justify the increased circuit complexity of the modulator. In addition, the impedances used to implement the PSK modulation induce a smaller available voltage at the rectifier input compared to ASK. This, in turn, results in a lower input sensitivity for PSK, and, hence, into a reduced operating range. Consequently, there is no clear advantage to implement PSK modulation over ASK modulation in passive ultralow power circuits. An excellent comparison among CMOS implementations of several backscatter PSK modulation schemes is carried out by De Vita, Battaglia, and Iannaccone [36], whereas Curty et al. [35] give a comprehensive treatment of both ASK and PSK modulation schemes.

8.7 Conclusions

Many issues in the design of radio frequency energy recovery systems have been discussed in this chapter. Most of the techniques discussed are commonly used in practical implementations of RFID tags, so they are targeted to passive devices with batteryless operations. In this scenario, the antenna plays a role of paramount importance—its impact on power transmission and backscattering modulation is fundamental and cannot be neglected. Energy harvesting in radio frequency systems depends on the combination of several techniques and on the ability to maximize power transfer between the antenna and the receiver.

The core element of the energy recovery system is the RF-to-DC rectifier; this circuit is capable of scavenging energy from an incoming RF signal and storing it in an internal storage system (i.e., a large capacitor). The choice of the best rectifier configuration is mainly driven by the target technology, the process options, the power efficiency, and by the distance from the transmitter. The voltage rectifier must supply power to the back-end circuits that must be kept simple to ensure batteryless operation and draw very little power from the supply. Deep submicron technologies with reduced supply swings are indispensable to comply with the power constraints set by the Friis transmission equation; however, deep submicron technology nodes are very sensitive to fluctuations of the supply voltage that could even lead to device breakdown. For this reason, a power conditioning circuit system with a clamping circuit is necessary to generate a stable and "clean" supply. The design of the power conditioning system is a challenging task because the circuit must provide a stable voltage while drawing minimum current from the supply.

To sum up, all the issues related to energy scavenging from RF signals are well understood and many ultralow power and cheap implementations are available. Nonetheless, the major limitation of energy harvesting from radio frequency signals is the limited amount of power that can be scavenged. This is a major shortcoming in complex and power-hungry systems for which RF scavenging is not a suitable approach, and limits the application domain of this technique to wireless sensor networks and passive RFID systems whose simple architecture requires little energy to operate.

REFERENCES

[1] Huang H., and Oberle, M., "A 0.5 mW Passive Telemetry IC for Biomedical Applications," *IEEE Journal of Solid State Circuits*, v. 33, n. 7, pp. 937–946, 1998.

[2] Marschner, C., Rehfuss, S., Peters, D., Bolte, H., and Laur, R. "Modular concept for the design of application-specific integrated telemetric systems." in *Proceedings of SPIE 4408*, 246–255, Cannes-Mandelieu, France, April 2001.

[3] Schuylembergh, K., and Puers, R. "Self tuning inductive powering for implantable telemetric monitoring systems," in *Proceedings of the 8th Conference of Solid-State Sensors and Actuators*, Stockholm, Sweden, June 1995.

[4] Sample, A., and Smith, J. R., *"Experimental results with two wireless power transfer systems,"* in Proceedings of the IEEE Wireless and Radio Symposium, San Diego, CA, 2009.

[5] Yan, H., Macias Montero, J. G., Akhnoukh, A., de Vreede, L.C.N., and Burghartz, J. N., "An integration scheme for RF power harvesting," in *Proceedings of the 8th Annual Workshop on Semiconductor Advances for Future Electronics and Sensors*, Veldhoven, The Netherlands, 2005.

[6] Karthus, U., and Fischer, M., "Fully Integrated Passive UHF RFID Transponder IC with 16.7 mW Minimum RF Input Power," *IEEE Journal of Solid-State Circuits*, v. 38, n. 10, pp. 1602–1608, 2003.

[7] De Vita, G., and Iannacone, G., "Design Criteria for the RF Section of UHF and Microwave Passive RFID Transponders," *IEEE Transactions on Microwave Theory and Techniques*, v. 53, n. 9, pp. 2978–2990, 2005.
[8] Curty, J.-P., Joehl, N., Dehollain, C., and Declercq, M. J., "Remotely Powered Addressable UHF RFID System," *IEEE Journal of Solid-State Circuits*, v. 40, n. 11, pp. 2193–2202, 2005.
[9] Umeda, T., Yoshida, H., Sekine, S., Fujita, Y., Suzuki, T., and Otaka, S., "A 950-MHz Rectifier Circuit for Sensor Network Tags with 10-m Distance," *IEEE Journal of Solid-State Circuits*, v. 41, n. 1, pp. 35–41, 2006.
[10] Bode, H. W., *Network Analysis and Feedback Amplifier Design*, 1st ed., Princeton, NJ: D. Van Nostrand Company, 1945.
[11] Fano, R. M., "Theoretical Limitations on the Broadband Matching of Arbitrary Impedances," DSc disser., Massachusetts Institute of Technology, Department of Electrical Engineering, May 1947.
[12] Yan, H., Popadic, M., Macías-Montero, J. G., de Vreede, L. C. N., Aknoukh, A., and Nanver, L. K., "Design of an RF power harvester in a silicon-on-glass technology," in *Proceedings of IEEE-STW PRORISC 2008*, Veldhoven, The Netherlands, pp. 287–290, 2008.
[13] Mandal, S., "Far Field RF Power Extraction Circuits and Systems," master's thesis, Massachusetts Institute of Technology, Department of Electrical Engineering and Computer Science, June 2004.
[14] Mandal, S., and Sarpeshkar, R., "Low Power CMOS Rectifier Design for RFID Applications," *IEEE Transactions on Circuits and Systems I*, v. 54, n. 6, pp. 1177–1188, 2007.
[15] Balanis, C. A., *Antenna Theory: Analysis and Design*, 3rd ed. New York: John Wiley & Sons, 2005.
[16] Johnson, R. C., and Jasik, H., *Antenna Engineering Handbook*, 3rd ed. New York: McGraw-Hill, 1993.
[17] Wheeler, H. A., "The radiansphere around a small antenna," *Proceedings of the IRE*, v. 47, n. 8, pp. 1325–1331, 1959.
[18] Friis, H. "A note on a simple transmission formula," *Proceedings of the IRE*, v. 34, pp. 254–256, May 1946.
[19] European Telecommunications Standards Institute (ETSI). Available online: http://www. etsi.org/WebSite/homepage.aspx
[20] Texas Instruments. UHF Gen2 STRAP. Available online: http://www.ti.com/rfid/docs/manuals/pdfSpecs/RI-UHF-STRAP DataSheet.pdf
[21] Lee, T. H., *The Design of CMOS Radio-Frequency Integrated Circuits*, 2nd ed., Cambridge, U.K.: Cambridge University Press, 2004.
[22] Neudeck, G., and Pierret, R., *The PN Junction Diode*, Vol. 2, Reading, MA: Addison-Wesley, 1989.
[23] Li, Q., Han, Y., Min, H., and Zhou, F., "Fabrication and Modeling of Schottky Diode Integrated in Standard CMOS Process," Auto ID Labs White Paper WP-HARDWARE-011, Massachusetts Institute of Technology, 2005.
[24] Milanovic, V., Gaitan. M., Marshall, J. C., and Zaghloul, M. E., "CMOS foundry implementation of Schottky diodes for RF detection," *IEEE Transactions on Electron Devices*, v. 43, n. 12, pp. 2210–2214, 1996.
[25] Cha, S. I., Cho, Y. H., Choi, Y. I., and Chung, S. K., "Novel schottky diode with self-aligned Guard Ring," *IET Electronics Letters*, v. 28, n. 13, p. 1221–1223, 1992.

[26] Mazzilli, F., Thoppay, P. E., Jöhl, N., and Dehollain, C., "Design methodology and comparison of rectifiers for UHF-band RFIDs," in *Proceedings of IEEE Radio Frequency Integrated Circuits Symposium*, pp. 505–508, Anaheim, CA, May 2010.

[27] Chatzandroulis, S., Tsoukalas, D., and Neukomm, P., "A Miniature Pressure System with a Capacitive Sensor and a Passive Telemetry Link for Use in Implantable Applications," *Journal of Microelectromechanical Systems*, v. 9, pp. 18–23, March 2000.

[28] Zhu, Z., Jamali, B., and Cole, P. H., "An HF/UHF RFID Analogue Front-End Design and Analysis," White Paper Series Edition 1. CD. Auto-ID Labs, Massachusetts Institute of Technology, September 2005.

[29] Zhang, L., Jiang, H., Sun, X., Zhang, C., and Wang, Z., "A passive RF receiving and power switch ASIC for remote power control with zero stand-by power," in *Proceedings of IEEE Asian Solid-State Circuits Conference*, Fukuoka, Japan, pp. 109–112, 2008.

[30] Che, W., Yan, N., Yang, Y., and Min, H., "A Low Voltage Low Power RF Analog Front-End Circuit for Passive UHF RFID Tag," *Journal of Semiconductors*, v. 29, n. 3, 2008.

[31] De Vita, G., and Iannaccone, G., "Ultra-low power series voltage regulator for passive microwave RFID transponders," in *Proceedings of NORCHIP Conference*, Oulu, Finland, 2005.

[32] Morales-Ramos, R., Vaz, A., Pardo, D., and Berenguer, R., "Ultra-low power passive UHF RFID for wireless sensor networks," in *Proceedings of EUROMICRO Conference on Digital System Design, Architecture, Methods, and Tools*, Parma, Italy, pp. 671–675, 2008.

[33] Yao, Y., Wu, J., Shi, Y., and Foster Dai, F. "A Fully Integrated 900-MHz Passive RFID Transponder Front End with Novel Zero-Threshold RF–DC Rectifier," *IEEE Transactions on Industrial Electronics*, v. 56, n. 7, pp. 2317–2325, 2009.

[34] Baker, R. J., *CMOS Circuit Design, Layout, and Simulation*, 3rd ed. New York: John Wiley & Sons, 2010.

[35] Curty, J.-P., Declerq, M., Dehollain, C., and Johel, N., *Design and Optimization of Passive UHF RFID Systems*, New York: Springer, 2007.

[36] De Vita, G., Battaglia, F., and Iannaccone, G., "Ultra-Low Power PSK Backscatter Modulator for UHF and Microwave RFID Transponders," *Microelectronic Journal*, v. 37, n. 7, pp. 627–629, 2006.

Chapter 9

Energy Scavenging for Magnetically Coupled Communication Devices

Mehrnoush Masihpour, Johnson I. Agbinya, and Mehran Abolhasan

Contents

9.1 Introduction

Energy scavenging dates back to the use of waterwheels and windmills to harvest the energy from the surrounding environment and convert it to useful forms of energy [1]. Energy scavenging techniques have been around for decades to provide electricity from different natural resources, such as wind power, water flow, heat, sunlight, and vibration. However, there has been a new interest in energy harvesting methods recently, due to the mass increase of wireless sensors and low power electronics [1]. Electronic devices, such as mobile phones, laptops, PDAs (Personal Digital Assistants), medical implant devices, and different types of sensor nodes, all require a power source to operate. Therefore, they are provided with either wires or disposable/rechargeable batteries to supply the power. Thanks to advanced fabrication technologies, very small sized batteries are introduced, which are small enough to be located on a chip to power up a sensor node or a small electronic handheld [1]. The problem with batteries is that they have limited lifetimes or they are required to be recharged on a regular basis. However, there have been technology advances to optimize the battery performance and its lifetime, such as dynamic optimization of voltage and clock rate, hybrid analog digital designs, and clever wake up procedures that allow a device to go to inactive mode when it is not required to operate a task [1].

All of those advances in technology allow the electronic devices to consume low power, but operate properly. Therefore, it is possible to power them up using energy sources already in the environment or even wasted from other systems. For example, a low power condition monitoring sensor, which is able to alert the failure of a machine, can be powered by the electromagnetic field created by surrounding working equipment in the workplace [2]. Another example is that the vibration of the floor or walls due to working machines and equipment in the environment can be used to create the required power for a low power sensor node to perform a task [1]. Even the heat from a human body can be used to provide the power for a low power wearable device, such as a wristwatch [1] [2]. There are a number of different energy sources in the environment that can be used to supply the required power for devices with different functionalities. Basically, there are two main types of energy scavenging devices: small scale and large scale. A large-scale energy harvester is a device that uses natural phenomena, such as wind, hydroelectricity, tidal, or geothermal to generate electric power [2]. However, small-scale energy scavenging devices are often less than microscale size [2]. Different types of small-scale energy scavenging generators are kinetic energy harvesters and electromagnetic radiations. Electromagnetic radiations, such as solar energy (photovoltaic), RF (radio frequency) radiation, and thermal energy (thermoelectric) convert the radiative energy, such as light, heat, or microwaves energy, to electric energy to provide DC or AC voltage. For example, to power up ID cards or passive RFID (radio frequency identification) tags, radio waves are used to supply the electric energy of the device in order to enable it performing a task [2] [3]. The difference between the body temperature and the ambient temperature is used to convert the thermal energy to a small amount of electric energy to activate a wristwatch [1]. Also the energy of sunlight can be converted to the electric energy using solar cells. The amount of energy harvested from sunlight depends on the area of the exposure and the size used in the solar panel. A small solar panel can be used to charge a couple of AA batteries; however, if larger panels are used, the produced electricity would be enough to run an industry site or to power a house.

This chapter focuses on energy scavenging techniques to power up low power magnetically coupled resonators. Passive RFID tags, biomedical monitoring devices, and some sensor networks might use the magnetic induction method for communication or actuation. The following section introduces the near field magnetic induction communications (NFMIC). In this chapter, we will discuss the possible energy sources to power NFMIC devices. The sequence of the chapter will be as follows. First, we introduce NFMIC and the applications. Then, we will discuss vibration-, solar-, and thermal energy-harvesting methods suitable to power up NFMIC devices. Because the output of these systems is DC voltage, we will introduce a DC/AC voltage converter, which is able to power up the devices with the AC voltage as required for NFMIC devices. Then, we will discuss an application of the NFMIC that is wireless power transmission for embedded medical devices and some techniques to improve the efficiency of the transmission.

9.2 Energy Harvesting for Magnetically Coupled Communication Devices

Near field magnetic induction communication (NFMIC) has been seen as a beneficial physical layer for the applications where communication occurs inside or in the close proximity of the human body, underground (such as tunnels), and underwater [4]-[14]. NFMIC uses near field magnetic flux for data transmission rather than radiating electromagnetic (EM) waves [15]. RF-based transmission systems radiate the EM waves, which are capable of travelling through the communication channel (often air) far from the source. Unlike transmitted EM waves, which do not come back to the source when transmitted (radiative transmission), MI (magnetic induction) waves can come back to the source due to the mutual coupling (reactive transmission) because the communication link is established through magnetic coupling between two inductive coils.

Although electromagnetic waves are suitable for long-range communications, they sometimes fail to address the requirements of short-range communications. However, unique characteristics of NFMIC make it favorable for short-range communication systems particularly where the communication channel is no longer air and contains humidity, soil, or body tissues. In such channel conditions, EM waves can be easily absorbed by the channel, which, for example, is body tissue in a BAN (body area network). In wearable BAN, using EM waves also has a multipath effect, which is a major problem due to the human body [16]–[18]. Therefore, reliability and QoS (quality of service) may be degraded. However, the effect of the human body on the MI waves is different. MI waves can easily penetrate the tissue with minimal losses, thus multipath effect and signal absorption due to the human body is not as severe as in EM-RF-based systems [4]–[6]. The most important factor affecting the magnetic wave is the magnetic permeability of the material within the channel. Therefore, since water, soil, and human body tissue have almost the same permeability as air, they have the same effect as air on the transmitted signal [12]–[14]. This characteristic of NFMIC can result in higher reliability and QoS, not only in BAN, but also in tunnels and rocky and costal regions.

NFMIC has a number of advantages over RF for short-range communication systems. For example, lower transmission power (a few milliwatts) is used to provide communication within a couple of meters [4] [5] [7]. According to Bansal [4], NFMIC is about six times more power efficient than the Bluetooth-enabled devices. Frequently, reuse also is well facilitated since NFMIC devices are not required to operate at the busy 2.4 GHz band. Because each user is assigned the frequency within his or her communication "bubble" [15], the same frequency can be used for multiple users with minimal chance of interference. The transmitted signal loses it power rapidly with distance, even more than RF signals. This characteristic of the NFMIC results in a well-sealed communication bubble. In other words, it is very difficult to be intercepted by an unauthorized party or to interfere with someone else's bubble. Moreover, NFMIC does not need to struggle with the fading due to multipaths because of the rapid power decay with distance [4]. Although the high path loss (due to the higher rate of power degradation with distance) offers some benefits, it makes the communication distance very short (centimeter range), hence, difficult to be used for the application where longer range is required.

Fundamentals of a magnetic induction communication system are discussed in the following section. However, to provide the primary power for such systems, often batteries are used, which results in larger-sized devices and limited lifetime.

On one hand, development of ultralow power micro systems, such as wearable devices, condition monitoring sensors, and implant devices is accelerating rapidly. On the other hand, advances in battery technology is not as fast as the growth of integrated circuit technologies [19] [20]. Batteries, which are used to supply the required power for micro systems, are still large, and have a limited lifetime, and they are required to be recharged or replaced frequently [19, 20]. Some devices, such as body implants, require no physical connection, such as wires and very small and long-lasting power sources. It is often very difficult and, in some cases, impossible to replace their batteries or even recharge them. Therefore, there is a great need for renewable power sources to replace batteries in such devices. Energy scavenging from the surrounding environment has been seen to be a promising solution for powering such devices.

9.2.1 Magnetic Induction Theory

In general, there are two main units in magnetic induction systems: passive and active units. The active unit is fed by an energy source to create a voltage and current in the circuit, which includes a magnetic coil. This coil creates an electromagnetic field and when another coil in the passive device is located near the created electromagnetic field, they resonate. The resonance results in inducing a voltage and current in the passive device, which has no individual power source of its own. By resonating at the same frequency, the energy resulting from coupling is maximized. Therefore, the induced power creates the circuitry in the passive unit and activates the unit to perform a function.

The block diagram of a magnetically inductive power transmission system is shown in Figure 9.1. The transmitter (active unit) and the receiver (passive unit) are composed of two coils each of radius r_T and r_R respectively, and separated from each other by a distance *d*. The transmission link between them is an inductive coupling *k*. The coupling between the two magnetic antennas can be estimated using the following equation [21]:

$$k = \frac{M}{\sqrt{L_T L_R}} \tag{9.1}$$

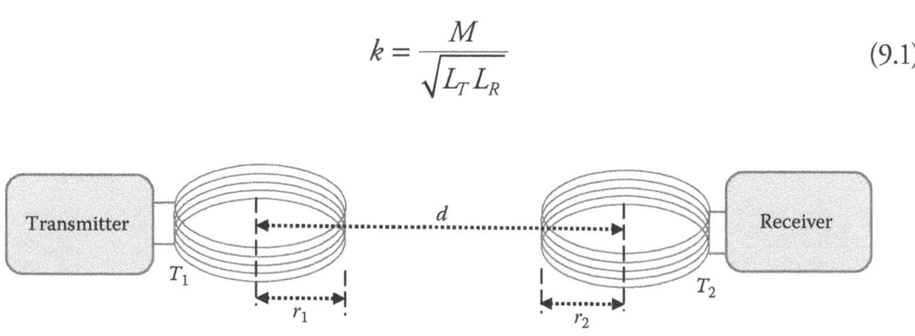

Figure 9.1 Magnetic induction system.

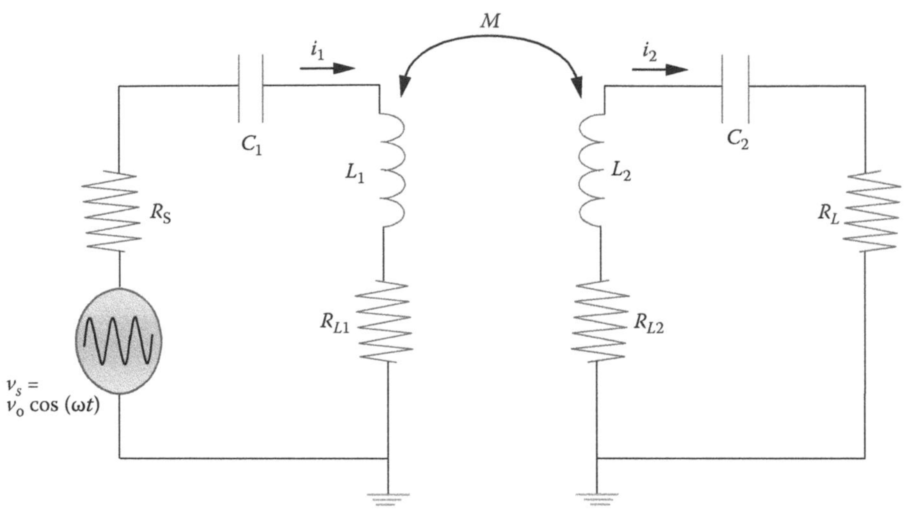

Figure 9.2 Circuit model.

M is the mutual inductance between the two inductive circuits and L_T and L_R are the self inductive values of the transmitter and receiver, respectively. The lump circuit model of Figure 9.1 is shown in Figure 9.2.

The efficiencies of the coils are given to describe how effective they are in transmitting power and, by definition, they are

$$\eta_T = \frac{R_S}{R_S + R_{LT}}; \eta_R = \frac{R_L}{R_L + R_{LR}} \tag{9.2}$$

Hence, wires with low paracitic resistances are desirable for achieving high coil efficiencies. Also, the quality factors are given by definition to be

$$Q_T = \frac{\omega_0 L_T}{R_{LT} + R_S}; Q_R = \frac{\omega_0 L_R}{R_{LR} + R_L} \tag{9.3}$$

The power delivered to the receiver load is given by the expression [15]:

$$\frac{P_L(\omega)}{P_S} = \frac{V_C}{V_D}\eta_T\eta_R Q_T Q_R = \eta_T\eta_R Q_T Q_R k^2(d) \tag{9.4}$$

$$k^2(d) = \frac{V_C}{V_D} \tag{9.5}$$

where

$$P_L = \frac{|i_R|^2 R_L}{2} = \frac{P_S \eta_T \eta_R Q_T Q_R k^2}{\left(1 + Q_T^2 \frac{(2\Delta\omega)^2}{\omega_0^2}\right)\left(1 + Q_R^2 \frac{(2\Delta\omega)^2}{\omega_0^2}\right)} \tag{9.6}$$

At resonance, this expression simplifies to:

$$P_L = P_S \eta_T \eta_R Q_T Q_R k^2(d) \tag{9.7}$$

The coupling coefficient for air-cored coil is

$$k^2(d) = \frac{V_C}{V_D} = \frac{\mu_0 A_R^2 \mu_0 r_T^4}{4 L_T L_R \left[\left(r_T^2 + d^2\right)\right]^3} \tag{9.8}$$

When the magnetic flux is enhanced by using a ferrite core, the coupling coefficient is increased to be [21]

$$k^2(d) = \frac{V_C}{V_D} = \frac{\mu_T \mu_0 A_R^2 \mu_R \mu_0 r_T^4}{4 L_T L_R \left[\left(r_T^2 + d^2\right)\right]^3} \tag{9.9}$$

The self inductances themselves can be estimated with the following expressions:

$$L = \frac{\mu_0 \pi r^2 N^2}{l + 0.9r} = \frac{\mu_0 \pi r^2}{l + 0.9r}; \quad N = 1 \tag{9.10}$$

In practice, the radius of the coil is far less than the length: $r << l$ and $l = 2\pi r$. Hence:

$$L_T \approx \frac{\mu_0 r_T}{2}; L_R \approx \frac{\mu_0 r_R}{2}; \quad A_R = \pi . r_R^2; N = 1 \tag{9.11}$$

Finally, the expression for the coupling coefficient is obtained:

$$k = \frac{r_T^2 r_R^2}{\sqrt{r_T r_R}\sqrt{\left(r_T^2 + d^2\right)^3}} \tag{9.12}$$

where, $r_R << r_T$. From the equivalent circuit model of Figure 9.2, the relationship for resonance frequency between the antennas is $\omega_0 = \frac{1}{\sqrt{L_T C_T}} = \frac{1}{\sqrt{L_R C_R}}$. Hence, the two coils are chosen so that they resonate at the same frequency and, thus, permit optimum mutual coupling to be established between them.

Finally, power at the receiver becomes [21]:

$$P_R = P_T Q_T Q_R \eta_T \eta_R \frac{r_T^3 \mu_0 \mu_T r_R^3 \mu_0 \mu_R \pi^2}{\left(r_T^2 + d^2\right)^3} \tag{9.13}$$

9.3 Vibration to Electric Energy Conversion Using Magnetic Induction

Magnetically coupled sensors are suitable for short-range communications and an indoor environment. Therefore, energy sources existing in an indoor environment can be harvested to power up such devices. Vibration, radiative, solar, and thermal energy are some examples of power sources that can be used to provide magnetic induction nodes with required power. Among the possible solutions, vibration is claimed to be a suitable and clean energy source to power up low power, small size electronics because of its abundance in many environments [19]–[22].

Conversion of vibration to electric energy can be done through three different methods using piezoelectric, electrostatic, and magnetic induction [20]. In the piezoelectric method, piezoelectric material is stressed by vibration, which leads to the generation of some voltage [22]. Created voltage can simply power up a low power microsystem. However, to convert the vibration to electricity through the electrostatic technique, two plates of a charged capacitor are pulled apart against the force of electrostatic attraction by vibration energy [22]. However, these two methods are out of the scope of this chapter. The third technique, which converts vibration to electric energy by a magnetic induction generator, is more suitable for this purpose. That is because, while piezoelectric and electrostatic techniques are useful for generation of a considerable amount of energy at lower volume and higher frequency, magnetic generators are suitable for large volume and lower frequency [22] and the vibration created from the environment is often very low frequency, usually less than 500 Hz [19] [20] [22]. However, the created electric energy is directly related to the frequency of vibration.

A magnetic energy harvester with a mass-spring method was proposed by Williams, et al. in 1996 for the first time [22]–[24]. The basic principle of magnetic induction generators for vibration conversion is based on Faraday's law of induction [19] [20] [22]. The system consists of a permanent magnet and a moving metal coil or vice versa [1]–[3] [19] [20] [22]. By moving either of the components (coil or magnet) relative to each other by using a vibration source, electromotive force is induced in the coil, which creates a current in the circuit and can power up electronics [19] [20] [22]. However, more power can be achieved at higher frequencies. Different methods have been proposed for increasing the vibration frequency harvested from the environment to improve the performance of such systems to collect more electric power [19] [22]–[25].

9.3.1 Fundamentals of Magnetic Generators

When vibration is applied to a flux generator as an input energy source, it causes motion of either a magnet or a coil relative to each other, which results in electromotive force around a closed loop. Electromotive force (EMF) is created by changing the magnetic flux of the loop in respect to time. If the coil has N turns, the electromotive force, according to Faraday's law, can be calculated by equation (9.14) [22].

$$emf = -N\frac{d\psi}{dt} \tag{9.14}$$

In this equation, $\frac{d\psi}{dt}$ is the flux changes with time of the magnet in Weber's theory. However, according to Lenz's law, the negative sign in equation (9.14) means that the induced electromotive force acts to appose the magnetic flux changes [22]. In other words, the current created by the electromotive force tends to act against the flux changes and not the flux itself [22]. To enhance the performance of the system, some influential properties of the electromagnetic microgenerators, such as output power, acceleration, and resonance frequency, can be altered [22]. To design an optimum electromagnetic generator, researchers mainly have been focusing on magnetic, spring-mass, and coil property, which will be discussed in the following section

9.3.1.1 Magnetic Property

Equation (9.14) implies that the induced electromotive force varies with the magnetic flux; therefore, it can be derived that the flux density has an important role to play in magnetic generator design. There are different types of magnets, which can be used in different magnetic generator designs according to the applications [22]:

- Rare earth magnets: These can provide very strong magnetic flux density and are suitable for use in micromagnetic generators.
- Alnico magnets: This conventional type of magnet is about five times less strong than the rare earth magnets.
- Neodymium Iron Boron (NdFeB): This type is categorized as the most powerful magnet type. It is capable of operating at up to 140°C. They are used for many microgenerators because they can provide strong magnetic flux density.
- Samarium Cobalt: If a higher temperature is required (up to 300°C), Samarium Cobalt, which is from the rare earth magnet family, can be used. However, they are less expensive and less powerful than NdFeB magnets.

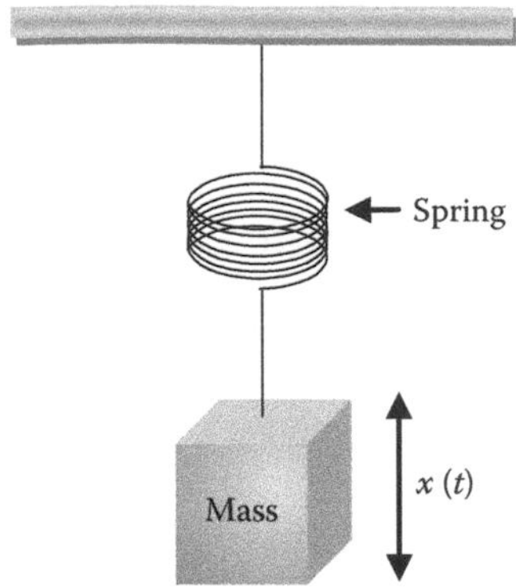

Figure 9.3 Simple harmonic motion of spring-mass system. (Adapted from W. C. Chye, et al., in *IEEE Symposium on Industrial Electronics & Applications (ISIEA), 2010*, 2010, pp. 376–382. With permission.)

9.3.1.2 Spring-Mass Property

Using spring-mass, as can be seen in Figure 9.3, oscillation for a magnet is achieved to provide a flux-cutting mechanism and to generate a specific amount of electromotive force [22]. To achieve higher electromotive force, spiral-shaped springs with low spring constant and lower stress concentration can be used [22]. These characteristics result in larger displacement, which refers to the distance of mass *x(t)* in vertical direction with respect to time (horizontal axis), and consequently, more flux cutting and higher electromotive force [22].

Displacement of the mass at time *t*, is expressed as:

$$x(t) = A\cos(\omega t + \phi) \tag{9.15}$$

where A is the amplitude of vibration in vertical direction, ω is the angular frequency, and φ is the phase. Phase shows the starting point on the wave. However, the velocity of the mass, which is derivative of the displacement, is

$$v(t) = \frac{dx}{dt} = -A\sin(\omega t + \varphi) \tag{9.16}$$

Similarly, the acceleration of mass can be computed using equation (9.17):

$$a(t) = \frac{dv}{dt} = \frac{d^2x}{dt} = -A\omega^2\cos(\omega t + \varphi) \tag{9.17}$$

To provide reliability and a higher power generation, copper springs are said to have a better performance than silicon [22] [25]–[27]. Copper also is more cost effective than titanium and 55-Ni-45-Ti, and therefore, it is used in many micromagnetic generator designs [22] [26]–[28].

9.3.1.3 Coil Property

There are two main types of coils; each has a different purpose. The first, which is used for macrosized prototype design, is wire-wound copper coil and the second is electrodeposited copper coil, which is useful for smaller designs, such as microgenerators [22]. Copper is used in coil designs because it is cost effective and highly conductive.

9.3.2 Micromagnetic Generator Design

In this section, a simple vibration power generator using magnetic induction is discussed. In this system, when there is vibration input, the coil will move and it cuts through the magnetic flux created by a permanent magnet. It then creates a sinusoidal electromotive force in the coil and, therefore, transfers the mechanical energy into electrical energy [29]. A simple structure of the magnetic generator is illustrated in Figure 9.4.

Such design is capable of providing about 1 mW to 20 mW power, which can be used to power up low-power small devices, such as sensor nodes [29]. According

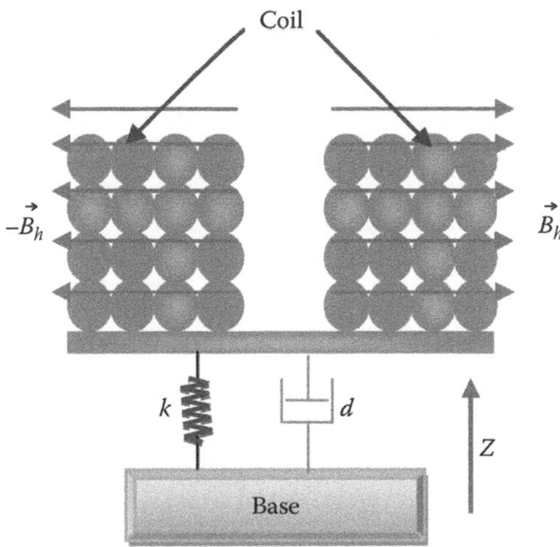

Figure 9.4 The simplified structure of the electromagnetic power generator. (From C. Xinping and L. Yi-Kuen, in *2006 IEEE International Conference on Information Acquisition,* 2006, pp. 91–95. With permission.)

to Xinping and Yi-Kuen [29], the generated force (F_e) is proportional to the flux density (B) in the coil and the induced current (i):

$$F_e = -\frac{B}{i} \tag{9.18}$$

where z is the velocity of the coil; electromotive force (*emf*) created in the circuit is

$$emf = \frac{B}{z} \tag{9.19}$$

To create a closed loop, the two ends of the coil are connected together using a resistor with value R. However, the damping ratio due to electromotive force is shown by ξ_t, c is the damping constant, Z_{max} is the maximum displacement, and ω_n is the angular frequency; the maximum achievable power (P_{max}) will be [29]:

$$P_{\max} = m\xi_t\omega_n^3 Z_{\max}^2 \tag{9.20}$$

$$\xi_t = \frac{c}{2m\omega_n} \tag{9.21}$$

$$c = \frac{(B_h l)^2}{R} \tag{9.22}$$

In equation (9.22), B_h is the average value of the horizontal vector of magnetic field and l is the total length of the inductor. Based on the typical design parameters, achievable power can be calculated using equation (9.23) [29].

$$P = \frac{(\omega B_h l Z)^2}{2R} = \frac{kB_h^2 l^2 Z^2}{2mR} = \frac{B_h^2 l^2 Za}{2R} \tag{9.23}$$

The size of coil (D) is defined as:

$$D \approx 10Z \tag{9.24}$$

where a is the peak value of the acceleration of mass, ρ is the resistivity of the metal used to create the coil, and the maximum power is [29]

$$P_{\max} = \frac{\pi B_h^2 Z^4 a}{20\rho} \tag{9.25}$$

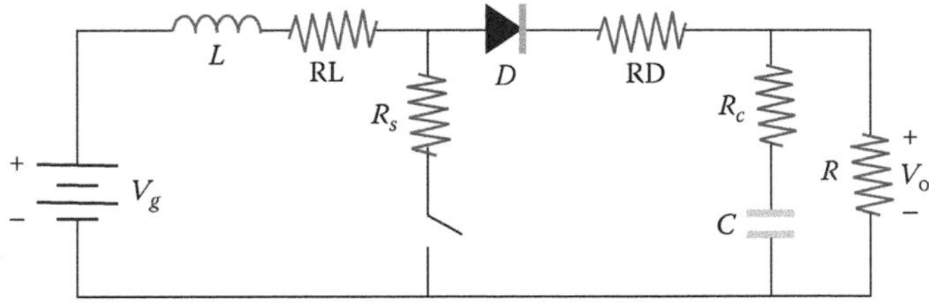

Figure 9.5 A typical boost converter. (From C. Xinping and L. Yi-Kuen, in *2006 IEEE International Conference on Information Acquisition,* 2006, pp. 91–95. With permission.)

The size of the generator can be determined by using equation (9.24) and equation (9.25).

9.3.2.1 Energy-Harvesting Circuit Design

Power storage elements require a DC voltage to operate, while the output voltage of an electromagnetic power device creates an AC voltage. Therefore, an energy harvester requires an AC-to-DC converter [29]. The output of AC-DC rectifier might be less than the required voltage of the storage element; hence, a step-up DC-to-DC converter is required to increase the harvested energy [29].

For the circuit shown in Figure 9.5, the following relationship holds [29]:

$$M = \frac{V_o}{V_g} = \frac{1}{1-D} \tag{9.26}$$

Conversion efficiency, which can be defined as a function of the duty ratio of the load resistance, is

$$\eta = \frac{1}{1 + \frac{R_d}{D'R} + \frac{DR_s + R_l}{D'^2 R}} \tag{9.27}$$

The total performance of such a converter is not optimum because, on one hand, the typical vibration input is random and, on the other hand, the duty ratio of a DC-DC boost converter is constant and directly proportional to the DC input voltage [29]. Therefore, the rectifier has different output voltage, which makes it very difficult to achieve high efficiency [29]. To improve the performance of a DC-DC boost converter, a feed-forward control circuit (DPBC) is proposed in Xinping and Yi-Kuen [29], which is shown in Figure 9.6.

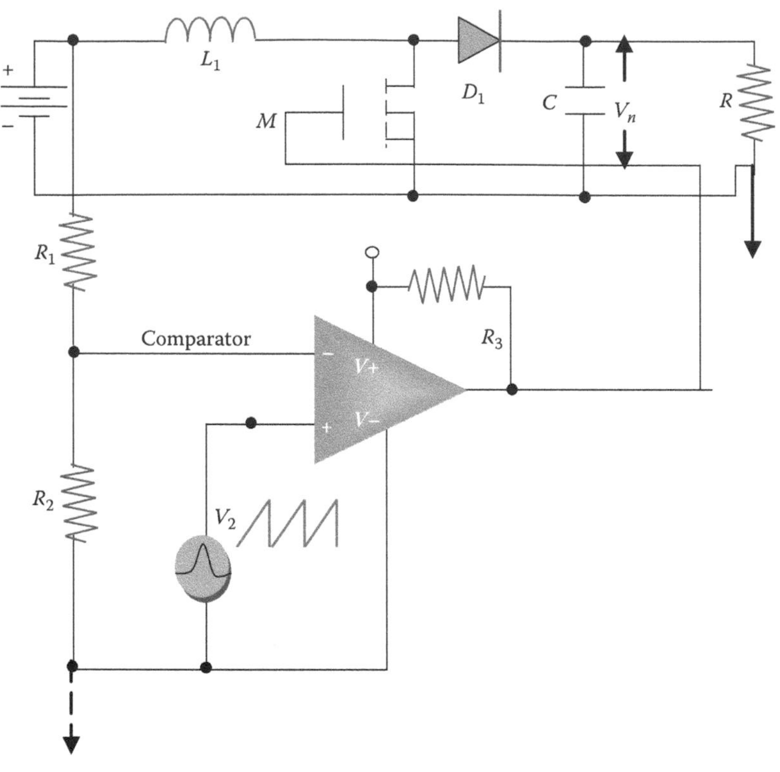

Figure 9.6 Feed forward control of DC-DC PWM boost converter. (From J. C. Xinping and L. Yi-Kuen, in *2006 IEEE International Conference on Information Acquisition,* 2006, pp. 91–95. With permission.)

The following voltage relationship (input and output voltage) holds for the circuit:

$$V_o = (R_1 / R_2 + 1)V_{Tm} \tag{9.28}$$

In this equation, V_{Tm} is the peak value of sawtooth voltage. The voltage of energy storage unit might vary between different values, thus, the output voltage of the converter also should vary to be compatible with the energy storage element [29]. To achieve the compatibility, a feedback control circuit is added to the feed forward control unit of a DPBC circuit, which is shown in Figure 9.7.

According to the modified circuit, the sawtooth voltage is defined to be [29]:

$$V(t) = V_o \frac{R_4}{R_3 + R_4}(1 - e^{-t/RC}) \tag{9.29}$$

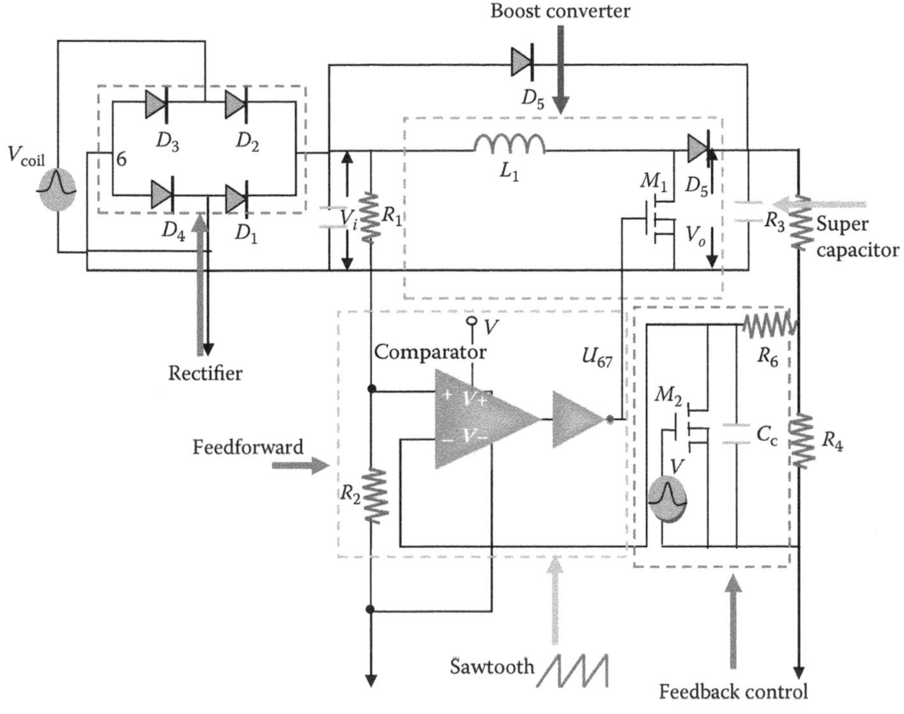

Figure 9.7 Energy harvesting circuit with feed forward and feedback control of DPBC. (From J. C. Xinping and L. Yi-Kuen, in *2006 IEEE International Conference on Information Acquisition,* 2006, pp. 91–95. With permission.)

Assuming that $t>>RC$, equation (9.29) will be simplified to:

$$V(t) \approx V_o \frac{R_4}{R_3 + R_4}\left(\frac{t}{RC}\right) \tag{9.30}$$

The sawtooth voltage value, which is directly proportional to the storage unit voltage, obtains its highest value when the switch is on and is [29]:

$$V_{Tm} \approx V_o \frac{R_4}{R_3 + R_4} \frac{(1 - D_p)T}{R_c C} \tag{9.31}$$

$$t = (1 - D_p)T \tag{9.32}$$

In the above equations, T is the period of the duty cycle of boost converter, D_p is the on-duty cycle, and V_o is the voltage of energy storage element. However, to attain higher efficiency of conversion, the output voltage is required to be higher than the voltage of the energy storage unit [29]. To achieve this, equation (9.33) should hold.

$$V_o \frac{R_4}{R_3 + R_4} \frac{(1 - D_p)T}{R_c C} \geq \frac{R_2}{R_1 + R_2} \tag{9.33}$$

The discussed design keeps the duty ratio as small as possible for better performance of the conversion system. However, the duty cycle is highly affected by the output voltage when high input voltage and small load resistance is considered. It is reported that, by applying a minishaker as an input to such design, 35 mW has been scavenged [29]. This amount of power would be enough to power up devices, such as a microsensor network with four commercial accelerometers [29].

9.4 Solar Energy Harvesting

Solar energy harvesting has been seen to provide the highest power density compared to other sources and is suggested to be the most effective source of power for wireless sensor nodes [30]. A solar system basically consists of a solar panel, DC/DC converter to boost the generated DC power, and a battery (Figure 9.8). The amount of electric power generated by a solar system depends on a number of factors, such as solar panel characteristics, the type of battery that is charged by solar energy, and the power management system.

The energy from sunlight is converted to electricity by means of a solar panel through photovoltaic effect. Photoelectric effect refers to the emission of electrons from

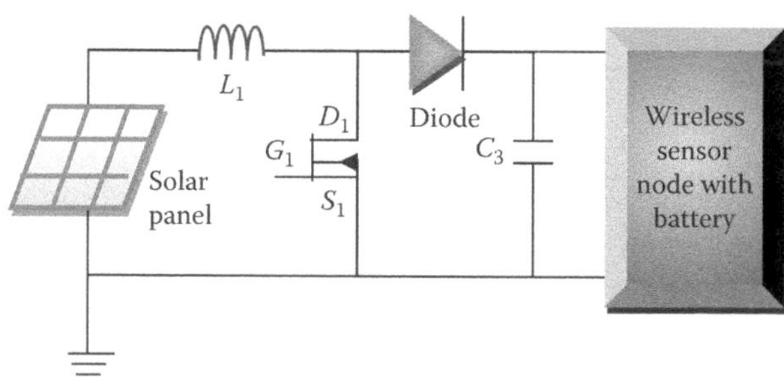

Figure 9.8 Schematic diagram of the solar energy harvester for wireless sensor node. (From W. Ko Ko, et al., in *2010 IEEE International Conference on Communication Systems (ICCS)*, 2010, pp. 289–294. With permission.)

matter (metals and nonmetallic solids, liquids, or gases) as a result of their absorption of energy from electromagnetic radiation of very short wavelength, such as visible or ultraviolet light. There are two common types of solar panels, known as polycrystalline and amorphous solar panels [30]. Polycrystalline solar panels provide about 25 percent more power than amorphous solar panels under the same conditions [30].

However, to improve the efficiency of a solar system, selection of an optimum battery is very important. Among the available batteries, SLA (sealed lead acid) and NiCd (nickel cadmium) are less efficient due to low energy density, while lithium-based and NiMH (nickel metal hybrid) batteries perform better for such systems due to a longer lifetime and cost effectiveness.

The output of a solar panel varies based on different loading and solar density circumstances. To provide the maximum power at the output of the panel and enhance efficiency, the system is required to perform around MPP (maximum power point). MPP refers to the operation point in which the panel is loaded correctly to optimize the output power. Most solar energy scavengers use two stages of DC/DC converter; one is used for MPPT (maximum power point tracking) implementation and another stage for the regulation of the output voltage [30]. Output power regulation is required in a solar system because the panel output is not constant and is often less than the required input voltage of the battery. However, each stage of a DC/DC converter is associated with some power losses. Therefore, it is optimal to implement one DC/DC converter to perform both tasks to minimize the power losses. In Ko Ko et al. [30], a solar harvester design is introduced with only one stage of a DC/DC converter to enhance the total efficiency of system.

To perform MPPT, there are two main techniques, known as buck and boost. Ko Ko et al. [30] suggest that boost is more suitable because the output of the panel is often less than the required input of battery; thus, panel output voltage boosting is needed. However, another advantage of the boost method over the buck is that it requires a nonisolated gate driver circuit, while buck needs an isolated gate deriver circuit.

To retain the output voltage of a solar panel at a constant value, even at different solar intensity levels, the panel should operate close to the MPP. To achieve a fixed voltage, a control system is required to regulate the input voltage of the battery (Figure 9.9). By controlling the duty cycle of the boost converter, the input voltage of the battery is maintained at a certain value [30].

V_o is the panel output voltage (clamped at battery voltage V_B), V_i is the controlled input voltage (MPPT voltage), and $f(D) = \frac{1}{1-D}$ for any DC/DC converter:

$$V_o = V_i \cdot f(D) \tag{9.34}$$

$$V_i = v_c \cdot k; \qquad k = V_B = V_o and\ v_c = \frac{1}{f(D)} \tag{9.35}$$

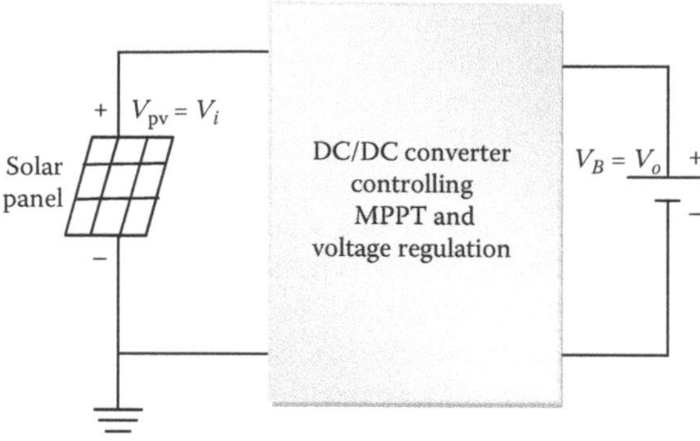

Figure 9.9 Schematic of the implemented system. (From W. Ko Ko, et al., in *2010 IEEE International Conference on Communication Systems (ICCS)*, 2010, pp. 289–294. With permission.)

Function *f*(*D*) varies for different DC/DC converters. According to equation (9.35):

$$D = f^{-1}\frac{1}{v_c} \tag{9.36}$$

In accordance with the above equations, if nonlinear feedback realization block ($f^{-1}(\frac{1}{u})$) is implemented in the control loop, the result is a linear control loop and to achieve the expected control performance, a simple proportional-integral (PI) controller can be implemented [30]. Figure 9.10 and Figure 9.11 show the whole control loop diagram and a simplified version of the diagram, respectively.

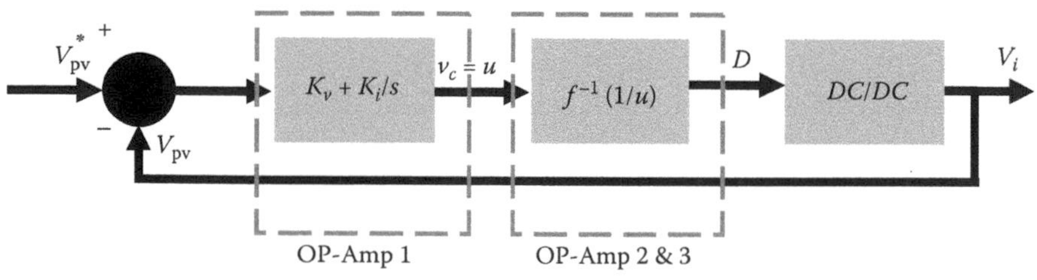

Figure 9.10 Implementation block diagram. (From W. Ko Ko, et al., in *2010 IEEE International Conference on Communication Systems (ICCS)*, 2010, pp. 289–294. With permission.)

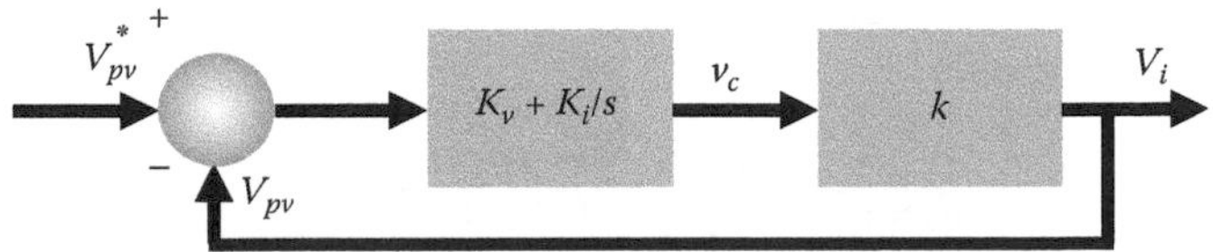

Figure 9.11 Simplified block diagram. (From W. Ko Ko, et al., in *2010 IEEE International Conference on Communication Systems (ICCS),* 2010, pp. 289–294. With permission.)

Figure 9.12 shows the control circuit diagram of an MPPT controller. As seen from the diagram, it consists of three parts [30]:

1. *Op-amp1 (PI control):* the first op-amp accomplishes the PI controller. First, it receives the solar input voltage (V_{pv}) and then compares it with a reference value (V_r), which is generated by the built-in reference generator of the op-amp. The ratio of the two resistances—R_3 and R_4—determines the proportional gain, and integral gain is determined by the capacitor C_1. The operating voltage of the solar panel (V_{pv}) is set by a potentiometer (R_p). By using only one op-amp in the design of the PI controller, the PI control action and the reference feed forward are added to implement the PI controller. This design enhances the power efficiency by decreasing the power consumption in the control circuit. It also makes the control loop faster by

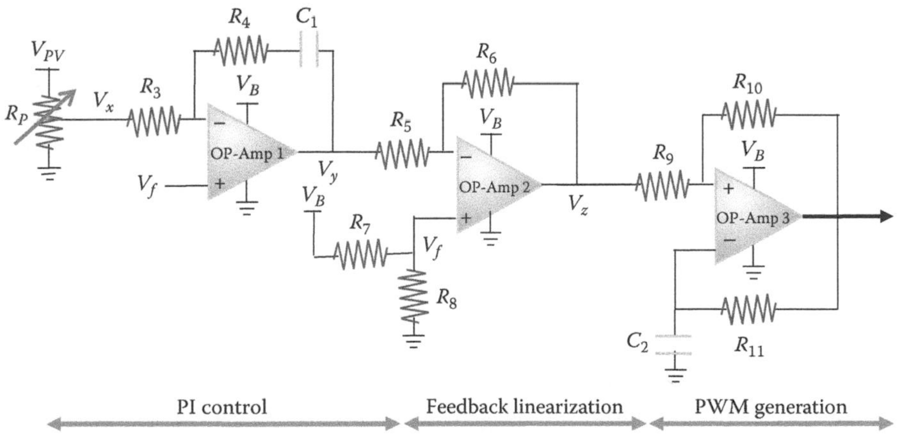

Figure 9.12 Control circuit diagram of MPPT controller. (From W. Ko Ko, et al., in *2010 IEEE International Conference on Communication Systems (ICCS),* 2010, pp. 289–294. With permission.)

facilitating the reference feed forward [30]. The following relationship holds for the PI controller:

$$v_c = v_y = \frac{sR_4C_5 + 1}{sR_3C_5}(-V_x + V_r) + V_r \tag{9.37}$$

2. *Op-amp2 (feedback linearization):* This section of the MPPT controller is a nonlinear block and the following equation represents the operation of the feedback linearization part:

$$v_z = (2 \times v_f) - v_y; \quad v_f = \frac{V_B}{2} \tag{9.38}$$

3. *Op-amp3 (PWM generation):* Pulse width modulation (PWM) signal generation is implemented using the third op-amp. It creates the signal with duty cycle proportional to the input voltage (V_z). However, the frequency of the signal is decided by R_{11} and C_2.

The discussed solar energy harvester is a useful, clean, and cheap method to power up magnetically coupled sensors. However, the output power from this system is a DC voltage and, to be used as an input to an inductively coupled system, it must be converted to AC voltage. A DC/AC conversion system is described in section 9.6.

9.5 Thermal Energy Harvesting

Heat is another form of energy that can be harvested from our surrounding environment. The heat generated by radiators, air conditioners, industrial machinery, household appliances, and even the human body can be used to run electrical devices. For example, a sensor network in a warehouse can be powered by using the heat from working machinery or a radiator can be the source of energy to power up some low-power electronics within a house or office. Thermic wristwatches produced by Seiko have been the first customer product using body heat to run the watch and also charge an internal battery [31].

Electric energy generation from heat is achieved thanks to the Seebeck effect [31] [32] by means of a thermoelectric generator (TEG). The Seebeck effect, named after Tomas Seebeck in 1821 [31], describes thermoelectric phenomena by which temperature differences between two dissimilar metals in a circuit convert into an electric current. However, the generated power through the Seebeck effect depends on different parameters, such as the thermoelectric material, temperature difference between the cold and the hot sides, and the structure and compactness of the TEG. More temperature gradient and compact TEG results in higher electric power generation and higher system efficiency. Also, using material with higher heat conductance capability leads to a better system performance.

Using thermal energy to power electronics, such as sensor nodes, is advantageous over using batteries due to its being more reliable, simpler, with no moving parts, and a longer lifecycle [31]. However, the limiting factor of such a system is very low efficiency. The existing TEGs in the market usually provide 5 to 6 percent efficiency and some have increased the efficiency to 10 percent [31] [33] [34]. Xin and Shuang-Hua [31] claim that their TEG provides 15 percent efficiency. Therefore, in this section, this design will be discussed in detail. However, we first describe a general thermal energy harvester system.

9.5.1 Thermal Energy Harvesting System

An equivalent circuit model of a general thermal energy harvester is illustrated in Figure 9.13. The energy resulting from the heat source of temperature T_H, is channeled through the TEG through a thermally conductive material, such as silver grease, which provides maximum electrical and thermal conductivity between surfaces, while providing protection from moisture and corrosion. A heat sink accumulates the heat and releases it to the surrounding ambient air at a lower temperature T_C [32]. The temperature difference across the junction of TEG (ΔT_{TEG}) is lower than the temperature gradient ($\Delta T = T_H - T_C$) due to the thermal contact ($R_{con}(H), R_{con}(C)$) and thermal grease resistances ($R_g(H), R_g(C)$) on the hot and cold sides. If the thermal resistance of TEG (R_{TEG}) is designed to be maximum or

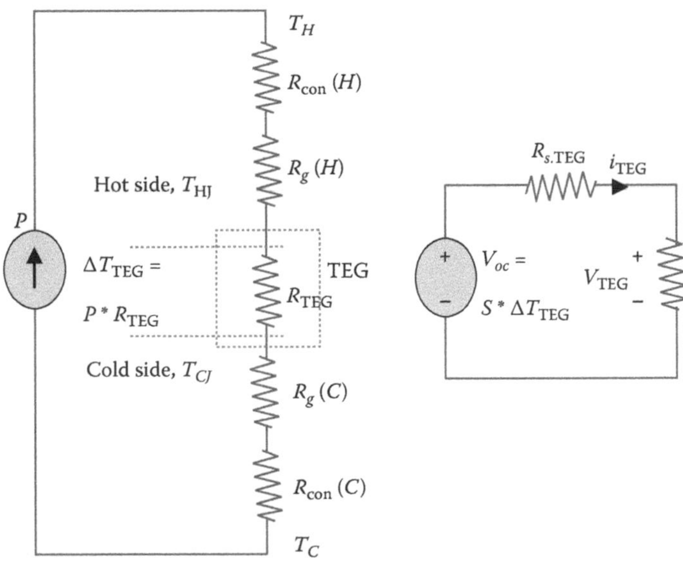

Figure 9.13 Equivalent circuit model of the thermal energy harvester. (From Y. K. Tan and S. K. Panda, *Industrial Electronics, IEEE Transactions on,* vol. 58, 9, pp. 4424–4435, 2010. With permission.)

the other resistances are designed to be as small as possible, this undesirable effect will be minimized [32]. To optimize the thermal scavenger, most of the heat has to be channeled through the TEG.

Based on the Seebeck effect, open circuit voltage V_{oc} of the TEG consists of n thermocouples connected in series electrically and in parallel thermally, and is achieved when α and S are the Seebeck's coefficients of thermocouples:

$$V_{oc} = S * \Delta T = n * \alpha(T_H - T_C) \tag{9.39}$$

By connecting a load resistance (R_L) to the circuit, a current I_{TEG} flows through the circuit. Where $R_{s,TEG}$ is the internal electrical resistance of TEG, the current in TEG is

$$I_{TEG} = \frac{V_{oc} - V_{TEG}}{R_{s,TEG}} = \frac{n * \alpha(T_H - T_C) - V_{TEG}}{R_{s,TEG}} \tag{9.40}$$

Therefore, the total power generated by the TEG is

$$P_{TEG} = V_{TEG} * I_{TEG} = \frac{V_{TEG} * n * \alpha(T_H - T_C) - V_{TEG}^2}{R_{s,TEG}} \tag{9.41}$$

To design a thermal energy harvester, different considerations need to be taken into account. Three important factors that need to be determined carefully are (1) the electrical characteristics of the target system, (2) thermal energy density available in the environment, and (3) the transfer efficiency of the harvester [31]. In this regard, a typical and efficient thermal energy scavenger for wireless sensor nodes is discussed below.

9.5.2 Thermal Energy Harvester for Wireless Sensor Networks

As mentioned earlier, to design a thermal scavenger, it is critical to be aware of the target system properties. Therefore, in this section, a TEG is designed for low power sensor nodes. However, one is required to determine the characteristics and behavior of the target system, such as input voltage, working current, sleeping current, working duration, and sleeping duration of the system. As described by Xin and Shuang-Hua [31], this TEG design is capable of providing the required power for ZigBee-enabled-type devices that are comparable with short-range, magnetic induction communication systems in terms of power consumption. The electrical characteristics of the target device are described in Table 9.1.

Table 9.1 Electrical characteristics of ZBRVA (a ZigBee-based home automation system)[31]

Parameter	*Value*
Supply voltage	2.3-3.7V
Working current	60mA
Sleeping current	10 uA
Working duration	20 seconds/hour
Sleeping duration	3579 seconds/hour

Source: L. Xin and Y. Shuang-Hua, in *2010 IEEE International Conference on Systems Man and Cybernetics (SMC),* 2010, pp. 3045–3052. With permission.

In this design, a radiator is considered as a heat energy source providing about 50°C at its surface and the room temperature is about 21°C. Based on Fourier's law, the energy density provided by the radiator is 1.6 KW per cubic meter [31]. TEG is attached to the radiator to provide electric energy by using the temperature difference between the radiator and the ambient air. Figure 9.14 shows the function diagram of this thermal energy harvesting system.

As can be seen from the diagram, the entire system consists of three main parts of an energy harvesting unit, which scavenges the heat energy and converts it to the electric energy; a DC/DC converter, which boosts the output voltage of the harvester unit; and a power management unit, which supports the efficient distribution of the generated energy to the target [31].

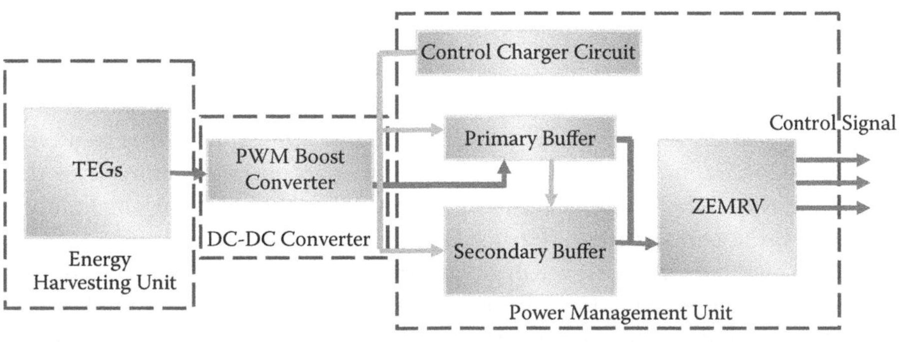

Figure 9.14 Function diagram of the thermal energy harvesting system. (From L. Xin and Y. Shuang-Hua, in *2010 IEEE International Conference on Systems Man and Cybernetics (SMC),* 2010, pp. 3045–3052. With permission.)

9.5.2.1 TEG Unit

Because the efficiency of a thermal energy harvester is quite critical, an efficient design is key to success of the design. The efficiency of a thermal energy scavenger is affected by the efficiency of thermoelectric module and heat temperature level. Therefore, by using highly thermally conductive material, the temperature difference can be increased and high thermal flow through the thermoelectric module is possible; consequently, the system efficiency is enhanced [31]. The best materials to use as heat conductors in the module are bismuth (Bi), anthimony (Sb), tellurium (Te), and selenium (Se) [31]. They are capable of operating in room temperature and up to 200°C [31]. However, Bi_2Te_3 is the most common material used in Pelteir coolers due to cost efficiency [31].

Module construction also can influence the maximum power conversion and electrical characteristics of the system. A thermoelectric module consists of a number of thermoelectric elements (often hundreds). These elements are built by either P-type or N-type semiconductors in series electrically and in parallel thermally between two ceramic layers [31]. The output voltage varies with the number of elements (series combination), also the length of element legs as well as the surface area of TEGs. By using more elements with shorter legs, the output voltage increases. The surface area of the TEG and the output power are also proportional.

As mentioned earlier, the temperature difference plays an important role in achieving an efficient design. Therefore, to maintain a large temperature difference, two parts must be added to the harvester. A heat exchanger, which is a thermal conductor like metal, is required at the hot side and an efficient heat dissipation, or a heat sink, is needed at the cold side (Figure 9.15) [31]. A heat sink is used to remove the energy from the cold side. There are three common types of heat sinks: copper, aluminum, and heat pipe-type heat sink. According to Xin and Shuang-Hua [31], a heat pipe-type heat sink is better in transferring the heat from the cold side.

Ambient air temperature influences the heat conductivity of the heat sink. Therefore, to minimize the air temperature surrounding the heat sink, some actions can be taken. By increasing the distance between the heat source and the sink, the sink's ability in heat transfer increases. One approach is to attach the thermoelectric modules in a stack, but there is an optimum number of the modules in the stack, which in this design is four [31]. By increasing the number of modules, the temperature difference between each module will be too small and not enough electric energy will be generated [31]. Also, the location of the TEG with respect to the radiator is important. Locating the TEG at the bottom of a radiator will maximize the temperature difference. However, to achieve larger thermal flow through the module, the heat from the source should be isolated from the ambient air. In this regard, some sponge or cotton can be used to isolate the module from the environment [31].

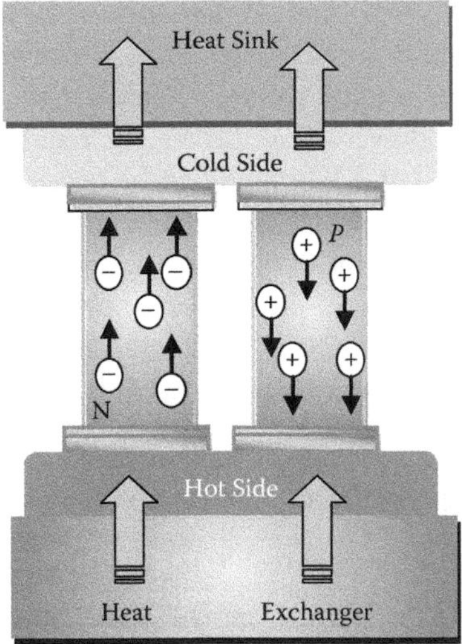

Figure 9.15 Thermal energy harvesting unit. (From L. Xin and Y. Shuang-Hua, in *2010 IEEE International Conference on Systems Man and Cybernetics (SMC)*, 2010, pp. 3045–3052. With permission).

9.5.2.2 DC-to-DC Converter

The output DC power from the TEG unit is often insufficient to power up the target device; therefore, it needs to be boosted by a step-up DC/DC converter. A DC/DC converter is shown in Figure 9.16. It is used in this design to amplify the output power of the thermal energy harvester unit [31].

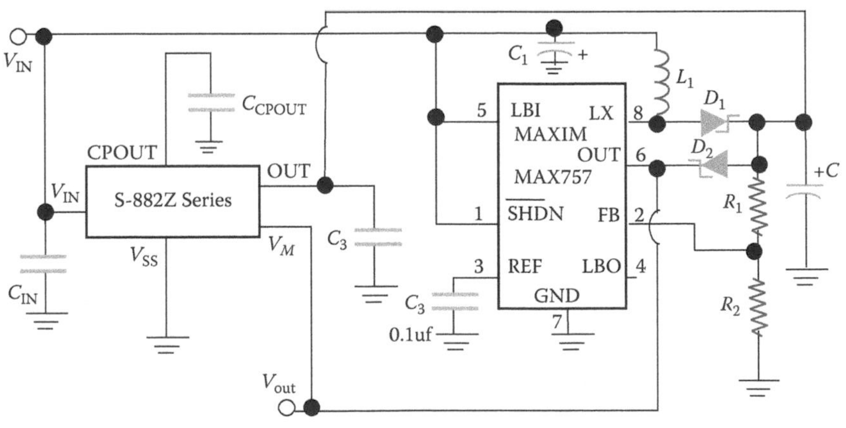

Figure 9.16 Schematic diagram of DC-DC converter unit.

Most of the step-up DC/DC converters require at least 0.7 V, while the generated power by the harvester is usually less. Thus, a conventional charge pump is used to boost the harvester output voltage. For example, S-8827 IC from Seiko can be used to increase the power up to 2.2 V. The required input voltage for this IC (integrated circuit) is only 0.3 V, which can be generated by the harvester [31]. Hence, the output power from the IC is used as input to the DC/DC converter to boost it up so it can power the target device. S-8827 and Max757 are combined to form a proper DC/DC step-up converter for this design (Figure 9.16) [31]. However, another function of a DC/DC converter unit is to isolate reverse current flow from the reservoir to the thermoelectric module [31].

9.5.2.3 Power Management Unit

The power management unit is responsible for distribution of power to each part of the system and to store the available energy in storage [31]. There are two storage buffers and a control and charge unit in the power management subsystem. Because the generated voltage from the harvester varies with time, there is a need for energy storage to accumulate the available energy from the DC/DC converter. The energy storage element or buffer can prevent an excessive power burst when the system starts up [31]. The primary buffer is charged by the thermal energy harvesting unit directly and powers the target device. When the first buffer is full, the second battery starts to charge, which will be used when the energy in the primary buffer is not enough to power the target device [31]. While the secondary buffer is charging, the primary buffer also can power up the target. However, the control and charge circuit supports the optimization of the harvested power for the entire system [31].

9.6 DC-to-AC Power Conversion

Since magnetic induction devices require AC power to function, there is a need to convert the DC output of discussed energy scavengers to an AC power. Thus, the DC output of the energy harvester needs to enter a DC/AC power converter. AC output then can be used to feed a magnetically coupled system. There are different DC/AC converter designs with different capabilities that are used for different purposes, such as high or low voltage conversion and power conversion/amplification. However, to design a DC-to-AC converter, size, weight, and efficiency are some critical factors that limit their use for some particular applications.

Conventional power converters are often composed of two or more elemental converters and they have a stiff DC voltage/current link between the elemental converters. Hence, to eliminate the fluctuation of the power difference between two elemental converters, a large smoothing capacitor or inductor is used for energy storage and results in larger sized converters [35]. However, a soft, switching boost,

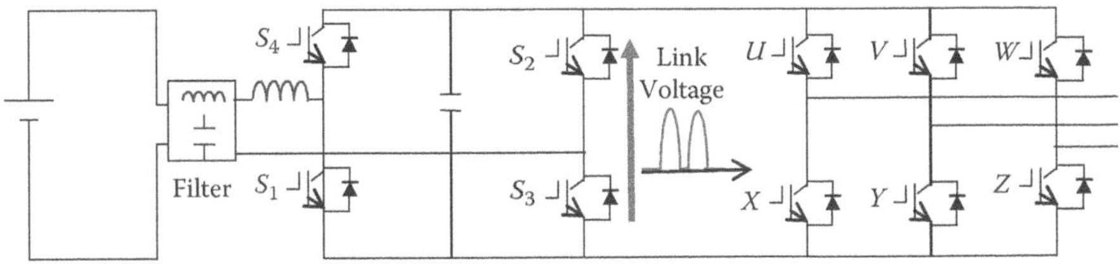

Figure 9.17 Circuit configuration of soft switching step-up DC-AC converter. (From T. Isobe, et al., in *2010 International Power Electronics Conference (IPEC)*, 2010, pp. 2815–2821. With permission.)

DC/AC converter without smoothing capacitor proposed by Isobe, et al. [35] is discussed in this section to provide the AC power for magnetic induction devices. This design is advantageous over the conventional power converters because it is more compact and efficient due to no need of smoothing capacitor and achieving soft switching for both boost-up chopper and DC/AC converter.

The power converter configuration, as shown in Figure 9.17, is made up of two elemental converters: a step-chopper and DC/AC converter. Four semiconductor switches (S1 to S4), an inductor, and a small capacitor form the chopper, while the DC/AC converter is formed by six semiconductor switches [35]. Pulsed DC voltage is generated in the capacitor by the chopper and the DC/AC converter inverts the DC output to AC by switching the pulsed voltage.

- *Step-up chopper operation principle* [35]: Two switches are used at the chopper to achieve positive power flow and, meanwhile, the other two switches are turned off (represented by a diode in Figure 9.18). The two active switches are controlled at the same time. When the switches are on, the source and capacitor voltages are applied to the inductor and result in boosting up the input current very fast (Figure 9.18b) until the capacitor voltage becomes zero. This results in the division of current in two paths depending on the semiconductors' properties. However, the input current increases by the source voltage applied to the inductor. When the switches are turned off, input current declines, while charging the capacitor until it becomes zero (Figure 9.18c). Due to the load current, the capacitor voltage decreases while the input current remains zero and, depending on the semiconductors' characteristics, voltage is divided between them.
- *DC/AC converter operation principle* [35]: As mentioned earlier, the DC/AC converter includes six active switches that distribute the pulsed link voltage to phases of an AC load. As an example, Figure 9.19 illustrates a diagram of the switching method of a DC/AC converter with PDM (pulse density modulation). To perform the modulation, only some of the pulses can be used because the switching needs to be performed in the zero voltage period

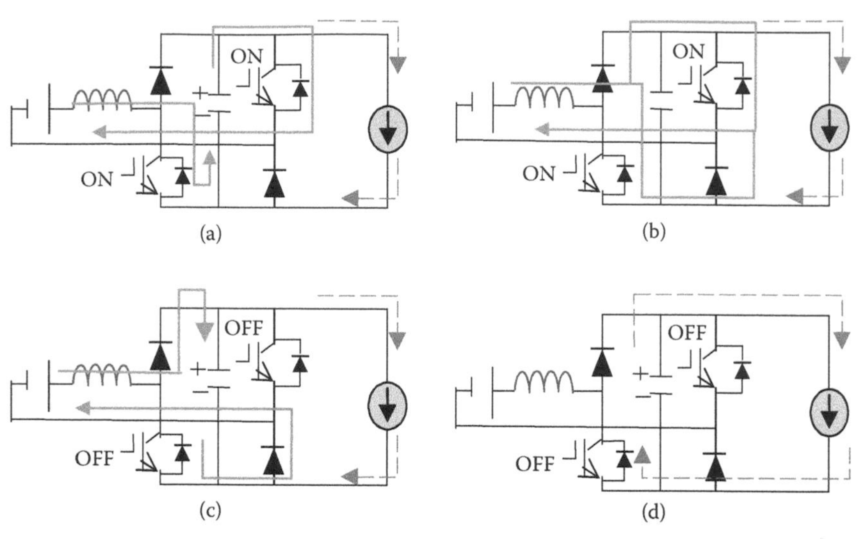

Figure 9.18 Possible current paths of the step-up chopper for positive power flow. (From T. Isobe, et al., in *2010 International Power Electronics Conference (IPEC)*, 2010, pp. 2815–2821. With permission.)

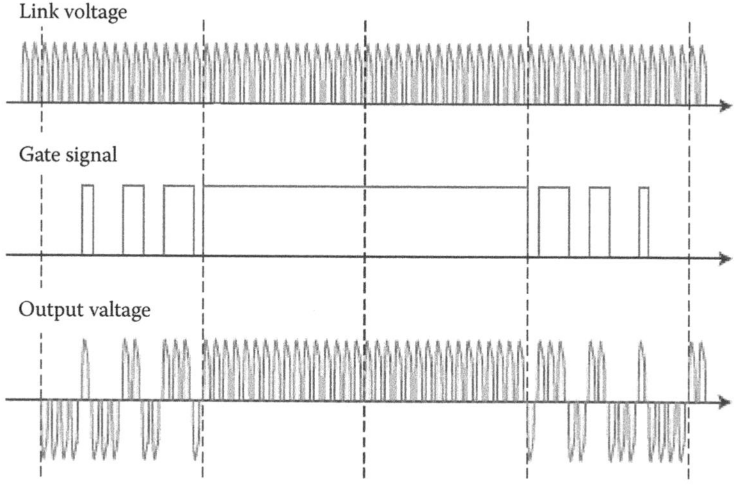

Figure 9.19 Diagram of switching method of DC-AC converter with PDM. (From T. Isobe, et al., in *2010 International Power Electronics Conference (IPEC)*, 2010, pp. 2815–2821. With permission.)

of pulsed link voltage to achieve soft switching. By controlling the magnitude of each pulse through the chopper, high control over the waveform can be fulfilled.

- *Step-up chopper control*: Although the need for an energy storage capacitor is mitigated in this design, precise power control is required to enhance the efficiency. The reason is the DC/AC converter input current i_{dc}, varies due to load of chopper, which is assumed to be a constant current [35]. The variation leads to the fluctuation of the peak voltage of pulsed link voltage and efficiency drops. Therefore, to overcome the undesirable fluctuation, a strict control system needs to be implemented. The control is done by defining a voltage threshold and by preventing the pulse voltage with peak value more than the threshold in each switching cycle [35]. However, the only available parameter to be directly controlled in such a system is the turn-off timing.

The following relationships hold for input current i_{in} and the capacitor voltage v_c with the other circuit parameters [35]:

$$i_{in}(t) = I_{dc} + E\sqrt{\frac{C}{L}}\sin\omega t + (I_{off} - I_{dc})\cos\omega t \tag{9.42}$$

$$v_c(t) = E - E\cos\omega t + \sqrt{\frac{L}{C}}(I_{off} - I_{dc})\sin\omega t \tag{9.43}$$

In the above equations, I_{off} is the initial value of i_{in}, I_{dc} is i_{dc} in this switching cycle and is considered a constant value. E is the source voltage, C and L are parameters of the chopper, and $\omega = 1/\sqrt{LC}$ is the resonance frequency. However, according to the above equations, the peak voltage value is depicted to describe the relationship between the input current at turn-off and resulting peak value [35]:

$$V_P = E + \sqrt{E^2\frac{L}{C}(I_{off} - I_{dc})^2} \tag{9.44}$$

Therefore, where V_P^* is the given peak voltage set-point, the input current set point to turn the switches off is described as [35]:

$$I_{off}{}^* = \sqrt{\frac{C}{L}(V_P^{*2} - 2V_P^*E)} + I_{dc} \tag{9.45}$$

The control diagram of step-up chopper for output peak voltage control is shown in Figure 9.20.

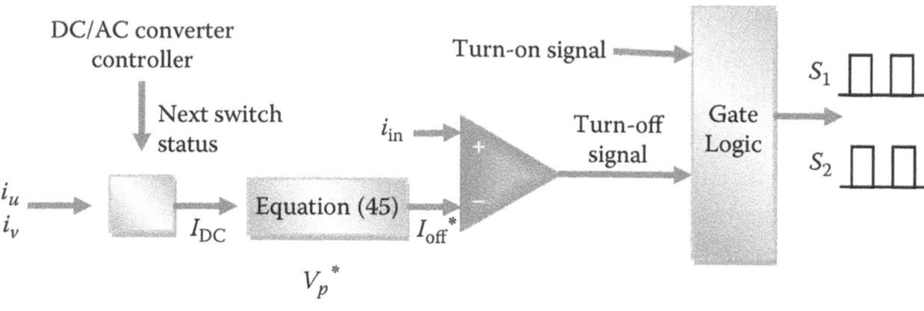

Figure 9.20 Control diagram of the step-up chopper for output peak voltage control.

9.7 Wireless Power Transmission System to Transfer the Harvested Energy to the Target

When the scavenged power is converted to AC, it is ready to feed a magnetic induction transmission system. As soon as the AC power is applied to a resonant circuit, it creates electric voltage and current through the circuit that is capable of powering up a matched resonant circuit as described in section 9.3.1. This concept is known as wireless power transmission, which has been employed in some communication systems, such as medical monitoring devices, passive RFID tags, electric cars, and some autonomous machinery. To describe this concept, wireless power transmission for passive RFID tags is discussed in the following section.

9.7.1 RFID Power Transmission System

RFID technology has gained great interest in commerce, industry, and academia as a technique for short-range communication [3] [36]. This technology has been used for identification, tracking objects in a supply change, monitoring the object status, and many more applications [36]. An RFID communication system consists of one or more readers and one or a number of tags, and they can communicate data with each other at a given frequency and up to a certain distance. There are two types of RFID tags, known as active and passive tags. Active tags are supplied with a power source (battery) that enables the tag to perform defined functions. However, passive tags have no battery and they receive their power from the reader whenever they are required to send information to the reader. The required power for passive tags is obtained through mutual magnetic inductive coupling. An unnecessary individual power source has made the passive tags much cheaper than active tags [36].

The antenna coil at the reader creates an RF field with a mainly magnetic component [37]. The coils in a passive tag and reader are then inductively coupled and a power link between them is established, which is not only used to power up the passive tag, but also a clock and a data signal can be transmitted through the RF field of the reader.

A typical passive tag includes an antenna coil and an IC with an internal capacitor. A resonant circuit is created by the antenna coil and capacitor, which delivers higher voltage to the resonant circuit than the voltage generated in the coil. However, the circuit is designed to resonate at the same or very close frequency as the frequency of the reader RF field [37]. The AC voltage generated in the circuit is required to be converted to DC voltage. The conversion process is simply done by means of a rectifier in the circuit and then the DC voltage is ready to activate the passive RFID tag. Figure 9.21 shows a typical setup of an RFID system.

The achievable delivered power to the tag is a function of the field strength, which is highly influenced by the coil specification and characteristics. Magnetic field strength (H) is computed using equation 9.46, where r is the radius of the reader coil, d is the distance of coupling, and N is the number of turns of the reader coil [37].

$$H = \frac{I \cdot N \cdot r^2}{2\sqrt{(r^2 + d^2)^3}} \tag{9.46}$$

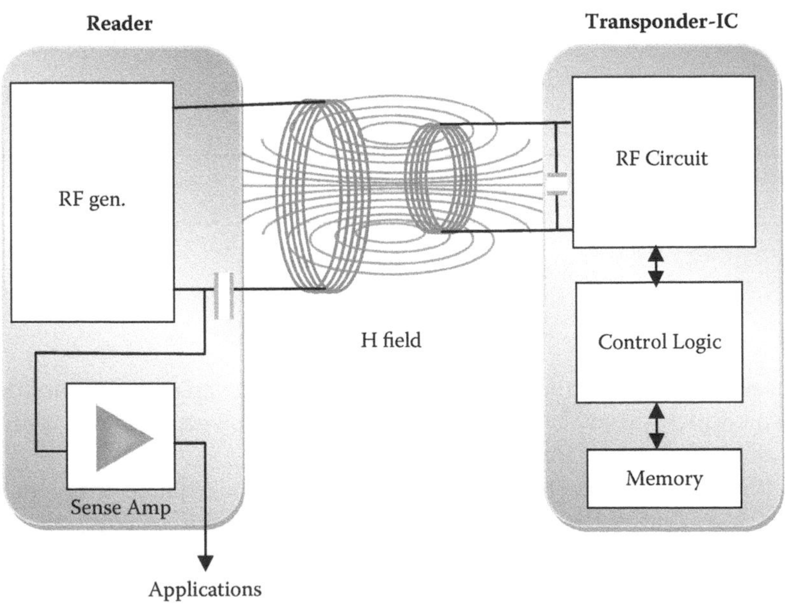

Figure 9.21 Typical setup of RFID system. (From S. Cichos, et al., in *2nd International IEEE Conference on Polymers and Adhesives in Microelectronics and Photonics, POLYTRONIC*, 2002, pp. 120–124. With permission.)

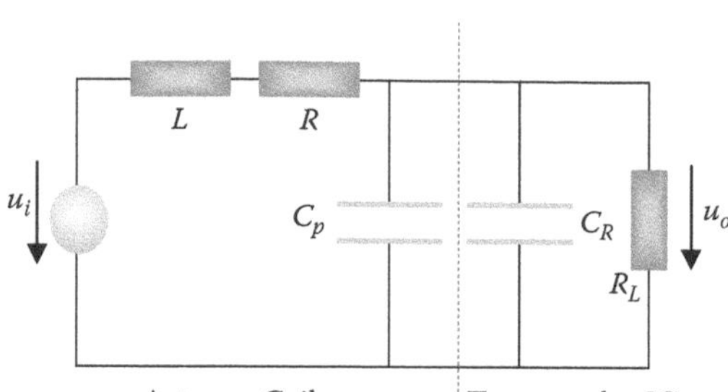

Figure 9.22 Equivalent circuit of the tag. (From S. Cichos, et al., in *2nd International IEEE Conference on Polymers and Adhesives in Microelectronics and Photonics, POLYTRONIC*, 2002, pp. 120–124. With permission.)

An equivalent circuit model of the antenna coil at the passive tag is shown in Figure 9.22. The induced voltage at the circuit is shown as a voltage source (u_i). R shows the ohmic losses, and it varies by the type of the material used to form the coil. The parasitic capacitance of the coil is shown by C_p, and C_R represents the internal resonance capacitance of the IC [37]. Voltage u_o, which is the voltage across the load resistance (R_L), can be described as [37]:

$$u_o = u_i \frac{1}{j\omega\left(\frac{L}{R_L} + RC\right) + \left(1 - \omega^2 LC + \frac{R}{R_L}\right)} \tag{9.47}$$

Summarizing the capacitors C_p and C_R to a common capacitor C, it is obtained using the following equation:

$$C = \frac{L}{R^2 + \omega^2 L^2} \tag{9.48}$$

One of the critical parameters to be considered for tag design is the ohmic losses (R) of coil, which affects the resonant frequency [37]. However, the load resistance has no effect on the resonant frequency.

For the tag to operate properly, there is a minimum required magnetic field strength. According to the law of induction, minimum required voltage to actuate the passive tag is [37]:

$$|u_i| = \mu_0 \cdot A \cdot N \cdot \omega \cdot H_{eff} \tag{9.49}$$

In this equation, A is the averaged coil area and H_{eff} is the effective value of a sinusoidal magnetic field. Where u_o is the minimum required voltage of the tag and H_m is the minimum operating field strength, the following relationship holds:

$$|H_m| = u_o \cdot \frac{\sqrt{\left(\frac{\omega L}{R_L} + \omega RC\right)^2 + \left(1 - \omega^2 LC + \frac{R}{R_L}\right)^2}}{\mu_0 \cdot A \cdot N \cdot \omega} \tag{9.50}$$

Equation (9.50) describes the tag properties independent from the reader [37]. Knowing the minimum operating field strength, the number of turns of the coil, inductive value, and the radius of antenna coil, maximum powering range (d_{max}) can be estimated.

$$d_{\max} = \sqrt{\left(\frac{I \cdot N \cdot r^2}{2|H_m|}\right)^{\frac{2}{3}} - r^2} \tag{9.51}$$

$d_{\max}$ is the maximum distance between the tag and reader, in which the reader can actuate the passive tag to be able to communicate with it.

Since the coupling between the reader and tag is highly dependant on the mutual inductance (M) between them, the power delivered to the tag can be maximized by tuning the tag and reader circuits to where they resonate at the same frequency [38]. If the reader and tag are tuned to a common frequency ω, the current in the tag circuit (I_t) is

$$I_t = V_{in} \frac{\omega M}{\{R_t R_r - \omega^2 M^2\}} \tag{9.52}$$

where R_r is reader load resistance, R_t is source resistance of tag, and V_{in} is the input voltage at the reader. However, tag current is maximized when [38]:

$$\omega M = \sqrt{R_t R_r} \tag{9.53}$$

Therefore, the optimum coupling is achieved. If the value of coupling is smaller than the optimum value, then the current responses of tag and reader imitate a single peaked resonance curve due to increasing isolation of the tag and reader as a result of reduction in coupling value [38]. On the other hand, if the value of coupling exceeds the optimum value, coupling increases. Therefore, it decreases the peak in reader current frequency response and results in a small peak on either side of the resonance frequency [38]. At the same time, the tag current is maximized while reader current decreases due to "humps" developing on either side [38]. Therefore, to achieve optimum coupling, it is critical to meet the condition described in equation (9.53).

9.7.2 Wireless Power Transmission Using Magnetic Induction for Medical Implants

Although RF energy harvesting using magnetic induction theory is different in terms of configuration, specifications, and required elements for different application types, the basic principle almost remains the same, as described in previous sections of this chapter. RF energy can be harvested by a device to generate electric energy through magnetic coupling, in order to perform a specific task. Similar to passive RFID tags, this concept is used in medical implants to power up the electronic device inside a patient's body, often to monitor the patient condition, diagnose, or even cure a disease. Implantable devices need to be small enough to be able to operate inside the body; therefore, implementing a battery in them results in a larger size and limits the use of such devices. They also might be supplied by means of wires that leave the patient uncomfortable. However, to bring comfort to the patient's life and provide ease of movement and energy saving as well as small embedded devices, RF energy can be scavenged from an external unit by the embedded unit (wireless power transmission).

The general operating principle of the embedded devices is the same as RF energy scavenging or wireless power transmission in passive RFID tags, but to enhance the performance of implant devices, different techniques are proposed. Some of these techniques will be discussed in the following sections.

9.7.3 Multivoltage Output System

With regard to providing electric energy for an implantable device using an RF energy harvesting system, it is often required to have different voltage levels [39]. High voltage is needed to provide the power for the stimulators and regular voltage to power up the other analog circuits and digital blocks [39]. There are two main types of multivoltage systems. The first method is to drive a resonant circuit using a voltage oscillator and then rectifying and converting it to other required levels. However, the second approach is to use both tapped and non-tapped coils, which is complex and requires larger size of coils since it increases total inductance [39]. The first is simpler to implement, but it suffers from lower efficiency. To address this problem, a cascaded resonant tank circuit is used in [39], which improves the efficiency of the powering system (43 percent power efficiency) for multivoltage wireless power transmission systems without the need for larger coils. The simplified circuit model is shown in Figure 9.23. The

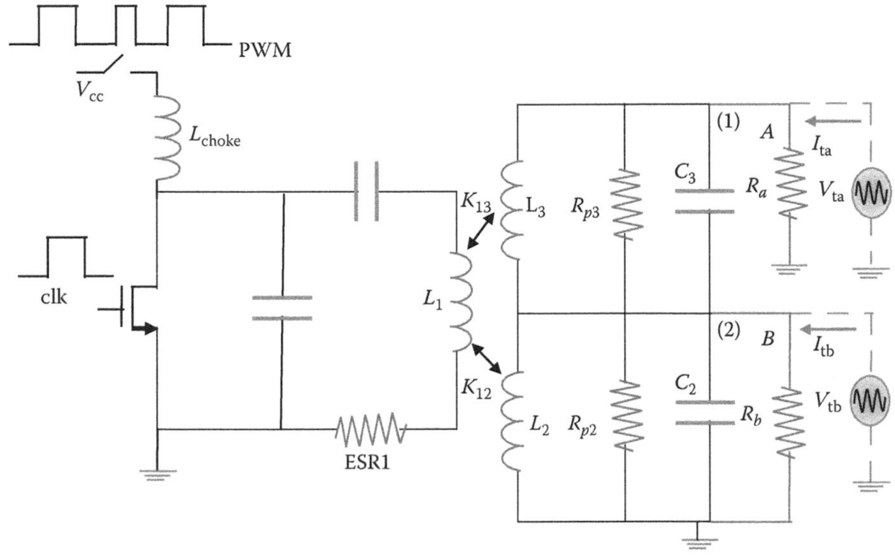

Figure 9.23 Simplified model of multiple voltage power telemetry system. (From M. Sawan, et al., *Biomedical Microdevices,* vol. 11, pp. 1059–1070, 2009. With permission.)

receiving circuit has two terminals T_1 and T_2 that provide the high voltage and regular voltage, respectively [39].

There are two resonant frequencies (ω_1, ω_2) when the resonant tank is resonating [39]. To calculate the value of ω, R_a and R_b are replaced by testing voltage source V_a and V_b with corresponding current i_{ta} and i_{tb}, respectively [39]. The first resonant frequency can be seen between point b and the ground $(di_{tb}/d\omega = 0)$ and the second between points a and b $(di_{ta}/d\omega = 0)$. By assigning the two resonant frequencies at the receiver to the resonant frequency of the transmitter, two conditions can be used to compute the design components. The two resonant conditions are as follows [39]:

$$\frac{1}{\omega_1^2} \approx L\gamma^2 \left(\frac{1}{\gamma} - 1\right)^2 C_3 \tag{9.54}$$

$$\frac{1}{\omega_2^2} \approx L(1-\gamma)^2 C_3 + L\gamma^2 C_2 \tag{9.55}$$

where L is the total inductance between the two terminals and $\gamma = V(regular)/V(high)$.

By using equation (9.54) and equation (9.55), system parameters can be computed.

However, the link efficiency is highly affected by the quality factor (Q) and coupling coefficient (k). Coupling coefficient has a value in the range 0.01 and 0.1 and is a function of distance and coil size [39]. For having an optimum design, the optimum Q needs to be determined. According to Sawan et al. [39], a coil can be equivalent to a lumped capacitor in parallel with an inductor at the frequency f and is derived using the following equation:

$$Q(f) \approx 2\pi f L\left(1-\frac{f^2}{f_{self}^2}\right) \Big/ R_{DC}\left(1+\frac{f^2}{f_h^2}\right) \tag{9.56}$$

In this equation, f_h represents the quantified impact of the proximity effect or so-called skin effect and is defined as [39]:

$$f_h = \frac{2\sqrt{2}}{\pi r_s^2 \mu_0 \sigma \sqrt{N_t N_s \eta A}} \tag{9.57}$$

Parameters in equation (9.57) are defined as follows: r_s is the wire radius used at the coils, μ_0 is magnetic permeability, σ shows magnetic conductivity, N_t is the number of turns, N_s shows the number of strands per turn, A is the coil cross-section area, and η is a value between 0.2 and 1 depending on the coil's geometry.

f_{self} in equation (9.58) is the coil's self-resonant frequency and is computed using the following equation:

$$f_{self} = \frac{1}{2\pi\sqrt{LC_{self}}} \tag{9.58}$$

where $C_{p,k}$ is the parasitic capacitance between turn p and k:

$$L = N_t^2 L_i \ and\ C_{self} = \sum_{p<k} C_{p,k}(k-p)^2 / N_t^2 \tag{9.59}$$

However, the frequency at which the coil has maximum Q (f_{peak}) can be derived by:

$$f_{peak}^2 \approx f_h^2 \left\| \frac{f_{peak}^2}{3} \right. \tag{9.60}$$

By using equation (9.54) to equation (9.60), not only the optimum Q is computed, but also the maximum efficiency for a multiple-voltage power telemetry system can be determined [39].

9.7.4 RF Energy Scavenging for Embedded Medical Devices Using Spiral Coils

One of the common types of inductive coils is the cylindrical configuration. This configuration provides a good quality factor and simple modeling of the inductance and associated losses [40]. However, they are not the optimum choice for inductive coupling of implantable devices because they cannot be further optimized in terms of power efficiency [40]. There is another type of inductive coil known as spiral coils that can be optimized in terms of quality factor and coupling link strength and efficiency [40]. Spiral coils consist of a number of conductive rings and the strongest coupling can be achieved when they are coaxial and placed on the same plane [40]. The most important design parameters in spiral coils are the impact of internal and external radius, separation between the windings, and the width of metal strips [40]. Spiral coils can be seen as a group of concentric rings. In the following, a wireless power transmission model for implantable devices using spiral microcoils will be discussed.

9.7.4.1 Transmitter Coil Modeling

The model description in this section is based on a conventional model that has an inductance in series with the conductor resistance [40]. The self inductance of concentric rings, forming the spiral coil, is computed using equation (9.61). It is a function of average ring radius (b) and wire radius (R) and magnetic permeability of the material used to form the coil. However, in free space, the wire radius is far smaller than the average ring radius [40].

$$L_P(b) = \mu_0 b\left(\ln\left(\frac{8b}{R}\right) - 2\right) \tag{9.61}$$

The mutual inductance between the windings of the coil (transmitting or receiving coil) is formulated as [40]:

$$Ma_1a_2 = \pi\mu_0\sqrt{a_1a_2}\left(\frac{2}{\pi}\sqrt{\frac{a_1}{a_2}}\right)\cdot\left[K\left(\frac{a_2}{a_1}\right) - E\left(\frac{a_2}{a_1}\right)\right] \tag{9.62}$$

where α_1 and α_2 are the mean radius of the two concentric rings, and $K(x)$ and $E(x)$ are the complete elliptic integrals of the first and the second order [40]. Therefore, the total inductance of coil is the result of summation of inductance of each ring and mutual inductance between different windings [40].

However, the resistance of a circular conductor is also formulated as [41]:

$$R_1(\omega) = \frac{\sum_{k=1}^{\infty} \frac{R_k}{R_k^2 + \omega^2 L^2}}{\left(\sum_{k=1}^{\infty} \frac{R_k}{R_k^2 + \omega^2 L^2}\right)^2 + \omega^2 \left(\sum_{k=1}^{\infty} \frac{L}{R_k^2 + \omega^2 L^2}\right)^2} \tag{9.63}$$

In this equation, $R_k = \xi_k^2/4\pi\sigma R^2$, and L is the self-inductance of the coil and is $L = \mu_0\mu_r/4\pi$. μ_r is the relative permeability and σ is the conductivity of the material used to form the coil windings. However, $\xi_k = (2k-1)\pi/2 + \pi/4$ [40].

Mutual coupling between the transmitting and receiving coils depends on different parameters, such as the size and shape of coils, and the side and angular alignment of them. Inductive link is highly affected by the side and angular misalignment of the transmitting and receiving coils and such effect can decline the link strength significantly even if by small misalignment. On the other hand, in medical-embedded devices, transceivers are not fixed and often they move and change their orientation; therefore, efficiency is highly influenced by motion or changing the location of the devices. The mutual inductance, taking into account all the mentioned functional parameters, can be calculated by:

$$M(r_T, r_R, \Delta, d) = \pi\mu_0\sqrt{r_T r_R}\int_0^{\infty} J_1\left(x\sqrt{\frac{r_T}{r_R}}\right)\cdot J_1\left(x\sqrt{\frac{r_R}{r_T}}\right)\cdot J_0\left(x\frac{\Delta}{\sqrt{r_T r_R}}\right)\cdot e^{\left(-x\frac{d}{\sqrt{r_T r_R}}\right)} dx \tag{9.64}$$

In this equation, Δ is the side misalignment, r_T, r_R are the transmitting and receiving coils, respectively, d is distance between the two coils, and:

$$J(r) = \begin{cases} \frac{1}{\omega} f(r) \quad when & \left(a - \frac{\omega}{2}\right) \leq r \leq \left(a + \frac{\omega}{2}\right) \\ 0 & otherwise \end{cases} \tag{9.65}$$

where, $\int f(r) = \omega$ is the radial distribution [40].

However, to improve the quality of the link and to decrease the sensitivity of the link to misalignment, an approach has been proposed in Lihsien et al. [40]. In this method, an array of spiral microcoils are used in implanted devices as the receiving antenna. However, the transmitting unit only has one microcoil to transmit the power to a multicoil receiver. Figure 9.24 illustrates this configuration [40].

The following equation holds for the system shown in Figure 9.24 [40]:

$$V_1 = Z_1 I_1 + j\omega M I_2 + j\omega M I_3 + j\omega M I_4 + j\omega M I_5 \tag{9.66}$$

$$0 = Z_i I_i + j\omega M I_i; \qquad i = 1,2,3,4,5 \tag{9.67}$$

Therefore, the current in the transmitting circuit is

$$|I_1| = \frac{|V_1|}{R_1 + \frac{(\omega M)^2}{Z_2} + \frac{(\omega M)^2}{Z_3} + \frac{(\omega M)^2}{Z_4} + \frac{(\omega M)^2}{Z_5}} \quad \textit{where} \quad (R_1 = Z_1) \tag{9.68}$$

However, when the two coils are coaxial, they provide the maximum inductive coupling. Coupling coefficiency and, consequently, the link quality decline because of displacement. Thus, as a result of misalignment, there is a decrease in received power. To increase the sensitivity of the receiver, a favorable configuration of receiving coils is proposed by Lihsien et al [40]. As shown in Figure 9.25, this topology consists of four similar spiral coils that are connected in a diagonal manner in series [40]. In case of displacement, while the received power increases at half of the receiving coil (L_1 or L_2), it decreases at the remaining half section [40]. Therefore, the tolerance of inductive link between the external power transmission unit and the embedded receiving array of microcoils to misalignment will be higher and less affected by both side and angular displacement [40].

9.8 Conclusion

In this chapter, an overview of the energy scavenging techniques has been discussed. Large- and small-scale energy harvesting methods are briefly introduced. However, the focus of this chapter has been on energy scavenging in small scale to power up magnetic induction transmission systems. In this regard, it is explained how energy from existing vibration in the environment is converted to electrical energy using magnetic induction. Solar and thermal energy scavenging also has been discussed. How to convert the harvested DC power to AC power to be usable in magnetic induction-enabled-type devices, and, finally, a wireless power transmission concept for inductively coupled devices were discussed. Passive RFID tags and medical implant power transmission systems have been introduced to describe the wireless power transmission principles. Some techniques are discussed as well that improve the performance of such systems. Multiple output voltage systems and using spiral coils in embedded devices are described as methods to enhance the efficiency of the system to be less prone to side and angle misalignment, and also to decrease the device size.

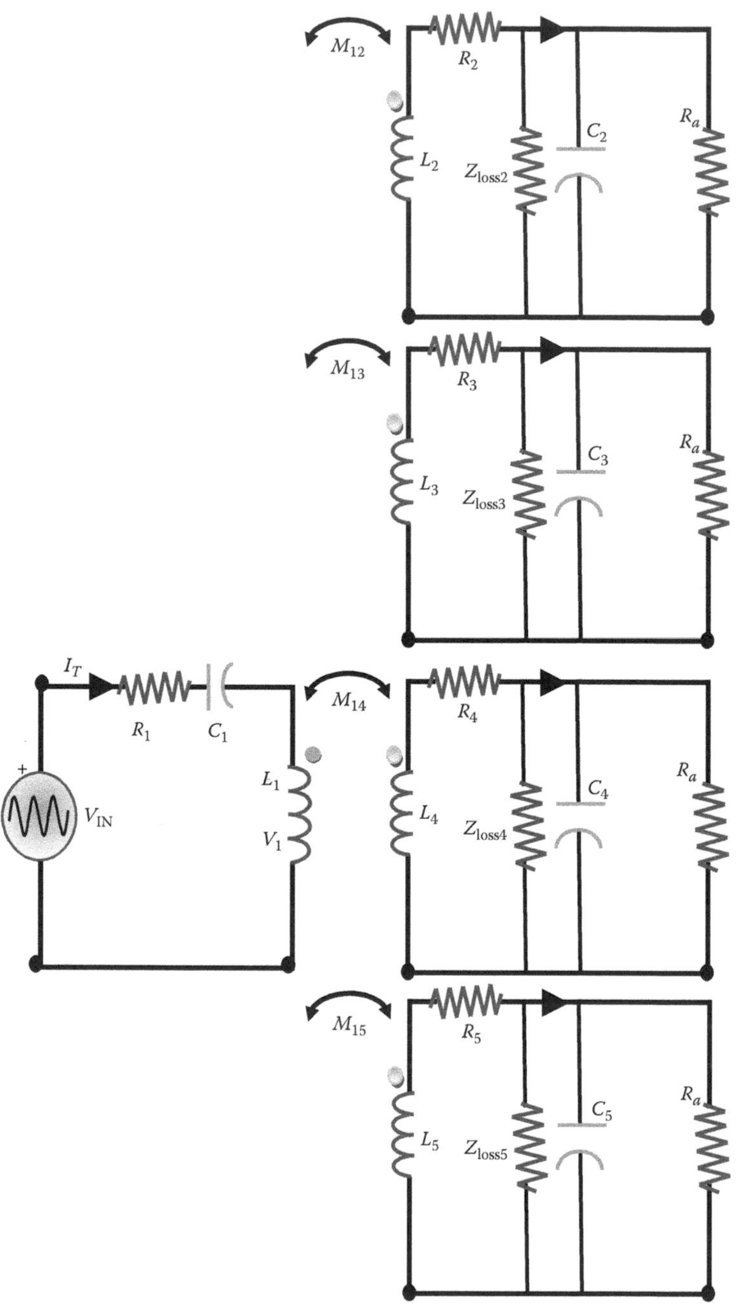

Figure 9.24 Equivalent circuit model of inductive link with four receivers. (From W. Lihsien, et al., in *Biomedical Circuits and Systems Conference, 2008. BioCAS 2008. IEEE*, 2008, pp. 101–104. With permission.)

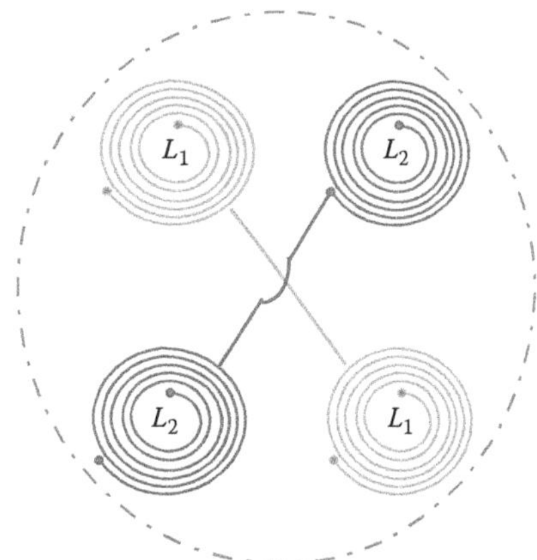

Figure 9.25 Schematic of diagonal connected coils in a multicoil structure. (From W. Lihsien, et al., in *Biomedical Circuits and Systems Conference, 2008. BioCAS 2008. IEEE*, 2008, pp. 101–104. With permission.).

REFERENCES

[1] H. Pei, et al., "Efficient solar power scavenging and utilization in mobile electronics system," in *International Conference on 2010 Green Circuits and Systems (ICGCS)*, 2010, pp. 641–645.

[2] O. Bertoldi and S. Berger, "Observatory NANO, report on energy," ed, 2009. http://www.observatorynano.eu/project/filesystem/files/WP2_5Energy_7EnergyHarvesting.pdf (accessed November 29, 2011).

[3] J. Bing, et al., "Energy Scavenging for Inductively Coupled Passive RFID Systems," in *Proceedings of the IEEE Instrumentation and Measurement Technology Conference, 2005. (IMTC)* 2005, pp. 984–989.

[4] R. Bansal, "Near-field magnetic communication," *Antennas and Propagation Magazine, IEEE*, vol. 46, pp. 114–115, 2004.

[5] C. Bunszel. (2001). Magnetic induction: A low-power wireless alternative. Available online at: http://rfdesign.com/mag/radio_magnetic_induction_lowpower/index1.html (accessed November 11, 2010).

[6] C. Evans-Pughe, "Close encounters of the magnetic kind [near field communications]," *IEE Review*, vol. 51, pp. 38–42, 2005.

[7] N. Jack and K. Shenai, "Magnetic Induction IC for Wireless Communication in RF-Impenetrable Media," in *IEEE Workshop on Microelectronics and Electron Devices, 2007. (WMED)* 2007, pp. 47–48.

[8] V. Palermo. (2003) *Near Field Magnetic comms emerges*. Available online at: http://www.eetimes.com/design/other/4009227/Near-field-magnetic-comms-emerges/ (accessed December 21, 2010).

[9] J. J. Sojdehei, et al., "Magneto-inductive (MI) communications," in *OCEANS, 2001. MTS/IEEE Conference and Exhibition*, 2001, pp. 513–519, vol. 1.

[10] R. R. A. Syms, et al., "Low-loss magneto-inductive waveguides," *Journal of Physics D: Applied Physics*, vol. 36, pp. 3945–3951, 2006.

[11] R. R. A. Syms, et al., "Magneto-Inductive Waveguide Devices," *IEE Proceedings Microwaves, Antennas and Propagation,* vol. 153, pp. 111–121, 2006.
[12] S. Zhi and I. F. Akyildiz, "Underground Wireless Communication Using Magnetic Induction," in *IEEE International Conference on Communications, 2009. (ICC '09)* 2009, pp. 1–5.
[13] S. Zhi and I. F. Akyildiz, "Magnetic induction communications for wireless underground sensor networks," *IEEE Transactions on Antennas and Propagation,* vol. 58, pp. 2426–2435, 2010.
[14] S. Zhi and I. F. Akyildiz, "Deployment Algorithms for Wireless Underground Sensor Networks Using Magnetic Induction," in *2010 IEEE Global Telecommunications Conference (GLOBECOM),* 2010, pp. 1–5.
[15] J. I. Agbinya, et al., "Size and characteristics of the 'cone of silence' in near-field magnetic induction communications," *Battlefield Technology,* vol. 13, 2010.
[16] H.-B. Li and R. Kohno, "Body area network and its standardization at IEEE 802.15. BAN," in *Advances in Mobile and Wireless Communications.* vol. 16, F. István, et al., eds., Berlin/Heidelberg: Springer Publishing, 2008, pp. 223–238.
[17] S. F. Heaney, et al., "Fading Characterization for Context Aware Body Area Networks (CABAN) in Interactive Smart Environments," in *Antennas and Propagation Conference (LAPC), 2010 Loughborough,* 2010, pp. 501–504.
[18] K. Y. Yazdandoost and R. Kohno, "UWB Antenna for Wireless Body Area Network," in *Asia-Pacific Microwave Conference, 2006. (APMC)* 2006, pp. 1647–1652.
[19] H. Kulah and K. Najafi, "Energy scavenging from low-frequency vibrations by using frequency up-conversion for wireless sensor applications," *Sensors Journal, IEEE,* vol. 8, pp. 261–268, 2008.
[20] W. Peihong, et al., "A Microelectroplated Magnetic Vibration Energy Scavenger for Wireless Sensor Microsystems," in *5th IEEE International Conference on Nano/Micro Engineered and Molecular Systems (NEMS), 2010,* pp. 383–386.
[21] J. Agbinya and M. Masihpour, "Power equations and capacity performance of magnetic induction communication systems," *Wireless Personal Communications,* pp. 1–15, 2011.
[22] W. C. Chye, et al., "Electromagnetic Micro Power Generator: A Comprehensive Survey," in *IEEE Symposium on Industrial Electronics & Applications (ISIEA),* 2010, pp. 376–382.
[23] C. B. Williams and R. B. Yates, "Analysis of a Micro-electric Generator for Microsystems," in *The 8th International Conference on Solid-State Sensors and Actuators, 1995 and Eurosensors IX. Transducers,* 1995, pp. 369–372.
[24] C. Shearwood and R. B. Yates, "Development of an electromagnetic microgenerator," *Electronics Letters,* vol. 33, pp. 1883–1884, 1997.
[25] S. Turkyilmaz, et al., "Design and Prototyping of Second Generation METU MEMS Electromagnetic Micro-Power Generators," in *International Conference on Energy Aware Computing (ICEAC),* 2010, pp. 1–4.
[26] W. J. Li, et al., "Infrared Signal Transmission by a Laser-Micromachined, Vibration-Induced Power Generator," in *Proceedings of the 43rd IEEE Midwest Symposium on Circuits and Systems,* 2000, pp. 236–239, vol. 1.
[27] J. M. H. Lee, et al., "Vibration-to-Electrical Power Conversion Using High-Aspect-Ratio Mems Resonators," presented at the *Power MEMS, 2003.*
[28] P.-H. Wang, et al., "Design, fabrication and performance of a new vibration-based electromagnetic micro power generator," *Microelectronics Journal,* vol. 38, pp. 1175–1180, 2007.

[29] C. Xinping and L. Yi-Kuen, "Design and Fabrication of Mini Vibration Power Generator System for Micro Sensor Networks," in *2006 IEEE International Conference on Information Acquisition*, 2006, pp. 91–95.

[30] W. Ko Ko, et al., "Efficient Solar Energy Harvester for Wireless Sensor Nodes," in *2010 IEEE International Conference on Communication Systems (ICCS)*, 2010, pp. 289–294.

[31] L. Xin and Y. Shuang-Hua, "Thermal Energy Harvesting for WSNs," in *2010 IEEE International Conference on Systems Man and Cybernetics (SMC)*, 2010, pp. 3045–3052.

[32] Y. K. Tan and S. K. Panda, "Energy harvesting from hybrid indoor ambient light and thermal energy sources for enhanced performance of wireless sensor nodes," *IEEE Transactions on Industrial Electronics*, vol. PP, pp. 1–1, 2010.

[33] A. J. Minnich, et al., "Bulk nanostructured thermoelectric materials: Current research and future prospects," *Energy and Environmental Science*, 2009.

[34] J. Tervo, et al., "State-of-the-art of thermoelectric materials processing," presented at the VTT Technical Research Centre of Finland, Oulu, Finland, 2009, vol. 2, No. 5, pp. 446–479.

[35] T. Isobe, et al., "A Soft-Switching Boost DC to AC Converter without Smoothing Capacitor Using a MERS Pulse Link Concept," in *2010 International Power Electronics Conference (IPEC)*, 2010, pp. 2815–2821.

[36] J. Bing, et al., "Energy scavenging for inductively coupled passive rfid systems," *IEEE Transactions on Instrumentation and Measurement*, vol. 56, pp. 118–125, 2007.

[37] S. Cichos, et al., "Performance Analysis of Polymer-Based Antenna Coils for RFID," in *2nd International IEEE Conference on Polymers and Adhesives in Microelectronics and Photonics, 2002. (POLYTRONIC)*, 2002, pp. 120–124.

[38] G. D. Horler, et al., "Inductively coupled telemetry and actuation," in *The IEE Seminar on (Refl No. 2005/11009) Telemetry and Telematics*, 2005, pp. 5/1–5/6.

[39] M. Sawan, et al., "Multicoils-based inductive links dedicated to power up implantable medical devices: Modeling, design and experimental results," *Biomedical Microdevices*, vol. 11, pp. 1059–1070, 2009.

[40] W. Lihsien, et al., "An Efficient Wireless Power Link for High Voltage Retinal Implant," in *IEEE Biomedical Circuits and Systems Conference, 2008. (BioCAS)*, 2008, pp. 101–104.

[41] O. M. O. Gatous and J. Pissolato, "Frequency-Dependent Skin-Effect Formulation for Resistance and Internal Inductance of a Solid Cylindrical Conductor," *IEE Proceedings of Microwaves, Antennas and Propagation*, vol. 151, pp. 212–216, 2004.

Chapter 10

Mixed-Signal, Low-Power Techniques in Energy Harvesting Systems

N. Fragoulis, L. Bisdounis, V. Tsagaris, and C. Theoharatos

Contents

10.1 Introduction

Energy harvesting systems pose a new challenge in the domain of circuit design because they must operate with an extremely low-power budget. As is evident, a circuit functioning in such an energy-starving environment must be operated near the fundamental low-power limits and should be designed on the basis of very strict guidelines conforming to the most recent advances in low-voltage and low-power design.

Modern portable systems, which are the main application field of energy harvesting techniques, are mainly mixed-signal systems comprised of a digital core including, amongst others, a central processing unit (CPU) or digital signal processing (DSP) and memory, often surrounded by several analog interface blocks, such as I/O (input/output), D/A (digital-to-analog), and A/D (analog-to-digital) converters, RF (radio frequency) front ends, and more. Therefore, a mobile device is a characteristic example of a mixed-signal system, which is, namely, a system that combines, to some extent, analog and digital circuitry.

The evolution in complementary metal oxide semiconductor (CMOS) technology, which is the dominant technology in portable systems, is motivated by the decrease in the price-per-performance factor for digital circuitry in a pace dictated by Moore's Law, the main effect of which is the shrinking of the dimensions (feature size) of the devices. To ensure sufficient lifetime for digital circuitry and to keep power consumption at an acceptable level, this dimension-shrink is accompanied by lowering of nominal supply voltages. While this evolution in CMOS technology is by definition very beneficial for digital circuits, this is not the case for analog circuits. In addition, although low-power techniques for analog and digital circuits have nowadays sufficiently matured, there still remain some fundamental controversies regarding the design of a mixed-signal system that a designer must take under consideration.

The most efficient way to reduce the power consumption of digital circuits is to reduce the supply voltage, since the average power consumption of CMOS digital circuits is proportional to the square of the supply voltage. On the other hand, the reduction of the supply voltage is also mandatory due to dimension shrinking in order to maintain the electric field at an acceptable level.

The rules for analog circuits seem to be different than those applied to digital circuits. This is mainly due to the fact that the power consumption of analog circuits at a given temperature is basically set by the required signal-to-noise ratio (SNR) and the required bandwidth.

A very important technique that seems to bridge the controversies between analog and digital low-power techniques is based on the ability of CMOS transistor devices to work in the *subthreshold region* (*weak inversion*). CMOS transistors functioning in this region exhibit extremely low power consumption, as a result of extremely low operating current densities that are inherent in the subthreshold operation. Subthreshold operation is not suitable for applications where high performance is needed, but seems a very attractive solution in energy harvesting systems where simple systems are generally implemented.

In the rest of this chapter, the particularities of the low-power design in the digital and the analog domain will be analyzed in order for the reader to gain a deeper knowledge of the effects that specific design choices have on the power performance in the analog and the digital world. In addition, new design techniques will be analyzed and discussed that bridge the analog and digital world controversies toward a successful, mixed-signal, ultralow-power design, suitable for use in an energy harvesting application. Finally, a brief review of the power-aware electronic design automation (EDA) software tools available in the market will be conducted in order to give to the reader a brief guide of the available means for analog and digital low-power design.

10.2 Mixed-Signal Environment in Energy Harvesting Systems

Typical applications of energy harvesting systems are small, wireless autonomous devices, like those used in wireless microsensor networks and radio frequency identification Systems (RFIDs). These types of applications would benefit from unbounded lifetimes in an environment where changing batteries is impractical or impossible, since the concept of energy harvesting involves converting ambient energy from the environment into electrical energy to power the circuits or to recharge a battery. Microsensor nodes must keep average power consumption in the 10 to 100 μW range to enable energy harvesting [1]–[3]. Combining energy harvesting techniques with some form of energy storage can theoretically extend system lifetimes indefinitely. Clearly, this type of system will be much more effective when coupled with the significant power and energy savings made possible by applying power reduction design techniques to their individual components.

10.2.1 Microsensor Wireless Networks

A microsensor node refers to a system that provides sensing, computation, and communication functionality. The block diagram of a typical microsensor node is shown in Figure 10.1. Wireless microsensor networks consists of tens to thousands of distributed nodes that sense and process data and relay the results to the end

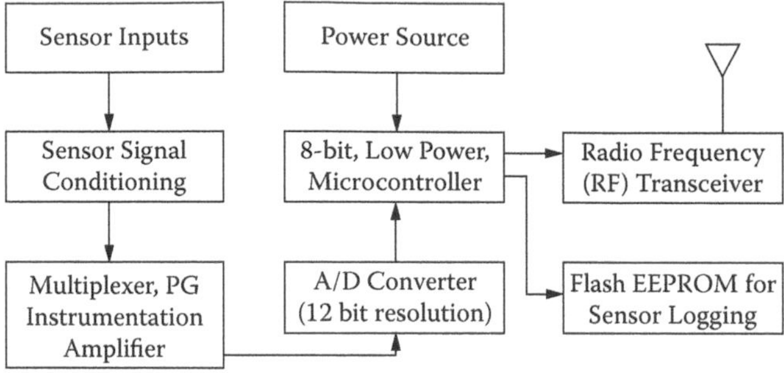

Figure 10.1 A block diagram of a typical wireless microsensor node.

user. Proposed applications for microsensor networks include habitat monitoring, structural health monitoring, and automotive sensing [3] [4].

The performance requirements for microsensor nodes in these applications are very low. The rate at which data changes for environmental or health monitoring, for example, is on the order of seconds to minutes, so the performance achieved even in subthreshold is more than adequate. A very common technique used in microsensor nodes is the duty cycle or shutdown of unused components whenever possible. Although duty cycling helps to extend sensor network lifetimes, it does not remove the energy constraint placed by the power source. Energy harvesting techniques are a necessity in these applications because, if a battery is used instead, it is not possible to recharge or replace batteries frequently. Thus, microsensor networks are a very interesting platform that showcases the need for new low-energy design techniques, which must be applied in the analog and digital domains. This is evident also from Figure 10.1, where a classical mixed-signal system can be easily identified.

10.2.2 Radio Frequency Identification (RFID)

RFID is another typical application that requires extremely low energy consumption [4] [5]. RFID is used to automatically identify objects through RFID tags that are attached to the object. The RFID tag is able to transmit and receive information wirelessly using radio frequencies. An RFID tag contains a limited amount of digital processing logic along with an antenna and communication circuits.

There are two main types of RFID tags. An *active* RFID tag communicates with the reader by transmitting data. Active tags frequently require a power source to supply the energy for transmission, and any extra energy saved due to application of low-power design techniques could be used for extended processing and longer range of communication. A *passive* RFID tag communicates with the reader by modulating the load that the reader sees. This means of communication requires less energy, so passive tags often operate on energy that is converted from the

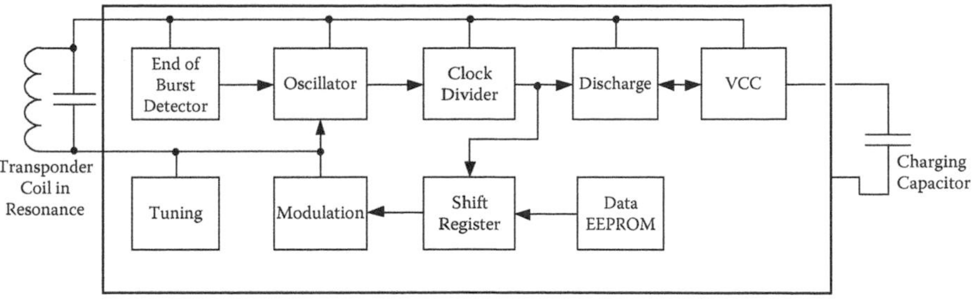

Figure 10.2 Typical block diagram of an RFID tag.

received signal. Passive nodes are usually smaller as a result, and their lifetimes are not limited by energy.

Reducing power consumption and using energy harvesting would benefit both types of tags. This is straightforward for active tags where minimizing the power consumption leads to both increased transmission range and/or longer battery lifetimes. For passive tags, the power is constrained by the ability to utilize the converted energy from the antenna. If the digital logic power dissipation could be reduced and the system could be assisted from an extra energy source, then the distance from the reader to the tag could increase because less transmitted power has to reach the tag.

As is indicated in Figure 10.2, an RFID tag is comprised of several analog and digital parts, which constitute a mixed-signal environment. To this end, if true low power is to be achieved, the reduction of power consumption of analog and digital parts of the system must be addressed.

10.3 Low Power Techniques in Digital Design

Average power consumption in digital CMOS circuits is more important than peak power as instances of peak power consumption when all the circuit components are on is rare. Ideally, CMOS gates consume power when the output node makes a switching transition. However, there are short circuit and leakage currents through the device that result in wasteful power dissipation. The average power P_{avg} consumed by a CMOS circuit can be represented mathematically as:

$$\begin{aligned} P_{avg} &= P_{switch} + P_{ShortCkt} + P_{lkg} \\ &= (C_L \cdot V_{SWING} \cdot V_{DD} \cdot f_{CLK} \cdot \alpha) + (I_{ShortCkt} \cdot V_{DD}) + (I_{lkg} \cdot V_{DD}) \end{aligned} \tag{10.1}$$

The first term on the right-hand side of equation (10.1), i.e., switching power P_{switch}, represents the power consumed by the switching capacitance or load capacitance C_L of the CMOS circuit. This is the power consumed in charging the load capacitance when the device makes a 0 to 1 transition. It represents approximately 60 to 70 percent [1] of total power consumed. Besides C_L, switching power is also a function of the supply voltage V_{DD}, the voltage difference between logic 1 and logic 0 V_{SWING}, the clock frequency f_{CLK}, and the node transition activity factor α. V_{SWING} usually equals the supply voltage V_{DD}, but for internal nodes, it could be less than V_{DD}.

The second term of equation (10.1), the short circuit power $P_{ShortCkt}$, refers to the power dissipated due to the direct path short circuit current $I_{ShortCkt}$, through the PMOS (p-type MOS) and NMOS (n-type MOS) transistors of static logic circuits during a switching transition. It accounts for 20 percent of total power dissipated in static circuits as there is no short circuit current for dynamic design because of precharging. The short circuit current is a function of the rise and fall time of the input and output signals, amount of capacitive load, size of the CMOS devices, and the gate capacitance, especially the equivalent gate to drain capacitance [6].

The last term in equation (10.1) corresponds to power dissipated due to leakage current I_{lkg}. Usually, leakage current accounts for approximately 2 to 3 percent of the total power. Though, ideally, no power is consumed when both PMOS and NMOS transistors are off, power due to I_{lkg} arises from inherent reverse biased diode currents I_{lkg} and subthreshold effects I_{subtkg} of the transistors. The leakage current is strongly a function of the fabrication technology.

Although power consumption is generally considered as a term identical to that of energy consumption, it is worthwhile to notice a fundamental difference in the case of digital circuits. The power consumed by a device is by definition the energy consumed by unit time. In other words, the energy (E) required for a given operation is the integral of the power (P) consumed over the operation time (T_{op}), hence:

$$E = \int_0^{T_{op}} P(t)\,dt \tag{10.2}$$

If we substitute $P(t)$ in equation (10.2) by the switching power of a digital circuit, which is the main component of the total power consumption P_{switch} and we assume that an operation requires n clock cycles, T_{op} can be expressed as n/f and so we get:

$$E = n \cdot C_L \cdot V_{SWING} \cdot V_{DD} \cdot \alpha \tag{10.3}$$

It is important to note that the energy per operation is independent of the clock frequency. Reducing the frequency will lower the power consumption, but will not change the energy required to perform a given operation. Since the energy consumption is what determines the battery life, it is imperative to reduce the energy rather than just the power. It is, however, important to notice that the power is critical for heat dissipation considerations.

10.3.1 Reducing Power in Digital Circuits

Because switching power accounts for the major portion of the power consumed, any attempt at low-power design should try to minimize it. To this end, low-power design methodologies [7] [8] at every level should aim at reducing the variables in that term, namely, C_L, V_{DD}, V_{swing}, f_{clk}, and a. However, significant reduction of the power consumption can occur through the following interventions:

1. *Supply Voltage V_{DD} Reduction:* Power consumed by a CMOS device is proportional to the square of the supply voltage V_{DD} and, hence, lowering the supply voltage would result in a quadratic reduction in power consumption, though device current reduces only linearly with V_{DD}. It can be proved that in this way the power can be practically reduced by one to eight times [9]. Supply reduction can be achieved through some special circuit manipulations and through feature size scaling, but the designer must be very careful because often these techniques impose serial limitations, such as circuit delay and degraded functional throughput. For computationally intensive functions, one of the effective ways to reduce power consumption while still operating at low voltage is to parallelize the computation by modifying the algorithm and the architecture. The key to architecture-driven voltage scaling is to exploit concurrency (pipelining and parallelism) in execution. Also, the combination of architectural optimization with threshold voltage reduction can scale down supply voltage to the sub-1 V range. To compensate for the loss in speed due to voltage scaling, it is possible to upsize the transistors that are in the critical delay path, or by transistor sizing, using fast logic structures [10].
2. *V_{swing} Reduction:* Power consumption of a CMOS logic gate with a fixed supply voltage V_{DD} also can be reduced by restricting the voltage swing V_{swing} at the output node [10] [11]. Usually, the output node of the gate will make rail-to-rail transitions (V_{DD} to 0 or 0 to V_{DD}). But, if an NMOS device has been used instead as a pull up, the output will limit the swing to $(V_{DD} - V_T)$. The power consumed for a 0 to $(V_{DD} - V_T)$ in such a case will be $C_L \cdot V_{DD} \cdot (V_{DD} - V_T)$, and the reduction in power consumption (over a rail-to-rail scheme) is proportional to $V_{DD} / (V_{DD} - V_T)$. However, there are a few drawbacks with such a design, such as the reduced noise margin and increased power consumption at the subsequent stage [12].
3. *Load Capacitance (C_L) Reduction:* An obvious way to reduce the load capacitance is to reduce the CMOS device size since scaling reduces the channel and parasitic capacitance [12]. Logic/circuit minimization through effective partitioning can also reduce the load capacitance.

4. *Node Transition Activity* α *Reduction:* Switching activity reduction can help to reduce power consumption in CMOS devices as power is consumed only during transitions. Various techniques range from simply powering down the complete circuit or portions of it, to more sophisticated schemes in which the clocks are gated or optimized circuit architectures are used that minimize the number of transitions [13]. An important attribute that can be used in circuit and architectural optimization is the correlation in the temporal sequence of data because switching should decrease if the data are slowly changing, i.e., highly positively correlated. Thus, knowledge about signal statistics can be used to reduce the number of transitions. The techniques for a reduction span all levels of the system design from the physical design level, to the logic level where logic minimization and logic level power down are the key techniques to minimize the transition activity [13].

Though switching power accounts for the major share of the total power dissipated, short circuit and leakage power usually amount to 20 to 30 percent. In order to reduce short circuit power, gate capacitances, device size, and the rise and fall time of the signals should be reduced. Leakage power, on the other hand, can be reduced by accurate device modeling and threshold control. The various power-reducing parameters discussed can be optimized at various design levels to a different extent.

10.4 Low Power Techniques in Analog Design

Power is consumed in analog signal processing circuits to maintain the signal energy above the fundamental thermal noise in order to achieve the required signal-to-noise ratio (SNR). A representative figure of merit of different signal processing systems is the power consumed to realize a single pole. The minimum power necessary to realize a single pole can be derived by considering the basic integrator presented in Figure 10.3 where an ideal 100 percent current efficient transconductor is used, in the sense that all the current pulled from the supply voltage is used to charge the integrating capacitor [14].

The power consumed from the supply voltage source V_{DD} that is necessary to create a sinusoidal voltage $V(t)$ across capacitor C having peak-to-peak amplitude V_{PP} and frequency f can be expressed as:

$$P = V_{DD} \cdot fCV_{PP} = fCV_{PP}^2 \cdot \frac{V_{DD}}{V_{PP}} \tag{10.4}$$

and the signal-to-noise ratio is given by:

$$SNR = \frac{V_{PP}^2/8}{kT/C} \tag{10.5}$$

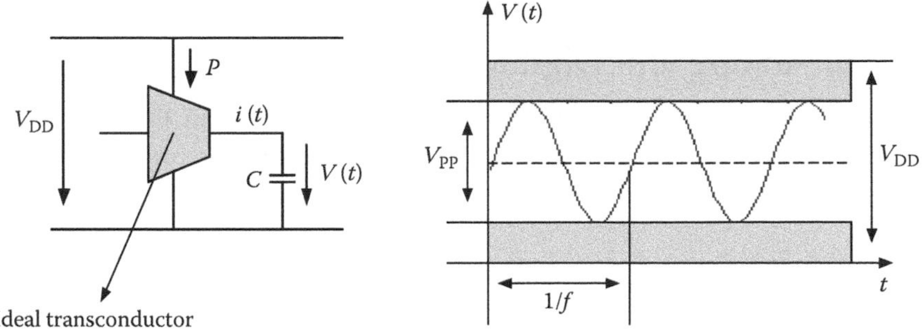

Figure 10.3 An ideal, single-pole analog processing filter.

Combining equation (10.4) and equation (10.5), we get:

$$P = 8kT \cdot f \cdot SNR \cdot \frac{V_{DD}}{V_{PP}} \tag{10.6}$$

According to equation (10.6), the minimum power consumption of analog circuits at a given temperature is basically set by the required SNR and the frequency of operation (or the required bandwidth). Since this minimum power consumption is also proportional to the ratio between the supply voltage and the signal peak-to-peak amplitude, power efficient analog circuits should be designed to maximize the voltage swing. The minimum power per pole for circuits that can handle rail-to-rail signal voltages ($V_{PP}= V_{DD}$) reduces to [15]–[18]:

$$P_{MIN} = 8kT \cdot f \cdot SNR \tag{10.7}$$

This absolute limit is very steep because it requires a factor 10 of power increase for every 10 dB of signal-to-noise ratio improvement. It applies to each pole of any linear analog filter (continuous and sampled data [19]) and is reached in the case of a simple passive RC (resistor–capacitor) filter, whereas the best existing active filters consume about two orders of magnitude more power per pole. High Q poles in the passband reduce the maximum signal amplitude at other frequencies and, therefore, increase the required power, according to equation (10.6).

Approximately the same result is found for relaxation oscillators, whereas the minimum power required for a voltage amplifier of gain A_v can be proved to be always larger or equal to:

$$P_{MIN} = 8nkT \cdot \Delta f \cdot \mathrm{A}_V \cdot SNR \tag{10.8}$$

which mean that is again proportional to SNR and is $n{\cdot}\times A_v$ times larger than the limit given by equation (10.7).

10.5 Comparison of the Power Consumption of Analog and Digital Circuits

The minimum power for an analog system can be compared to that of a digital system, which, if the transistors are considered ideal, corresponds to the switching component P_{Switch} of equation (10.1), i.e.:

$$\begin{aligned} P_{\min-digital} &= C_L \cdot V_{SWING} \cdot V_{DD} \cdot f_{CLK} \cdot \alpha \\ &= P_{\min-digital} = E_{tr} \cdot f_{CLK} \cdot \alpha \end{aligned} \tag{10.9}$$

In equation (10.9), each elementary operation requires a certain number of binary gate transition cycles, each of which dissipates an amount of energy E_{tr}. Due to the Nyquist theorem, f_{CLK} must be at least twice the signal bandwidth so f_{CLK} can be considered to be the signal bandwidth if we account a factor (1/2) into the constant E_{tr}.

The number m of transitions is only proportional to some power m of the number of bits N, and, therefore, power consumption is only weakly dependent on SNR (essentially logarithmically) [20]:

$$a = N^m \approx [\log(SNR)^m] \tag{10.10}$$

Comparison with analog is obtained by estimating the number α of gate transitions that are required to compute each period of the signal, which for a single pole digital filter can be estimated to be approximately:

$$a \cong 50 \cdot N^2 \tag{10.11}$$

From equation (10.9), $E_{tr} = C_L \cdot V_{SWING} \cdot V_{DD} = C_L \cdot V_{DD}^2$, which varies from 10^{-12} to 10^{-15} Joule.

Combining equation (10.9) to equation (10.11), we get:

$$P_{\min-digital} \cong E_{tr} \cdot f_{CLK} \cdot 50 \cdot [\log(SNR)]^2 \tag{10.12}$$

Therefore, the relationship between switching energy and signal-to-noise ratio (S/N) is logarithmic. Comparison of analog and digital fundamental limits is depicted in Figure 10.4, and clearly shows that analog systems may consume much less power than their digital counterparts, provided a small signal-to-noise ratio is acceptable. But, for systems requiring large signal-to-noise ratios, analog becomes very power-inefficient.

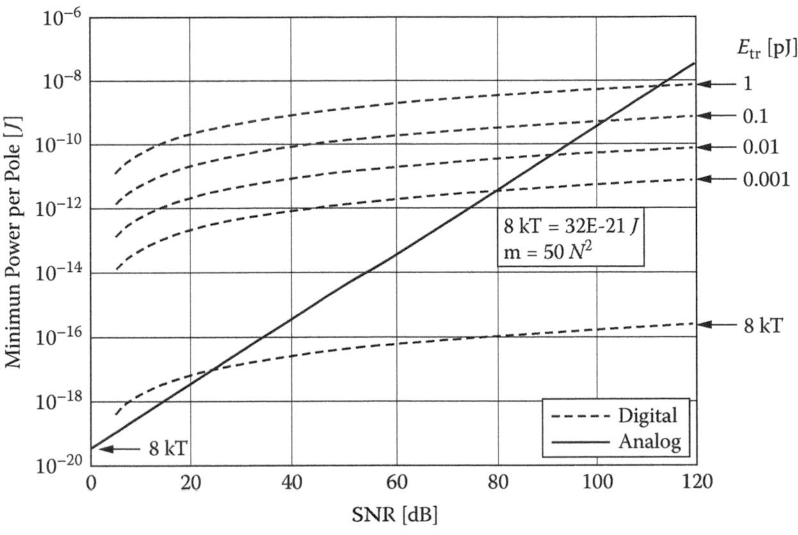

Figure 10.4 Minimum power consumption of analog and digital circuits.

10.6 Combination of Techniques Toward Low-Voltage, Mixed-Signal Design

Unlike digital circuits where, according to equation (10.3), the power consumption decreases with the square of the supply voltage, reducing the supply voltage of analog circuits while preserving the same bandwidth and SNR has no fundamental effect on their minimum power consumption. However, this absolute limit was obtained by neglecting the possible limitation of bandwidth BW due to the limited transconductance g_m of the active device. The maximum value of BW is proportional to g_m/C. Replacing the capacitor value C by g_m/BW in equation (10.5) and expressing the product of the SNR times the bandwidth yields:

$$SNR \cdot BW = \frac{V_{pp}^2 \cdot g_m}{8kT} \tag{10.13}$$

In most of the cases, scaling the supply voltage V_{DD} by a factor K requires a proportional reduction of the signal swing V_{pp}. Maintaining the bandwidth and the SNR, therefore, is only possible if the transconductance g_m is increased by a factor K^2. If the active device is a bipolar transistor (or a MOS transistor biased in subthreshold region), its transconductance can only be increased by increasing the bias current I by the same factor K^2 and, therefore, power $V_{DD} \cdot I$ is also increased by K.

The situation is different if the active device is a MOS transistor biased in strong inversion. Its transconductance can be shown to be proportional to I/V_P, where V_P is the pinch off or saturation voltage of the device. Because this saturation voltage also has to be reduced proportionally with V_{DD}, then increasing g_m by K^2 only requires an increase of current by a factor K and, hence, the power remains unchanged. However, even in this case, supply reduction has serious effects on the functionality of the circuit, since it affects the maximum frequency of operation. For a MOS transistor in strong inversion, the frequency f_{max}, for which the current falls to unity, is given by approximately:

$$f_{\max} = \frac{\mu \cdot V_P}{L^2} \tag{10.14}$$

Therefore, if the process is fixed (channel length L constant), a reduction of V_{DD} and V_P by a factor K causes a proportional reduction of f_{max}.

Reduction of the supply voltage also has an implicit effect on the dynamic range of the analog processor. The *dynamic range* (*DR*) of an ideal integrator, such as this of Figure 10.3, is given by [20]:

$$DR_{\max} = \frac{CV_{DD}^2}{8kT} \tag{10.15}$$

Therefore, implementation of an analog signal processing circuit with a specific dynamic range, in an environment of low-supply voltage, poses an additional challenge for the analog designer.

Unfortunately, in analog systems, low-voltage limitations are not restricted to power or frequency problems. For example, reducing V_P also increases the transconductance-to-current ratio of MOS transistors that, in turn, increases the noise content of current sources, decreasing SNR this way, while at the same time it drastically degrades their precision.

10.7 Optimum Combination of Analog and Digital Low-Power Techniques

As is evident from the above, the main tool that a designer has toward lowering the power consumption of a digital circuit, namely the voltage supply reduction, is not so effective in analog design because power consumption of analog circuits is mainly dependent on SNR, and voltage supply usually leads to an increase of power consumption.

10.7.1 Instantaneous Companding Technique

A possible way to maintain a sufficient dynamic range when reducing the supply voltage, without degrading the power consumption of analog signal processing circuits, is to use the instantaneous companding technique [21]–[23]. In this approach, the currents are compressed when transformed into voltages and expanded when transformed back to currents. The input current has to be predistorted in order to preserve a linear operation.

The basic idea is to ensure that the signal in the channel is always on a level that is significantly over the noise level. To achieve this, the signal is preamplified (predistorted) through amplifier *g*, but, in order to keep distortion at an acceptable level, it is important not to over-amplify the signal. For this reason, large signals are amplified by a smaller gain than small signals, in a way that amplified signals are always near the maximum dynamic range of the channel. After passing through the channel, signals must be recovered by undergoing the inverse amplification procedure.

As is evident from the above, the gain of the amplifier *g* depends on the signal level, and, of course, is nonlinear. In instantaneous companding (Figure 10.5), this can be achieved by using nonlinearity, which has small signal gain increases inversely proportional with the signal level, such as that indicated in Figure 10.6.

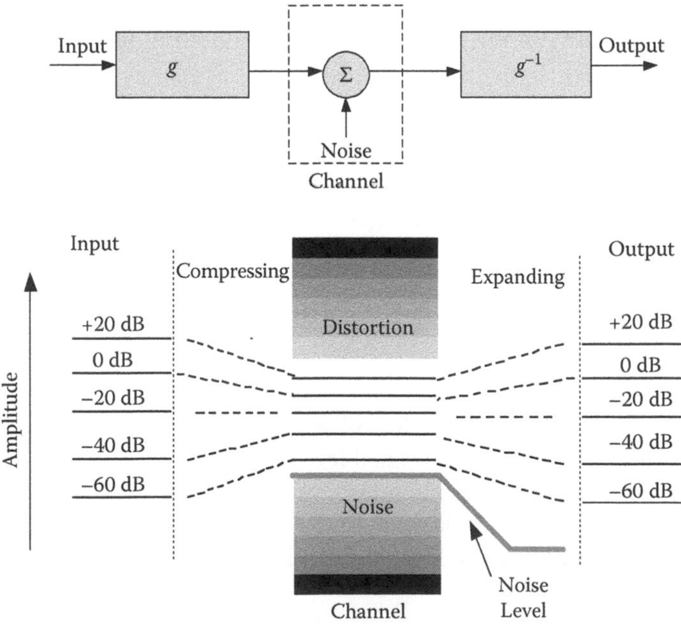

Figure 10.5 The instantaneous companding principle.

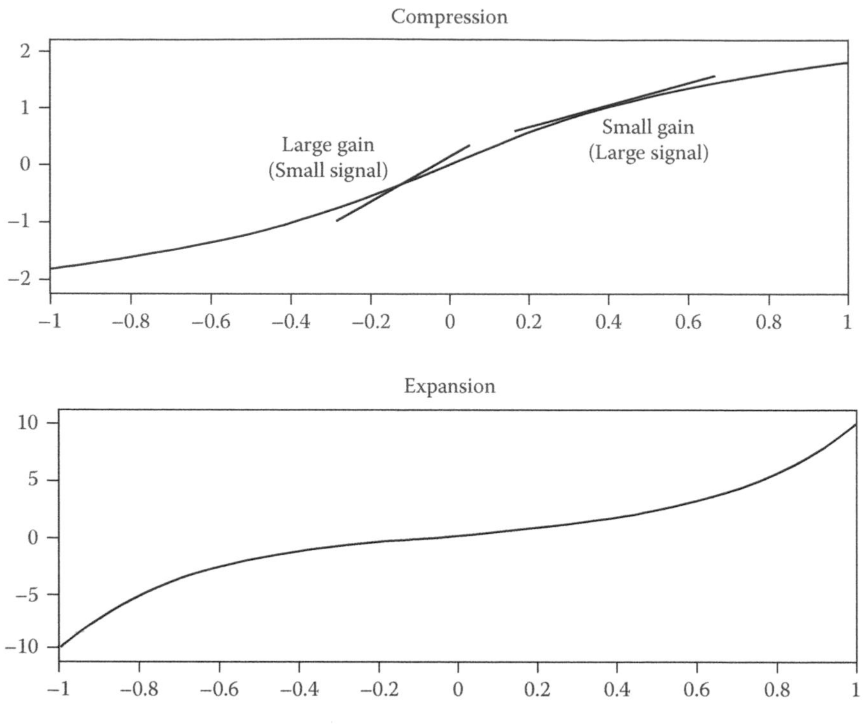

Figure 10.6 Appropriate, signal-dependent shapes for instantaneous companding.

The choice of the amplifying function is not restricted, in theory, but, since the predistortion of the signal requires the derivative of this expanding function, it is much easier to realize it using the exponential function because it is invariant to the differentiation operator and can be implemented either by the current-to-voltage characteristic of a bipolar transistor or a subthreshold-biased MOS transistor.

It can be shown that, for the instantaneous companding circuits, the following statements are valid:

1. SNR is constant and independent of signal level, in contrast to common circuits where SNR depends on signal level.
2. Dynamic range in companding circuits is larger than the maximum SNR, while in common circuits is equal to the SNR.
3. The maximum SNR does not change and, therefore, the power consumption remains unchanged.

Hence, in these circuits, power can be saved if the SNR can be reduced to the minimum required while maintaining the necessary dynamic range.

10.7.2 Subthreshold CMOS Design

A very important technique that seems to bridge the controversies between analog and digital low-power techniques is based on the ability of CMOS transistor devices to work in the subthreshold region (or weak inversion). CMOS transistors functioning in this region exhibit extremely low power consumption as a result of extremely low-operating current densities.

A CMOS transistor is considered to be in a strong inversion region of operation, when the voltage V_{GS} between its gate and source terminals is greater than a threshold voltage V_T. In this region, the drain current I_D of the transistor is considered to have a nonzero value, while for $V_{GS} < V_T$, it is usually considered to have a zero value. However, such an abrupt behavior is certainly not usual in nature, and actually the MOS transistor conducts a very low drain current even when $V_{GS} < V_T$. The main characteristic of this region of operation, which is called *weak inversion* or *subthreshold* region, is the extremely low drain current flow, which corresponds in turn to an extremely low drain current. This behavior is well characterized far into the subthreshold region and varies exponentially rather than quadratically with V_{GS}, according to the relation [24]–[26]:

$$I_D = \frac{W}{L} I_{DO} e^{(V_{GS}-V_T)/(nV_t)} \quad (10.16)$$

where

$$I_{DO} \cong \frac{K'2(nV_t)}{e^2} \quad (10.17)$$

In equation (10.17), K' is the transconductance parameter, $V_t = kT/q = 26\ mV$ is room temperature, and n is a constant between 1 and 2. Typical current densities, for I_D in this region of operation, lie in the pA range. Due to these very low current density levels, weak-inverted CMOSs are convenient for the design of ultra-low power applications.

In this region of operation, CMOS devices exhibit some very interesting features:

- Exponential I–V characteristics of subthreshold MOS devices provide the opportunity to implement analog current-mode circuits with very wide tunability. The possibility to change the bias current in a wide range especially provides appropriate bases to design wide frequency tuning range circuits.
- Exponential I–V behavior of the subthreshold MOS devices makes them suitable to be used for designing for analog log-domain, instantaneous companding circuits.

- CMOS devices in this regime exhibit maximum transconductance (g_m) to bias current (I_{DS}) ratio, i.e., g_m/I_{DS}, which means that the power efficiency of the MOS circuit can be maximized.
- Another very attractive characteristic of subthreshold MOS transistors is their ability to work under very low supply voltage. Therefore, it is possible to reduce the supply voltage of a CMOS inverter down to almost $4V_T$, while preserving sufficient gain for logic operation. Therefore, it is possible to use CMOS logic circuits deeply biased in subthreshold region. This means that if the speed of operation is not the premier design issue, it is possible to reduce the supply voltage and, hence, reduce the power dissipation of a system, which is mostly proportional to the dynamic power consumption.

Emerging new applications, such as energy harvesting systems, which require very low power consumption, has made subthreshold circuits very popular. Subthreshold operation is not suitable for applications where high performance is needed, but seems a very attractive solution in medium (1 Ms–10 Ms) or low (10 Ks–100 \Ks) data throughput systems, where energy consumption and cost are the most important parameters [25].

10.8 Power-Oriented Electronic Design Automation (EDA) Tools

As has been pointed out throughout this chapter, power consumption is a very critical parameter that has to be taken into account during the design of electronic circuits for energy harvesting applications, in order to provide the appropriate energy savings. To facilitate low-power design, electronic design automation (EDA) tools are required, which include efficient methods for fast and accurate estimation of energy dissipation as well as for the design of circuits and systems with certain power consumption constraints [27]–[34].

To date, power-oriented EDA tools have been developed in two main directions:

1. Analysis and modeling
2. Optimization (reduction) of the power consumption of circuits and systems

As shown in Figure 10.7, tools regarding both directions have been developed in several abstraction levels (i.e., level of the input design description), such as transistor level, gate (logic) level, register transfer (RT, architectural) level, and behavioral (algorithmic, system) level [28]-[35]. A netlist of interconnected transistors is at the transistor level, while a netlist of interconnected logic cells is at the logic level. At the register transfer level, designs are described in hardware description languages (e.g., VHDL), and at the behavioral level the

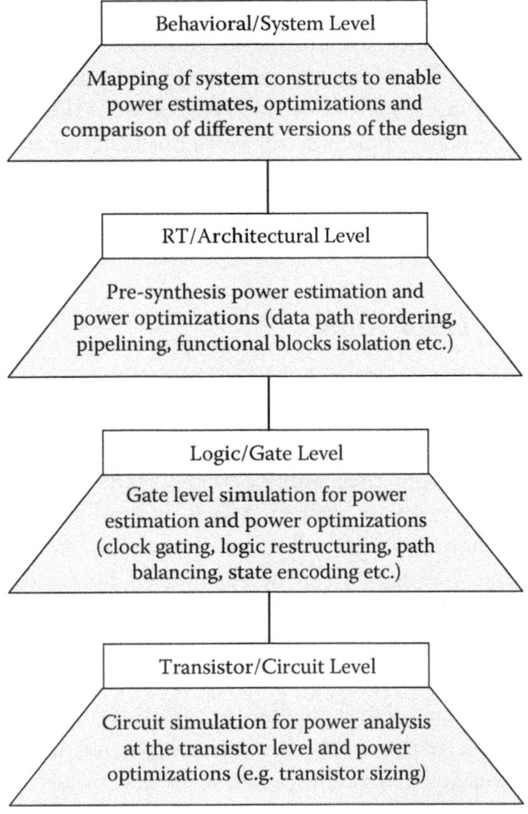

Figure 10.7 Power-oriented design flow with power analysis and optimization steps.

functionality of a design can be described by using hardware description languages (with more abstract functions) or high-level programming languages, such as C, C++, and SystemC.

While the evolution of analysis, synthesis, and optimization EDA tools for digital circuits and systems is fast and satisfactory, the analog portion of design automation has not been able to keep up with its demand. Despite that some efforts have been attempted in this area, there are not yet practical and efficient power-aware analysis and optimization tools that are generally accepted in the analog designers' community [36] [37]. The absence of analog power-aware EDA tools comparable to digital counterpart constitutes a serious bottleneck in designing mixed circuits and systems. Existing analog design automation methodologies are trying to optimize performance and power by using extensive circuit (SPICE-like) simulations based on precise transistor models (e.g., EKV subthreshold model [15]), in order to adjust transistor sizes. An alternative approach is the sizing of the circuit by using equation-based methods that are based on simplified device equations and approximations. An additional option is the use of analog/mixed-signal languages (AMS-HDL) [38] to speed up mixed-signal current and power simulation. While languages, such as Verilog-AMS and VHDL-AMS, hold a lot of promise, they are only really applicable for full-chip functional verification. They cannot be used for accurate transistor-level power and mixed-signal timing analysis because they cannot accurately model the analog circuitry and device-level effects that can cause leakage.

10.8.1 Transistor Level Tools

Power analysis tools at the transistor level are the most accurate, but also require the most time-consuming analysis. Their run time characteristics in combination with the fact that the whole transistor level description has to be available to the designer before their use limit their applicability to large circuits. Such tools are usually used for creating power models for relatively small elements (characterization) in order to use them at a higher level of abstraction. Transistor level circuit simulators (SPICE-like simulators) can be easily used for circuit power analysis. Their operation is based on detailed equations in order to model the transistors' behavior under various conditions, resulting in limited capacity and analysis speed. An advantage of the use of circuit simulators for power analysis is that they can be applied in either digital, analog, or mixed-signal circuits as well as for transmission line power analysis. An alternative approach to SPICE-like circuit simulators is the power analysis at the switch level that models each transistor as a nonideal switch considering several electrical properties, and leads to significant capacity and run-time improvements.

The process of circuit power characterization is provided by modeling tools that control the circuit simulation engine in order to produce the required power characterization data. These modeling tools use as input the transistor level netlist and the functional description for each cell that needs to be characterized, along with process and operating conditions, such as transition times, output loads, temperature, supply voltage, etc., while at the same time they perform stimulus generation in order to lead the simulation and to produce the power characterization data for each cell.

Power optimization at the transistor level is based on the transistor sizing concept, i.e., employing the smallest transistors to achieve low power while still satisfying the circuit's timing constraints as well as on the transistor reordering concept. The disadvantage of these approaches is that they can be applied only in custom designs. The input in such tools is the transistor netlist and the circuit's timing constraints, and the target is to reduce power consumption in circuit paths with positive timing margins. An additional category of power optimization tools at the transistor level concerns power grid analyzers that report the gradient of voltages along the power rails, as well as the current densities at different points on the power rails, in order to indicate electromigration violations.

Examples of transistor-level power analysis tools are HSPICE [39] by Synopsys, PSPICE [40] by Cadence (circuit simulators), NanoSim [41] by Synopsys (switch-level power analyzer), and SiliconSmart [42] by Magma Design Automation (transistor level modeler). Examples of transistor-level power optimization tools are AMPS [43] and RailMill [44] by Synopsys.

10.8.2 Gate-Level or Logic-Level Tools

In order to improve the speed and the capacity of transistor level tools, logic or gate-level EDA tools have been introduced, which are more compatible with application-specific integrated circuits (ASIC) design flows than transistor-level tools. However, gate-level tools are still limited in capacity and exhibit the disadvantage that they can be applied to a completed design (synthesized and simulated) before meaningful power results can be obtained.

Gate-level power analysis tools compute power consumption per logic element (logic gates, flip-flops, multiplexers, etc.) based on the computation of the nodal activities obtained by logical simulations. Their input is the structural netlist (HDL code) of the design, the power models of the logic elements, as well as the activity information for each logic element produced by logic simulation. The accuracy of gate-level power analysis tools is less than that of transistor-level tools.

Gate-level power optimization tools have been included in logic synthesis tools in order to improve power consumption at the same time as timing and area during the process logic synthesis. These tools search for power-saving opportunities and implement changes, such as clock gating, unit isolation, logic restructuring, path balancing, state encoding, retiming, dual-threshold voltage or dual-supply voltage cell swapping, in order to reduce dynamic as well as leakage power consumption. Reduction of power consumption up to 25 percent can be achieved without violating timing constraints.

An example of gate-level analysis tool is PrimePower [45] by Synopsys, while commercially available power optimization tools at this level are PowerCompiler [46] by Synopsys, and Low Power Solution [47] by Cadence.

10.8.3 Register Transfer-Level Tools

The register transfer (RT) or architectural level is the level of the design hierarchy at which the majority of the system's functional design is performed. Because of that, power analysis and optimization at this level becomes very important. RT-level, power-oriented tools are primarily used as design tools in contrast to gate- and transistor-level tools that are used mainly as verification tools because they can be applied when most of the creative part of the design process is completed. However, in custom circuit design (digital, analog, or mixed-signal), designers should use transistor-level tools for circuit design and characterization.

Architectural-level power estimation tools are less accurate compared to the other two levels defined earlier; however, they help designers make decisions at early stages of the design cycle and provide faster run time. In terms of speed, RT-level tools are about an order of magnitude faster than gate-level tools, which, in turn, are about an order of magnitude faster than transistor-level tools, while in terms of accuracy, RT-level power estimates are 20 percent of actual measurements.

Architectural power analysis tools estimate the power consumption of a design described at the RT level in a hardware description language. The estimation is performed prior to synthesis and the results are linked to the RT-level code to indicate the power contribution of the design portions. The input of power analysis tools at this level is a RT-level code, on which an inferencing procedure is applied that converts the RT-level code into a netlist of instances, such as adders, registers, decoders, and memories. After that, high-level power models are used to estimate the power consumption on an instance-specific basis. Elements not yet present in the design, such as clock distribution and wiring capacitances, are estimated according to specifications produced by subsequent design steps.

Architectural power optimization tools use as input the RT-level description and produce a power-optimized RT-level description. Optimizations, such as clock gating, data path reordering, pipelining, memory restructuring, functional blocks isolation and reduction, data path precomputation, and detection of idle conditions in functional blocks, are implemented.

An example of an RT-level analysis tool is PowerArtist by Apache Design Solutions [48], while commercially available power optimization tools at this level are PowerArtist, PowerCompiler [46] by Synopsys, and Talus PowerPro [49] by Magma Design Automation.

10.8.4 Behavioral-Level Tools and Power Emulation

Behavioral-level power analysis tools are used as input designs in a behavioral hardware description language or in high-level programming languages, such as C, C++, and SystemC. Their distinguishing feature is that the system designer can perform a power analysis at the very beginning of the design process. In addition, since power reduction opportunities are larger at the higher abstraction levels, the ability to evaluate power optimization trade-offs at this level can be very effective. In system level power analysis tools, a mapping between language constructs and hardware objects must be made in order to enable a power estimate. This is achieved by analyzing various combinations of scheduling, allocation, and binding, and then producing power consumption results, along with performance and area estimates.

Behavior-level power optimization is based on the comparison between different versions of the target design, which are analyzed for power in order to select the optimal one. Each of the design versions is mapped onto precharacterized (in terms of power consumption) objects. Optimizations at this level include rescheduling control and data flow, reducing the number of memory accesses, minimizing overall memory storage requirements, use of different data encodings, power-oriented memory partitioning, and hardware–software mapping.

PowerOpt [50] by ChipVision is a commercially available power analysis tool at the behavioral level, while Atomium [51] developed by IMEC is an example of a behavioral level, power optimization tool.

The increase of circuit sizes and test-bench complexities is straining the capabilities of power estimation tools. Power emulation [52] exploits hardware acceleration to drastically speed up power estimation. The adoption of power emulation is based on the observation that power estimation and analysis is typically performed by evaluating power models for different circuit components, based on the input values seen at each component during circuit simulation, and aggregating the power consumption of individual components to compute the design's power consumption. The functions performed during power estimation and analysis (power model evaluation, aggregation, etc.) can be implemented as hardware components. Therefore, any given design can be enhanced with "power estimation hardware," and mapped onto a prototyping platform (i.e., FPGA-based platform) in order to exercise it with any given test stimuli for obtaining power consumption estimates and supporting power optimization actions.

10.9 Conclusions

In this chapter, the particularities of the low-power design in the digital and the analog domain have been analyzed in order for the reader to gain a deeper knowledge of the effects that specific design choices have on the power performance of circuits in the analog and the digital world. In addition, new design techniques have been discussed that bridge analog and digital world controversies toward a successful, mixed-signal, ultralow-power design, suitable to be used in energy harvesting applications. Finally, the power-oriented EDA tools that are available in the market for the design of low-power analog and digital circuits have been reviewed in order to give the reader a guide to the available means for low-power analog and digital circuits design.

REFERENCES

[1] J. Kahn, R. Katz, and K. Pister, "Next century challenges: Mobile networking for smart dust," in *Proc. of the ACM International Conference on Mobile Computing and Networking (MobiCom)*, August 1999, pp. 271–278.

[2] M. Hempstead, N. Tripathi, P. Mauro, G. Wei, and D. Brooks, "An ultra-low power system architecture for sensor network applications," in *Proc. of the International Symposium on Computer Architecture (ISCA)*, June 2005, pp. 208–219.

[3] S. Priya, and D. J. Inman (Eds.), *Energy Harvesting Technologies*, New York: Springer, 2009.

[4] L. Schwiebert, S. Gupta, and J. Weinmann, "Research challenges in wireless networks of biomedical sensors," in *Proc. of the ACM International Conference on Mobile Computing and Networking (MobiCom)*, July 2001, pp. 151–165.

[5] R. Weinstein, "RFID: A technical overview and its application to the enterprise," *IEEE IT Professional*, vol. 7, no. 3, pp. 27–33, May–June 2005.

[6] L. Bisdounis and O. Koufopavlou, "Short-circuit energy dissipation modeling for submicrometer CMOS gates," *IEEE Trans. Circuits & Systems I: Fundamental Theory and Applications*, vol. 47, no. 9, pp. 1350–1361, September 2000.

[7] J. D. Carothers and R. Radjassamy, "Low-power VLSI design techniques—the current state," *Integrated Computer-Aided Engineering, IOS Press*, vol. 5, no. 2, pp. 153–175, April 1998.

[8] J. Rabaey and M. Pedram (eds.), *Low Power Design Methodologies*, Boston: Kluwer Academic Publishers, 1996.

[9] D. Liu and C. Svensson, "Trading speed for low power by choice of supply and threshold voltages," *IEEE Journal of Solid State Circuits*, vol. 28, no. 1, pp. 10–18, January 1993.

[10] A. P. Chandrakasan and R. W. Brodersen, *Low Power Digital CMOS Design*, Boston: Kluwer Academic Publishers, 1995.

[11] A. P. Chandrakasan and R. W. Brodersen, "Minimizing power consumption in digital CMOS circuits," *Proceedings of the IEEE*, vol. 83, no. 4, pp. 498–523, April 1995.

[12] S. Wolf, *The Submicron MOSFET*, Sunset Beach, CA: Lattice Press, 1995.

[13] M. B. Srivastava, A. P. Chandrakasan, and R. W. Brodersen, "Predictive system shutdown and other architectural techniques," *IEEE Trans. Very Large Scale Integration (VLSI) Systems*, vol. 4, no. 1, pp. 42–55, March 1996.

[14] E. A. Vittoz, "Low-power design: Ways to approach the limits," in *Proc. of the IEEE International Symposium on Circuits and Systems (ISCAS)*, June 1994, pp. 14–18.

[15] C. C. Enz and E. A. Vittoz, "CMOS low-power analog circuit design," in Designing Low Power Digital Systems, Emerging Technologies Tutorial, *Proc. of the IEEE International Symposium on Circuits and Systems (ISCAS)*, May 1996, pp. 79–133.

[16] M. Declercq and M. Degrauwe, "Low-power/low-voltage IC design: An overview," *Advanced Engineering Course on Low-Power/Low-Voltage IC Design*, Lausanne, Switzerland: École Polytechnique Fédérale de Lausanne (EPFL), June 1994.

[17] A. P. Chandrakasan, S. Sheng, and R. W. Brodersen, "Low-power CMOS digital design," *IEEE Journal of Solid-State Circuits*, vol. 27, no. 4, pp. 473–484, April 1992.

[18] T. G. Noll and E. de Man, "Pushing the performance limits due to power dissipation of future ULSI chips," in *International Solid-State Circuits Conference (ISSCC) Digest of Technical Papers*, February 1992, pp. 1652–1655.

[19] R. Castello, and P. R. Gray, "Performance limitation in switched-capacitor filters," *IEEE Trans. Circuits Systems*, vol. CAS-32, no. 9, pp. 865–876, September 1985.

[20] C. Toumazou, G. Moschytz, and B. Gilbert (eds.), *Tradeoffs in Analog Circuits Design, A Designer's Companion*, Boston: Kluwer Academic Publishers, 2003.

[21] Y. Tsividis, "Companding in signal processing," *IEE Electronics Letters*, vol. 26, no. 17, pp. 1331–1332, August 1990.

[22] Y. Tsividis, "Externally linear, time-invariant systems and their application to companding signal processors," *IEEE Trans. on Circuits & Systems II: Analog and Digital Signal Processing*, vol. 44, no. 2, pp. 65–85, February 1997.

[23] E. Seevinck, "Companding current-mode integrator: A new circuit principle for continuous-time monolithic filters," *IEE Electronics Letters*, vol. 26, no. 24, pp. 2046–2047, November 1990.
[24] R. L. Geiger, P. E. Allen, and N. R. Strader, *VLSI Design Techniques for Analog and Digital Circuits*, New York: McGraw-Hill, 1990.
[25] A. Wang, B. H. Calhoun, and A. P. Chandrakasan, *Sub-threshold Design for Ultra Low-Power Systems*, New York: Springer, 2006.
[26] A. Tajalli and Y. Leblebici, *Extreme Low-Power Mixed Signal IC Design*, New York: Springer, 2010.
[27] J. Rabaey, *Low Power Design Essentials*, New York: Springer, 2009.
[28] D. Soudris, C. Piguet, and C. Goutis (eds.), *Designing CMOS Circuits for Low Power*, Dordrecht, The Netherlands: Kluwer Academic Publishers, 2002.
[29] M. Santarini, "Taking a bite out of power: Techniques for low-power ASIC design," *Electronic Design, Strategy & News (EDN) Magazine*, vol. 10, no. 11, May 2007.
[30] L. Benini, and G. De Micheli, "System-level power optimization: Techniques and tools," *ACM Trans. Design Automation of Electronic Systems*, vol. 5, no. 2, pp. 115–192, April 2000.
[31] E. Macii (ed.), *Ultra Low-Power Electronics and Design*, Dordrecht, The Netherlands: Kluwer Academic Publishers, 2004.
[32] C. Piguet (ed.), *Low-Power CMOS Circuits: Technology, Logic Design and CAD Tools*, Boca Raton, FL: CRC Press, 2006.
[33] M. Pedram and J. Rabaey (eds.), *Power Aware Design Methodologies*, Dordrecht, The Netherlands: Kluwer Academic Publishers, 2002.
[34] D. Chinnery and K. Keutzer, *Closing the Power Gap between ASIC & Custom: Tools and Techniques for Low Power Design*, New York: Springer, 2007.
[35] S. Henzler, *Power Management of Digital Circuits in Deep Sub-Micron CMOS Technologies*, New York: Springer, 2007.
[36] C. Toumazou, and C. A. Makris, "Analog IC design automation, Part I—Automated circuit generation: New concepts and methods," *IEEE Trans. Computer-Aided Design of Integrated Circuits and Systems*, vol. 14, no. 2, pp. 218–238, February 1995.
[37] J. Lee, and Y. B. Kim, "ASLIC: A low power CMOS analog circuit design automation," *Integration, The VLSI Journal, Elsevier*, vol. 39, no. 3, pp. 157–181, June 2006.
[38] B. Geden, "Taking power analysis to the transistor level for a full chip," *Electronic Design, Strategy & News (EDN) Magazine* (guest opinion), December 2009.
[39] Synopsys Inc., *HSPICE Simulation and Analysis, User Guide*, Mountain View, CA, 2007.
[40] Cadence Design Systems Inc., *Cadence PSPICE A/D & PSPICE Advanced Analysis, Technical Brief*, San Jose, CA, 2010.
[41] Synopsys Inc., *NanoSim Datasheet*, Mountain View, CA, 2001.
[42] Magma Design Automation Inc., *SiliconSmart Datasheet*, San Jose, CA, 2009.
[43] Synopsys Inc., *AMPS Datasheet*, Mountain View, CA, 1999.
[44] Synopsys Inc., *RailMill Datasheet*, Mountain View, CA, 2000.
[45] Synopsys Inc., *PrimePower Datasheet*, Mountain View, CA, 2002.
[46] Synopsys Inc., *PowerCompiler Datasheet*, Mountain View, CA, 2007.
[47] Cadence Design Systems Inc., *Building energy efficient ICs from the ground up*, White Paper, San Jose, CA, 2009.
[48] Apache Design Solutions Inc., *RTL Design for Power Methodology*, White Paper, San Jose, CA, 2010.

[49] Magma Design Automation Inc., *Talus PowerPro Datasheet*, San Jose, CA, 2008.
[50] ChipVision Design Systems Inc., *PowerOpt Datasheet*. Available online: http://www.chipvision.com/products/index.php
[51] F. Catthoor, *Unified Low-Power Design Flow for Data-Dominated Multimedia and Telecom Applications*, Boston: Kluwer Academic Publishers, 2000.
[52] J. Coburn, S. Ravi, and A. Raghunathan, "Power emulation: A new paradigm for power estimation," in *Proceedings of Design Automation Conference (DAC)*, June 2005, pp. 700–705.

Chapter 11

Toward Modeling Support for Low-Power and Harvesting Wireless Sensors for Realistic Simulation of Intelligent Energy-Aware Middleware

Philipp M. Glatz, Leander B. Hörmann, Christian Steger, and Reinhold Weiss

Contents

11.1 Introduction

Low-power embedded system architectures with integrated standardized wireless transceivers and versatile CPUs enable previously impossible wireless and mobile devices. The quality of mobile entertainment platforms, sensing system, automation, and control constantly increase with the advent of new architectures. While the field of high performance and low-power embedded architectures has matured and offers many trade-offs today, the field of energy-harvesting wireless sensor networks has evolved from power aware ones.

Today, a number of optimizations, modeling approaches, and tools exist for various wireless embedded devices. However, there are still issues where there is a strong need for deepening the understanding of power-aware and energy harvesting technology. First, there is a lack of battery performance-aware, low-power modeling approaches. Second, there is only little support available for cost-efficient dimensioning of energy harvesting architectures. The reason for still having open issues and very limited support for design space exploration in these fields is the complexity of systems and the time it takes to validate models and assumptions in real world experiments.

This chapter provides an overview of existing devices and solutions for wireless embedded systems with a special emphasis on low-power solutions and energy harvesting opportunities. Modeling and simulation environments, including wireless communication simulation, will be briefly surveyed. A lack of integrated low-power chips, radios, and energy harvesting with suitable tool support will be identified. Discussion will then be narrowed down to state-of-the-art solutions and environments in the area of wireless sensor networks. As a case study, an energy harvesting and battery plug-in for a power profiling simulation and emulation environment will be developed. It then will be shown how the environment can be used for profiling the optimization of an energy storage-aware middleware for wireless sensor networks. Accuracy of power profiling and energy simulation results will be determined by a comparison with accurate energy harvesting and mote (sensor node) hardware measurements.

11.2 Energy Constraints for Mobile and Wireless Embedded Systems

Since the advent of pervasive computing [1] and the vision of smart dust [2], several different scientific fields and technical solutions for various application domains have positioned themselves in our everyday life.

Wireless sensor networks (WSNs) as surveyed in Akyildiz, et al. [3] and Yick, Mukherjee, and Ghosal [4], and mobile ad hoc networks (MANETs), as in Kahn, Katz, and Pister [2] and Giordano [5] have found their way into scientific, technical, and consumer applications. Especially, findings from experiments in related scientific fields pave the way for innovations when the acquired knowledge is used to implement novel optimizations in end-user applications. A striking example of how knowledge from scientific fields may foster innovation was given in a speech by Jeffrey Sharkey at the Google I/O Developer Conference 2009. The major topics of his speech, Coding for Life, can best be depicted by citing parts of its introductory outline:

> The three most important considerations for mobile applications are, in order: battery life, battery life, and battery life.

This reflects the generic nature of energy conservation issues. Unlike any other, it can critically impact mobile and distributed systems in general. The speech also shows how energy conservation schemes that have first been researched for WSNs and MANETs are being implemented for end-user applications after some time. Some of the issues like (1) using low-power modes, (2) estimating energy consumption from processing and transmission power, and (3) conserving energy with opportunistic behavior, or with (4) using event-driven paradigms seem pretty obvious. However, for the application domain under consideration, the developer's paradigm had not been shifted to incorporating these issues up to that time.

A similar situation can be found for the battery performance awareness of WSNs. Although there exist simulation environments that claim to be battery aware, and despite the fact that a number of battery performance models are around that are far better than linear ones, there is a lack of evidence that power profiling results from simulations have been compared to results from actual hardware measurements of battery performance. The situation for modeling energy harvesting and dimensioning cost-efficient hardware is similar. Wan Du, et al. [6] survey different WSN simulation environments. However, there is no implementation available that incorporates voltage dependencies of power profiles, rate discharge, and relaxation effects or temperature dependencies [7]. The modeling errors that come from neglecting these effects and neglecting the impact of energy harvesting efficiency models complicates matters when measuring energy conservation optimization methods [8].

11.2.1 Chapter Outline

The chapter authors make the point that mapping applicable means of optimization from one field or domain to another (e.g., from WSN/MANET research to smartphone implementations) and developments within single fields of research are mainly inhibited if there is something missing at the basis of the theoretical frameworks or tools that are supporting their implementation and simulation. For this reason, the chapter will start with outlining the state-of-the-art of low-power, energy-efficiency, and harvesting support in hardware; its measurement and its simulation techniques; and tool-chain availability of major players in the field. Summing up related work, we will identify a lack of battery modeling support in simulation as well as a lack of tools for harvesting hardware dimensioning in system modeling environments. Especially, seamless integration of simulation and profiling tools at different levels of abstraction is missing. There is no such tool that lets the system architect choose energy harvesting devices (EHDs) with suitable characteristics and, e.g., explore different low drop out (LDO) regulator analog designs at the same design step when designing an energy harvesting system (EHS) at system level for supplying a WSN mote. However, the efficiency of differently designed systems may vary drastically for different kinds of systems running or simulating different kinds of application settings, even if as basic as setting a duty cycle (DC) as evaluated in Hörmann, et al. [9]. As WSNs and MANETs have many similar characteristics, but WSNs tend to be a little more resource constrained, we choose WSN platforms as examples for showcasing novel tools and techniques. These will include battery model evaluation and its integration for WSN simulation as well as EHS and EHD characterization with energy efficiency modeling. Results of these energy storage-focused considerations then will be applied for simulating the appliance of novel optimization techniques for networking middleware design. WSNs provide good examples for constrained devices, especially when it comes to low-power, energy-efficiency, battery performance, and energy harvesting. Section 11.4 will discuss results of power profiling simulation and hardware measurements. The accuracy of these results will be discussed as well. This provides the basis for the two subsequent sections on the implementation and evaluation of battery models for WSNs and embedded systems in general. Finally, at the conclusion of the chapter, an outline of promising future directions will be discussed.

11.3 Industry Tool Chains for Energy Harvesting and Related Work

This section is split in two parts. First, tool chains and hardware platforms of some of the major players in the field of low-power and energy harvesting solutions are briefly surveyed. For these, the completeness of available approaches will be summed up and missing links will be identified. Second, related work will be introduced that provides some background and highlights relating to work for missing parts in the approaches presented in the first subsection. They will serve as a basis for the energy-aware considerations of battery and harvesting modeling and simulation.

11.3.1 Industry Platforms for Modeling, Simulation, and Implementation

The scope of this short outline on industry energy energy-aware tool chains and hardware is limited to technologies that are relevant for WSNs.

11.3.1.1 Texas Instruments

A state-of-the-art WSN platform, namely the TelosB [10], is implemented using a product from the MSP430 family of central processing units (CPUs). Instruction-level simulation and emulation engines are available. Furthermore, for WSNs, simulation tools exist that let one integrate embedded system (ES) platforms like the TelosB in simulations that span different levels of abstractions. TOSSIM [11] is a simulation environment that allows simulation with incorporating channel effects as well. For this simulation environment, there exist extensions like PowerTOSSIM that enable power profiling features. Power profiling is available for different versions of TOSSIM, where one example is shown in Perla, et al. [12]. Other options exist as well. If disregarding channel modeling, Avrora [13] and its power profiling extension in AEON [14] even enable power profiling at the instruction level.

11.3.1.2 Atmel

For the ARM-based Mica2 [15] platform, there are several well-integrated and hybrid simulation approaches available that consider different levels of abstraction. A well-known example is given by ATEMU [16]: the ATmelEMUlator.

For both types of platforms—TI and ARM—and their ES, there is a lack of integrated expressive battery models, although power simulation is a well understood aspect of these platforms. Extension points for possible integration of battery models exist as exemplified by the PAWiS environment [17], but a proper implementation is missing.

There are even existing approaches that try to combine several different techniques. COOJA [18] integrates and interconnects Avrora with TOSSIM and NS-2 to cover instruction-level aspects as well as operating system considerations and network-level issues.

11.3.1.3 Energy Harvesting Modeling and Simulation

With EHS-enhanced WSNs, an even more energy-constrained field of scenarios is being targeted. For both ARM and TI platforms, examples exist that integrate their most well-known motes in an EHS solution. While Heliomote and Prometheus are solely or mainly based on rechargeable batteries, other EHS solutions with ARM and TI platforms exist that are only supplied by energy storage structures made of double-layer capacitors (DLCs). The EHS in Glatz, et al. [19] supplies a Mica2 mote, and the RiverMote [20] platform, with integrated EHS, is implemented using the same processor as the TelosB platform.

Compared to power profiling simulation issues, things get a lot more complicated when it comes to designing systems with energy harvesting in mind. A sign for the need of coping with more complex WSN and harvesting demands can be found in NXP acquiring Jennic or by TI rolling out EHS design kits like the EZ430-RF2500-SEH. Over time, in particular for platforms appearing since 2009 and 2010, more platforms enable versatile trade-offs between low-power states and short high-speed operation. The TI CC430 and ATmega128RFA1 are very well designed for meeting high performance and low power for EHS requirements.

However, only a few tools exist for EHS simulation. Furthermore, the scientific community lacks an integrated approach with the combination of system-level simulation and networking. Approaches, as shown in Mateu and Moll [21], are limited to single hardware instance considerations. Most environments for network-level consideration of harvesting systems take a very abstract point of view with stochastic modeling, as in Seyedi and Sikdar [22]. The few integrated solutions, as in Merrett, et al. [23] and Castalia in De Mil, et al. [24], lack versatile tools for accurate hardware measurements for validating simulation results. Another approach, presented as Tospie2 in Glatz, Steger, and Weiss [25], integrates accurate profiling of power state models (PSMs) and EHS efficiency models (EEMs), but the architecture is a very loosely coupled system.

11.3.2 Battery Models and Applications to Wireless Sensor Networks

When modeling and simulating low-power, ES architecture power consumption, one needs a firm basis on which different power aware optimizations can be compared. For an often neglected issue for achieving accurate WSN power dissipation profiles (namely battery effects modeling), we have outlined related work for battery characterization setups, battery effects modeling, and networking optimizations that are applied to them.

Different types of energy storage devices exist where each type implies special characteristics that have to be considered when tuning a system. However, as existing simulation environments lack support for nonrechargeable batteries, we will limit our concept to these types. Rechargeable batteries and their effects will not be considered, but for extending the scope to battery-free systems, systems with DLCs will be considered.

11.3.2.1 Automated Battery Characterization Setups

The battery performance at a given point in time depends on several factors that are influencing the system, but it is also a system with memory. The different possibilities of temperature history, the discharge rate of the load, the current state-of-charge (SOC), and possible combinations of what these values have been in the past, result in a large number of different contexts that the battery may be in. Therefore, profiling performance characteristics—as shown for alkaline manganese types of Duracell primary cells [26]—may take some time and needs accurate measurements. In general, characterization may be based on SOC measurements that are chemical methods, current integration methods, or voltage measurement methods. However, for finding a profiling methodology that can be integrated on the WSN mote later on, chemical methods and current integration are not practical.

For profiling the SOC, voltage-based measurements are promoted in Pop, et al. [27]. They also discuss applicability to state-of-health (SOH) measurements, which are especially useful for extending ideas to rechargeable systems. Setups for profiling different types of batteries (charge and discharge performance) are presented by Schweighofer, Raab, and Brasseur [28] and by Abu-Sharkh and Doerffel [29]. Schweighofer, Raab, and Brasseur [28] provide simple equivalent circuit diagrams (ECDs) of batteries using internal resistance and two low-passes with different time constants for modeling a battery.

11.3.2.2 Battery Effects Modeling

ECD modeling of battery effects is quite common in the fields of electrical and computer engineering. Distinguishing between theoretical capacity and actual capacity, as in Rao, Vrudhula, and Rakhmatov [30], the actual battery performance in terms of the amount of charge that is extracted mainly depends on discharge rates, temperature conditions, and aging effects. Better actual performance means more extracted charge before the battery cutoff voltage is reached. For primary cells, recovery effects can be utilized for increasing the performance.

Different simple empirical models are given by the Thevenin battery model and linear electrical model with its ECDs given by Salameh, Casacca, and Lynch [30], Peukert's power law explained by Linden [31], and the model of Pedram and Wu [32]. What all of these models have in common are batteries that are modeled as nonlinear systems with memory. Despite the existence of model evaluation experiments with an error of one order of magnitude or below, there are a number of side effects that need to be considered for WSNs that cannot be coped with completely by using these standard models and variants.

First of all, a drain down to the cutoff voltage of batteries often cannot be achieved because of a limited voltage range of components, and the need for converters' capabilities and efficiencies. Depending on the type of simulation integration, motes have to be removed from simulation when the lower threshold first has been hit or they have to be shut off temporarily if proper functionality can be assumed after a relaxation period.

Furthermore, the evaluation of SOC-dependent effects that can be used in simulation environments cannot directly be mapped to real-world deployments. While the SOC models for simulations may utilize a lot of computational power, several difficulties arise at runtime when resources may be very limited. The accuracy of voltage-based SOC determination methods depends on measurement capabilities and ADC (analog-to-digital converter) accuracy. In addition, determining the SOC by applying battery models at runtime can become computationally demanding. In particular, gathering and memorizing exact and long-time series of the load's characteristics increases the complexity of the problem. Glatz, Hörmann, and Weiss [20] depict such measurement's accuracy for an EHS-enhanced WSN platform to be below 10 percent in relevant areas. Adding such a bias to battery model input may drastically limit model performance. Finally, timing measurements and memorizing their results may heavily impact PSMs of applications using sophisticated platforms.

Likewise, for approximation of battery performance alone, EHS energy storage structures usually cannot be modeled with curve-fitting techniques alone. Therefore, EEMs and leakage are used for dynamic and static energy budget calculation. This chapter will mainly focus on battery modeling considerations and static EHS effects.

11.3.2.3 Battery Technology and Networking Optimizations

Knowledge of energy storage characteristics allows applying power management techniques and optimizations. Transmission scheduling by Nuggehalli, Srinivasan, and Rao [33] and Ma and Yang [34] and traffic shaping techniques [35] especially apply. Dealing with energy storage performance modeling issues necessitates complex measurement, analysis, and algorithmic efforts. This fact—and the time it takes to prove novel energy storage technologies work—makes battery modeling and optimizations lag behind current technology capabilities and application requirements. Therefore, even though novel technologies may outperform older technologies, the considerations in this chapter's modeling section are based upon technologies as they can be compared to examples by YEG Components [36] and EPCOS [37] for ultracapacitors, and White and Beeby [38], Belleville, et al. [39], and Chu [40] for batteries and energy harvesting aspects. More recent developments include graphene-based ultracapacitors (Stoller, et al.) [41] and lithium ion batteries with carbon nanofibers (Fan, et al.) [42].

11.4 Power Profiling Accuracy and Conceptual Considerations

The conceptual part of this chapter starts with experimenting with related power profiling approaches. Expressiveness and possible exactness are quantified. Different aspects of power profiling that can still be optimized deserve consideration.

11.4.1 Expressiveness of WSN Power Profiling Based on Power State Models

Mica2 motes, the EHS in Glatz, et al. [19], RiverMote and the measurement setup from Glatz, et al. [8] are used for dimensioning, modeling, and measuring systems.

Figure 11.1 can be used for comparing simulation-based power profiles with results from hardware measurements. Although a modified Avrora in Tospie2 [25] allows the performing of decent profiling of power traces and the main components' power state switches, there are some issues that need to be considered when applying simulation-based approaches.

11.4.1.1 Issues with Simulation Environments

First, there are several aspects that are hard to model exactly when it comes to using the radio. This is mainly due to different output power and complex signal propagation issues and media access control (MAC) interaction among motes. Second, it is a nontrivial task to deduce correction values from these comparisons for the simulation-based approach and vice versa. This is what a developer using Tospie2 can try to do with the PSM database. It holds the power values of all different power states and allows setting a time span for transient effects.

11.4.1.2 Simulation Evaluation Results

Energy, power, and timing results that are shown in Figure 11.1 are summarized in Table 11.1. Radio activation and its power dissipation need careful consideration on nearly all WSN platforms. Targeting low-power or energy neutral operation (ENO)-capable WSNs, we evaluate low-power listening (LPL) because we expect long-term

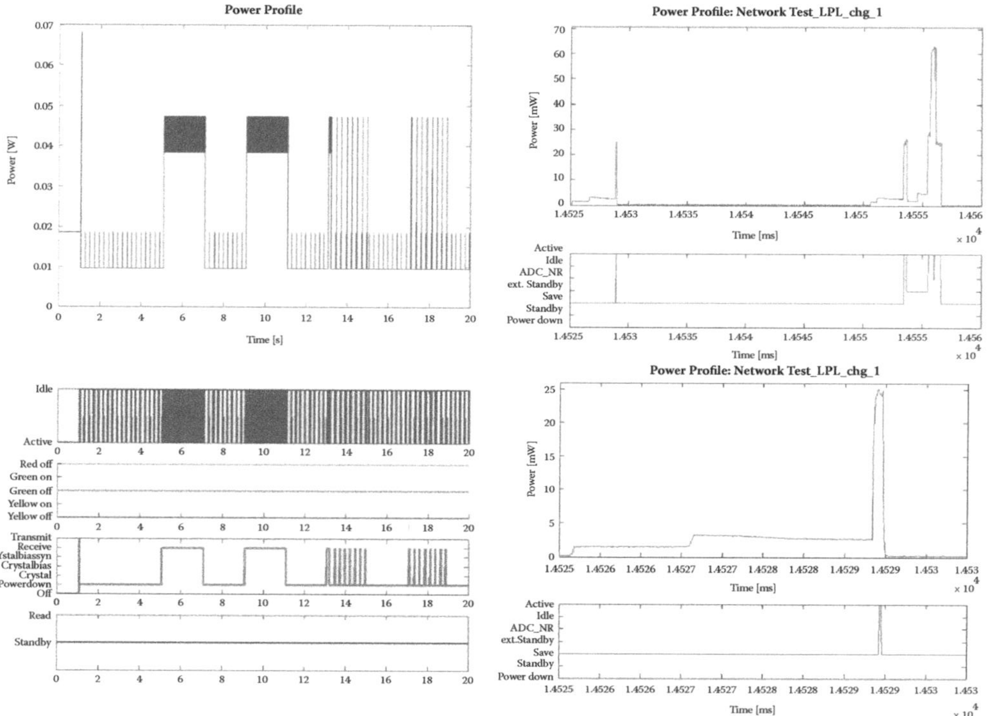

Figure. 11.1a-h. Although simulation-based approaches (Avrora power profile, 11.1(a); power states, 11.1(c); and LPL peak 11.1(e)) can give good estimates, different problems can be identified as far as hardware models and timing accuracy are concerned. Hardware measurements give far more accurate results when it comes to short-time radio activation as can be seen from the power profile and state switches, 11.1(g); the complete radio activation process, 11.1(b); its first peak from an interrupt, 11.1(d); its second peak including full radio activation, 11.1(f); and measuring a full LPL period, 11.1(h), of 255 msec.

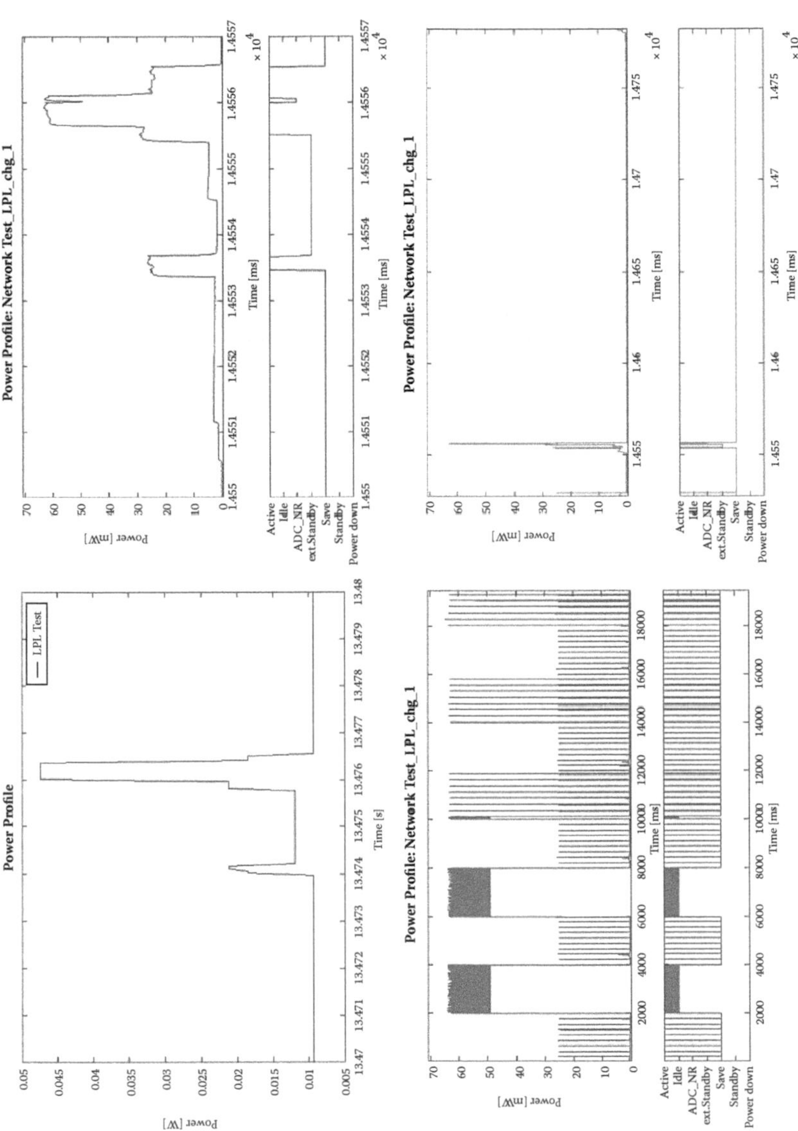

Figure 11.1 (Continued)

Table 11.1 Key results of comparing power profiles from measurement-based approaches with those that have been gathered from simulations using a standard PSM. LPL is used with a sleep interval of a quarter second.

Type of trace	*Avg. Power [mW]*	*Time [ms]*	*Energy[μJ]*
NI 1st peak	2.3322	5.0100	11.680
NI 2nd peak	9.6604	7.0100	67.720
NI both peaks	2.6162	32.0100	83.746
NI full period	0.4678	255.0100	119.286
Sim 1st peak	12.4000	0.1497	1.900
Sim 2nd peak	17.9000	2.9448	52.600
Sim both peaks	10.2000	28.2006	288.100
Sim full period	9.4000	252.1088	2372.400

deployments' applications to be asleep most of the time, unlike communication or multimedia networks, such that LPL plays an important role in their PSMs. The main focus is less on the sending process, but rather on the listening cost. This is also because sending power can be varied on most platforms while listening power dissipation is pretty much fixed most of the time. Summing up, LPL is a significantly important and good PSM part to compare quality of simulation-based, measurement-based, analytical, and hybrid power profiling models with annotations.

The test application wakes up every 4 sec for 2 sec, where LPL is used to check for incoming transmissions. During the first two activation periods, the noise floor needs to be estimated. Therefore, the radio is constantly listening. Exact profiled state switches from simulation are shown in Figure 11.1c and their hardware measurement counterpart in Figure 11.1g accordingly. An erroneous behavior of TinyOS2.1 is corrected. The latest community implementation keeps the serial peripheral interface (SPI) bus active while shutting down the radio for LPL. This keeps the CPU from being put to sleep.

Problems with Hardware Models—This is the first big problem of using simulation-based approaches alone: wrong machine or hardware models. Although both tests run exactly the same (SPI-corrected LPL) executable binary, it is a fact that Avrora cannot correctly profile power dissipation because SPI activity is out of its scope. While this over-estimation of power dissipation can be corrected by subtracting approximately 10 mW power dissipation on average, it makes the power profiling process more error prone. A possible solution can be to use an approach as it is supported by Tospie2: Do design space exploration with simulation-based approaches and validate important results with accurate measurements.

Limited Resolution of Simulated Time—The second issue is related to timing. While hardware measurements need to take into account Shannon's theorem, simulation environments have issues with sampling rates as well. A slightly extended Avrora that is used in Tospie2 takes 1 h time on an otherwise idling dual-core processor to simulate 10 sec of the LPL test at 100 kS to achieve the same sampling rate as with the hardware measurement setup. To get to a more acceptable performance simulation profile in this chapter, use 20 kS, and comparing Figures 11.1e and Figure 11.1f shows how accurate the simulation can get. However, the achievable accuracy is not accurate enough to accurately profile low-power networking optimization measures.

Profiling LPL Radio Activation—Figure 11.1b shows the process of LPL activation. This includes two timer interrupts and radio activation with the states "power down," "crystal activation," "crystal bias activation," "crystal bias synchronization," and "receive," whereas "receive" turns out to be an RSSI (received signal strength indicator) measurement. Table 11.1 tells us that simulation results are far from being accurate enough to trace a timer event's power dissipation. The time span between the two timer interrupts cannot easily be compared either due to the error in the machine model (time span between $t = 14.53$ sec and $t = 14.55$ sec in Figure 11.1b). Therefore, the discussion concentrates on the second timer interrupt and radio activation periods that are shown in Figure 11.1e and Figure 11.1f for simulation and measurement. Their evaluation in Table 11.1 (2nd peak) shows that both timing and power values vary a lot and result in the simulation having an error of more than 22 percent under the estimation.

Average Power Dissipation of LPL—We argue that, in a setting where the LPL alone (with no incoming messages) already makes up more than 70 percent of the overall energy needs, errors of more than 22 percent cannot be acceptable. It is of no use—from a power profiling point of view—to use such an approach for profiling power-optimization capabilities when developing new protocols. Although the simulation environment is an invaluable tool for characterization, other approaches like measurements, analytical measures, or hybrid approaches are needed as well. Further calculations of how much energy is needed will be based on the average power dissipation during a full LPL period if no message is received, which has been measured to be approximately $P_{LPL,period} = 468$ µW.

Average Transceiver Power Dissipation—The energy budget that is needed for the transmission of a single message depends on the preamble length, the time needed for transmission of the message with the payload itself, and the transmission power. For actually deciding whether the message could be received by another party, their antenna models, channel models, and MAC considerations have to be taken into account as well. Therefore, we do not evaluate a power profiling simulation environment's accuracy here because, for virtually any application, one might find a setting that leads to an accurate value of overall energy needs, but only with simulation settings tuned toward a specific application that might completely fail for another one. We want to note that it is out of the scope of this chapter, and remind the reader that not only is there a significant error of simulated power dissipation to be expected, but also that we are pretty uncertain what the error will be. Several environmental conditions, and especially water, human beings, and crops may render an accurate offline error analysis useless when it comes to real deployments. The authors believe that it is worth the time to set up design rules and best-practice guidelines for deployments (especially mote placement, temperature dependencies, and activation and reset policies). Then, still, one might be better off using maximum transmission power and simple slotted networking design with worst-case assumptions on power dissipation for offline analysis instead of power optimization for coverage efficiency and uncertainty and synchronization-based MAC protocols for possibly a little less energy needs. Characteristics of low-power embedded systems for long-term operation impose the need for error-aware and deterministic-driven approaches for their design process. So, these less deterministic approaches—despite their possible savings—are not considered here.

Worst case analysis is based on results from measurement-based power dissipation profiling. Although there is a significant mote-to-mote variation, it is valid to assume the power dissipation to be approximately 65 mW when the mote is sending. The actual average power dissipation during the transmission process also varies if the transmission power is kept constant. For a preamble length as in the LPL experiments above, it is further assumed that a message transmission takes 260 msec until the power state can be left again. Therefore, on average, one transmission will need 16.900 mJ. Motes that are listening to incoming messages have been measured to consume the 22,309th part of the sending process on average, which is approximately 7.575 mJ. Depending on the time offset of when the sender starts sending the preamble and when the receiver samples the RSSI, the energy needed at the receiver may vary from 0.119 mJ over the expected value of 7.575 mJ up to 15.032 when receiving a message. The authors, therefore, instead suggest introducing a random variable and evaluating networks for upper and lower bounds of how much energy may be needed. This may give a feeling for how much can be gained from tuning the networking mechanism and especially motes' synchronization.

Choosing a Suitable Preamble Length—Depending on how dense a network has been deployed, different LPL preamble lengths may be optimal when optimizing with subject to overall network energy conservation. This assumes that an energy-aware or energy management protocol is applied where motes balance their energy or their workload accordingly.

We compare two cases and show trade-offs for a nine-node network with full adjacency matrix in Figure 11.2: Eight motes are uploading their data to a root or cluster head in Figure 11.2a and one mote is downloading information into the network (e.g., for network control) in Figure 11.2b. Plots assume the energy budget of RiverMote's DLCs and further assume that load-balancing or energy management is available, such that the overall network's energy budget can be modeled instead of a single mote's energy. Modeling a single mote's energy budgets can be done the same

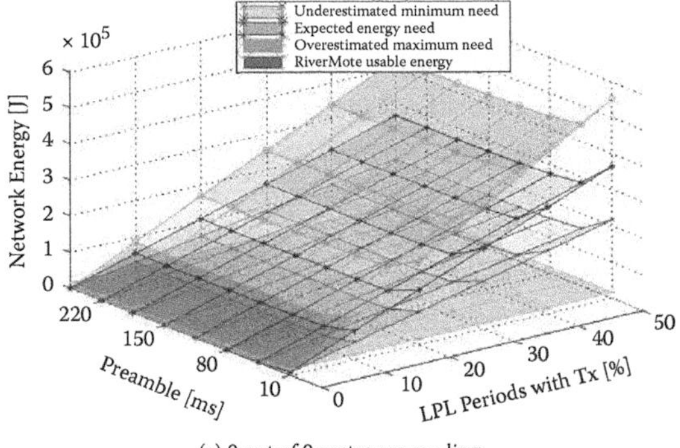

(a) 8 out of 9 motes are sending

One Day Network Energy Need - Different Synchronization and Error

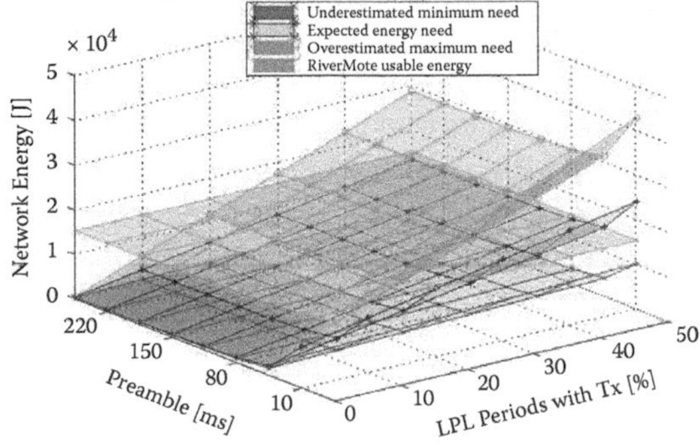

(b) 1 out of 9 nodes is sending messages

One Day Network Energy Need - Different Synchronization and Error

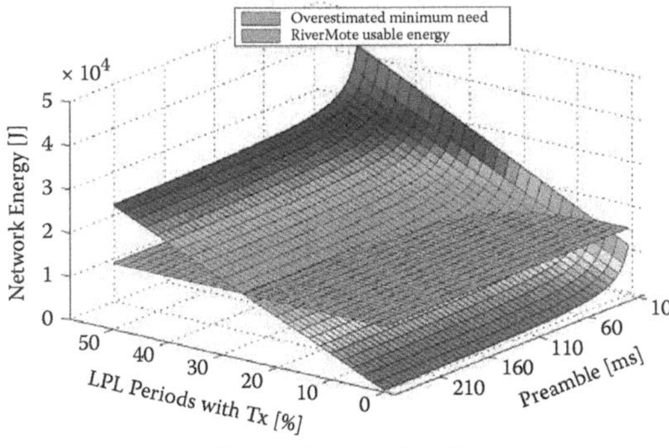

(c) Possible network settings for a given error

Figure 11.2 The subplots analyze the energy budget and needs of a nine-node network where all motes are in communication range of each other. Results are shown for different LPL settings, transmission rates, MAC synchronization, modeling error, and number of senders compared to the available energy budget.

way, but it only provides a little more insight while it would complicate the discussion at hand. The modeling results show that network connectivity or density heavily impact the energy model. Figure 11.2c shows different possible network settings for worst case MAC synchronization and the maximum simulation error profiled in Figure 11.1 in greater detail. We show that choosing an LPL preamble length, which may be constrained by scalability, real-time, or channel capacity constraints, impacts the overall network energy needs if short LPL periods have to be chosen. The network consumes over 50 percent more energy when using small preambles; for example, 10,100 J instead of 2,500 J; 13,310 J instead of 23,940 J; and 28,570 J instead of 42,620 J for the underestimated, expected, and overestimated energy needs for 5 msec and 255 msec LPL periods in Figure 11.2b. The maximum sustainable data rate increases linearly with the available energy; 0 to 50 percent of the LPL periods are used for transmitting data. In case eight out of nine motes are transmitting (Figure 11.2a), the maximum sustainable data rate for a usable energy of 1,686 J is bounded between 2 percent (overestimated) and 5 percent (underestimated).

Energy Needs of CPU, Sensors, and Data Logging—Figure 11.3 shows the process of sampling the photo sensor on a Mica2 mote. It is shown that the time span for reading the sensor with TinyOS2 takes 9 msec, while the actual conversion of the value at the ADC only needs 260 μsec in the end. The ADC is given the long time span to let its value settle down. The maximum value of power dissipation that is to be expected is approximately 48 mW at initialization time when the CPU

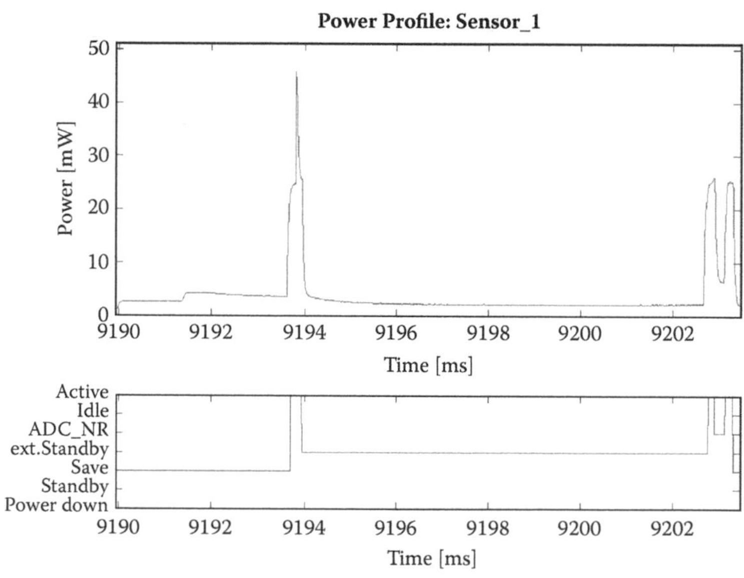

Figure 11.3 Read–ReadDone for the photosensor on the MTS300 sensor board with a Mica2 mote.

is active. It has to be taken into account that the time measurements include the software with hardware abstraction and access times as well.

Figure 11.4 shows power profiles taken with the hardware measurement setup similar to these taken for the radio before in Figure 11.1. What is left out is the plot of the second peak of activation that actually operates the sensor. It is shown separately in Figure 11.3. Finally, Figure 11.5 shows corrected measurements. Summing up the results in Table 11.2, it turns out that sampling a sensor is of similar cost (43 μJ) to sensing the channel (83 μJ) for LPL. We point out that it is pretty hard to model sensor readings' power consumption that way, while it might actually be even harder to give accurate estimates with simulation environments. It is presented here because current literature lacks detailed comparison of exemplary sensors. Other platforms and other sensors might behave in other ways, but the example presented here perfectly shows what it takes to accurately model and quantitatively discuss and evaluate WSN optimization measures. Other exact sensor and actuator profiles for the same platform can be found in Glatz, Steger, and Weiss [43] where they are put in the context of being optimized subject to implementing a low-cost and accurate ranging solution.

Other sensors exist that need significantly more time and power than what is needed for sampling sensors directly connected to the ADC with no special control circuitry. For example, in the datasheet of the accelerometer of the MTS300 sensor board, one can see that the setup time is 16.5 msec. TinyOS puts itself to sleep for 17 msec while the module comes up. Next, there exist modules where it depends on quality of service (QoS) demands to decide how long and which way a component should be used to find an optimal power dissipation to QoS trade-off. Glatz, Steger, and Weiss [43] give an overview of power awareness aspects and how they can be optimized when implementing localization with the MTS300 sounder and microphone.

11.4.2 Impact of Variable Battery Voltage: From Power State Models to Resistance Models

Unfortunately, PSMs alone are not sufficiently expressive for describing WSN motes' energy budgets. While errors from measurements, simulation, and modeling have to be coped with when dealing with PSMs, the concept of PSMs has to be rethought when it comes to different supply voltages [9]. Figure 11.6 shows experiments taken every 30 min. A battery pack with two Duracell PLUS batteries has constantly been drained. Figure 11.6 plots average power, current, and voltage values that are normalized to one over the first experiments' values. Within six hours, the average power dissipation drops more than 8 percent. This trend continues down to voltages where system components cannot be operated any more. Different components of the Mica2 mote—processor, radio, LEDs, sensor—have been profiled for their power dissipation and functionality at different voltage levels. Converting circuits allow operating these components down to battery voltage below their specified voltage range. Figure 11.7 shows results for profiling the same application as in Figure 11.6. This time, a laboratory power supply is used for

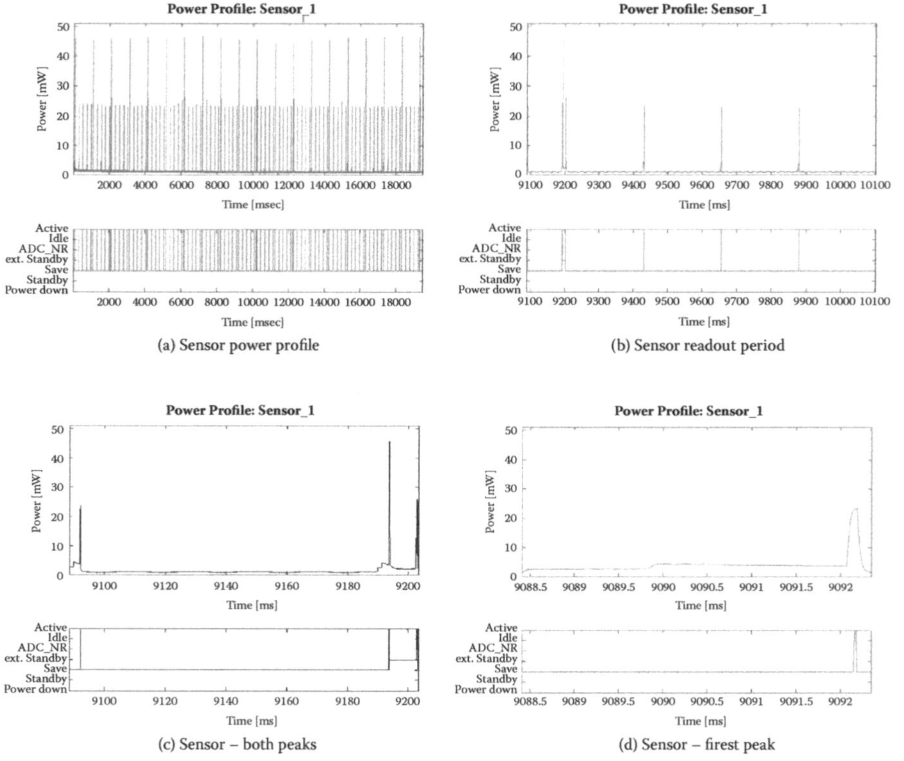

(a) Sensor power profile

(b) Sensor readout period

(c) Sensor – both peaks

(d) Sensor – firest peak

Figure 11.4 The plots in (a) show a full 20 sec, 100 kS measurement with Tospie2 for characterizing the power consumption of reading sensors that are directly attached to the ADC of the processor. It is an example for showing issues when running hardware measurements.

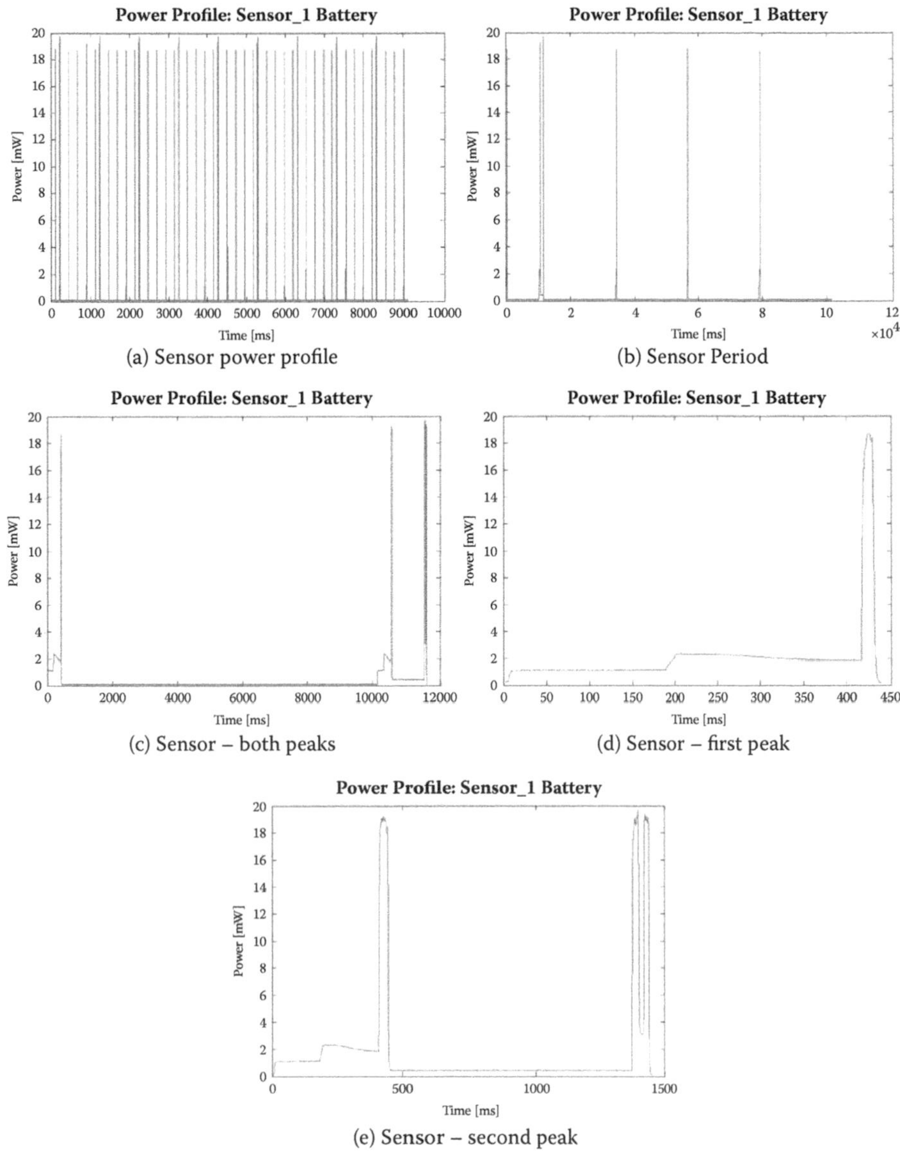

Figure 11.5 Corrected sensor profiles.

supplying the mote. Although no complete tests on component functionality could be performed, radio communication was still working in a small laboratory setup. Finally, it can be concluded that the error in measurement, simulation, and modeling can easily be exceeded by the error of PSMs if their variation at different supply voltages is not taken into account.

Table 11.2 Key results of power profiles from a sensor that is directly connected to an ADC. The sensor is sampled every second

Type of trace	*Avg. Power [mW]*	*Time [ms]*	*Energy[μJ]*
NI 1st peak	40.189	3.9500	15.875
NI 2nd peak	4.0471	13.5200	54.717
NI both peaks	1.4617	115.0500	168.164
NI full period	1.0727	1012.7500	1086.331
Sim 1st peak	2.0930	4.3900	9.1881
Sim 2nd peak	1.7959	26.0555	14.5100
Sim both peaks	0.3701	116.0400	42.9520
Sim full period	0.1362	1013.7100	138.0976

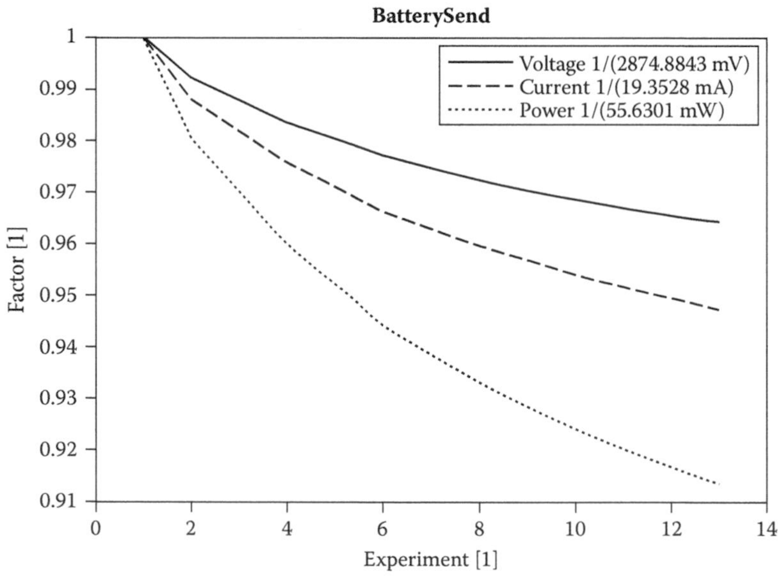

Figure 11.6 The RadioCount2Leds application from TinyOS2 is profiled at different voltages of a battery pack supply.

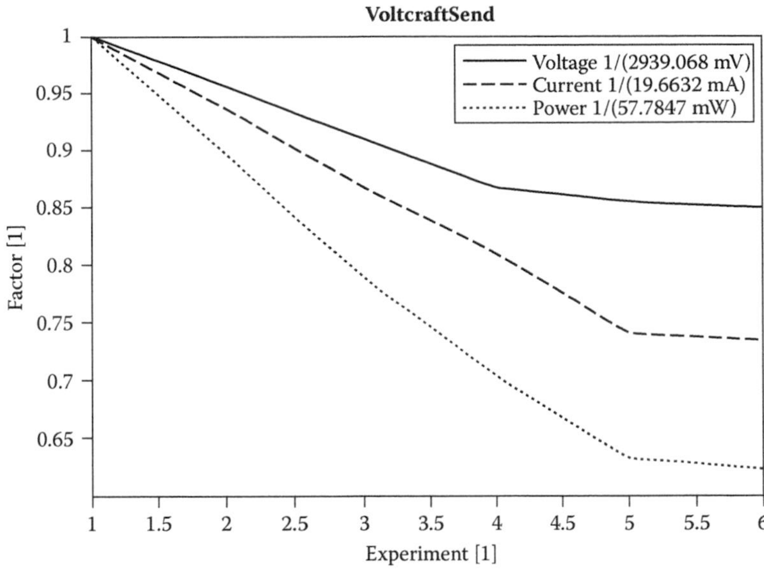

Figure 11.7 The RadioCount2Leds application from TinyOS2 is profiled at different voltage settings of a laboratory power supply.

11.4.3 A Battery Modeling Concept

For implementing battery awareness in a power profiling extension of Avrora in Tospie2, extended standard ECD models have been implemented in Matlab® with SimScape and a large number of hardware measurements have been performed for training these models. Figure 11.8 shows the overall performance of a Duracell PLUS battery pack. A Mica2 mote is used as load. It is running a modified RadioCount2Leds application with LPL and a 50 percent application DC. Figure 11.9 outlines core parts of the main modeling concept. Subcomponents are depicted in Figure 11.10. The idea is to set up SimScape models of different complexities. These implement battery and load models that can be compared with measurements shown in Figure 11.8. The load is to be represented as impedance value and the battery is represented by partly precharged capacitors. Their capacitance values have to be learned with gradient decent methods together with related resistor values to achieve correct time constants. Upper parallel capacitors are precharged while lower parallel capacitors contain no charge at all when the simulation starts, similar to the capacitors that are in series with the main capacitance and mote. The mote implements model parameters that can be profiled from hardware measurements. Parallel, first-order capacitors are mainly responsible for long-term rate discharge and relaxation effects. Second-order capacitors allow modeling voltage overshoots. Series capacitors can be used conveniently to model short-term deviation with different time constants.

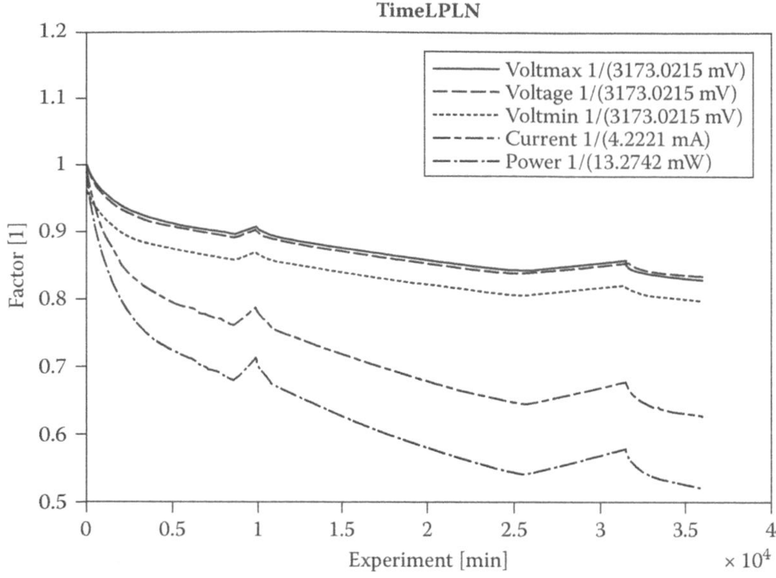

Figure 11.8 The battery performance has been profiled over 25 days with two periods of relaxation in between.

All states of the mote are automatically captured by a profiling system. These include exact measurements of the timing of processor states, LED states, and radio activation. An example is given in Figure 11.11. The ability to track down which component is responsible for a load change is essential for setting up correct impedance models. This also will reveal why PSMs cannot be as expressive and correct as impedance models. The power dissipation trace that is shown in Figure 11.11 is representative of the test applications being used here. The modified TinyOS2 implementation is used with a 50 percent DC. LPL is activated all the time. Half of the time, the LEDs are on and a message is being sent.

11.4.3.1 Impedance Models for Energy Budget Design

Figure 11.12 reveals the direct dependency of impedance values of the processor and LED's power states from the supply voltage. The impedance of the power-intensive sending process stays the same over 25 days.

11.4.4 Harvesting Modeling Concept

EHSs are best described by their efficiencies, their blackout sustainability (BOS) [44] and voltage, current, and energy thresholds that have to be met. A detailed

Battery Simulation Engine
Environment Representation – C-Code
Environment Representation - Mathematica
Environment Representation - SimScape
Battery Model
Shunt
Measurement U/I
Solver Configuration
Load Model (Mote)
Model Design – Laplace Transfrom – Differential Equation Solver – Code Generation
Simulation Trace Generation – Analytic Model Comparison – Executable Validation

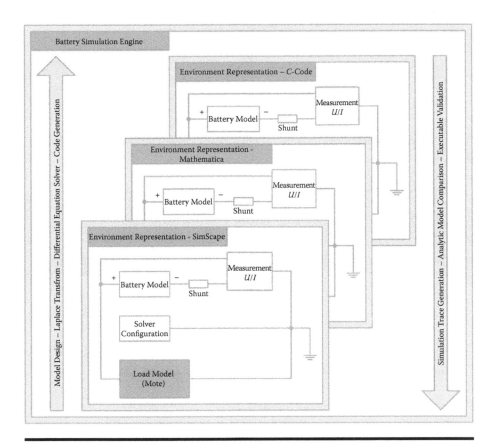

Figure 11.9 Setting up a flow for battery model compilation and testing.

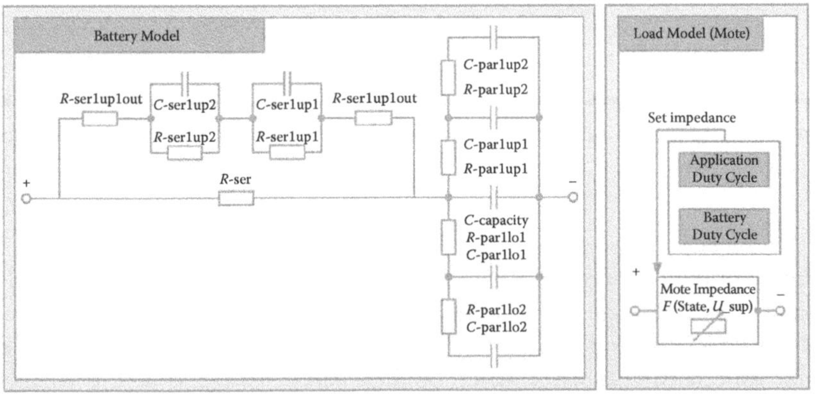

Figure 11.10 Inner parts of the battery model and the load model.

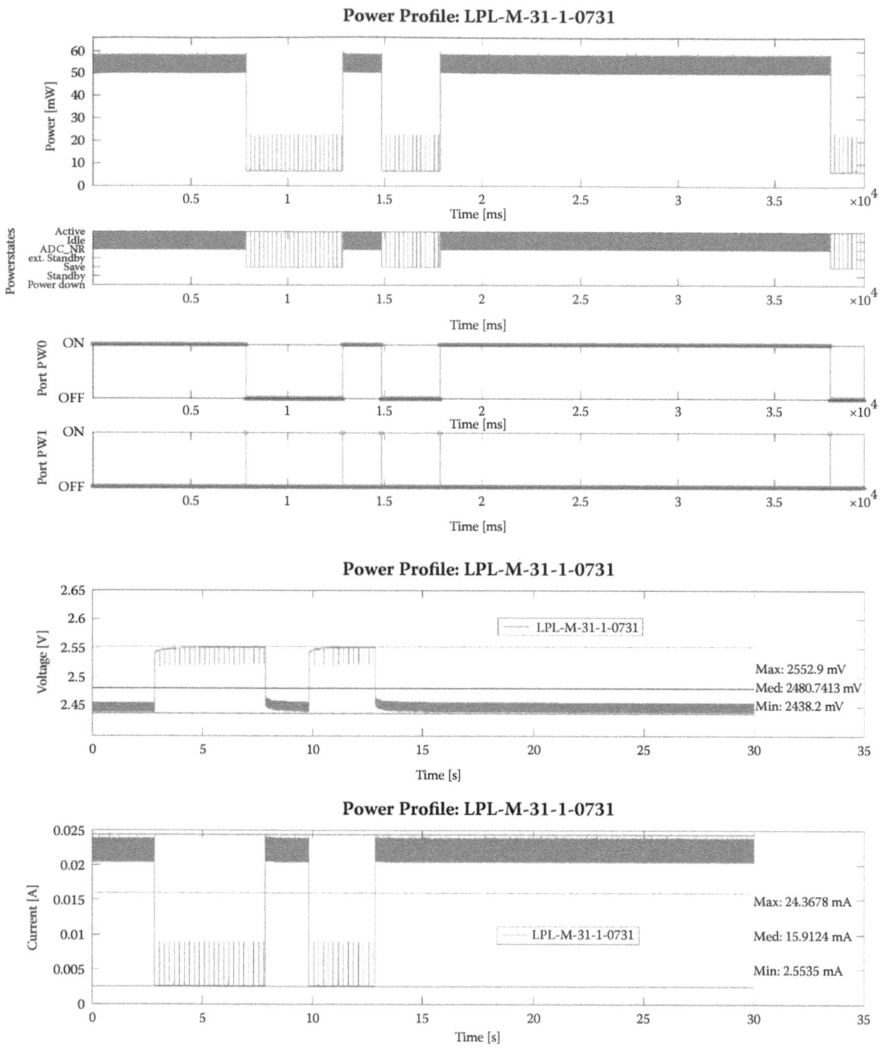

Figure 11.11 Detailed output of the measurement unit for tracking power states and resulting load.

description of the basics of the RiverMote platform that have to be considered for BOS-aware system design can be found in Glatz, et al. [44].

BOS characteristics depend on DLC leakage and the EHS components' efficiencies. Especially, the performance of the energy storage structure balancing mechanism that prevents the system from being physically damaged has to be taken into account. Figure 11.13 shows results of the balancing current evaluation.

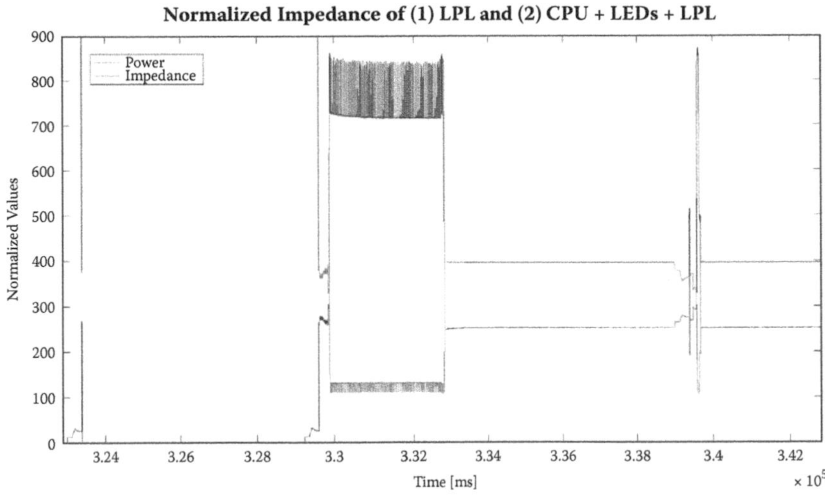

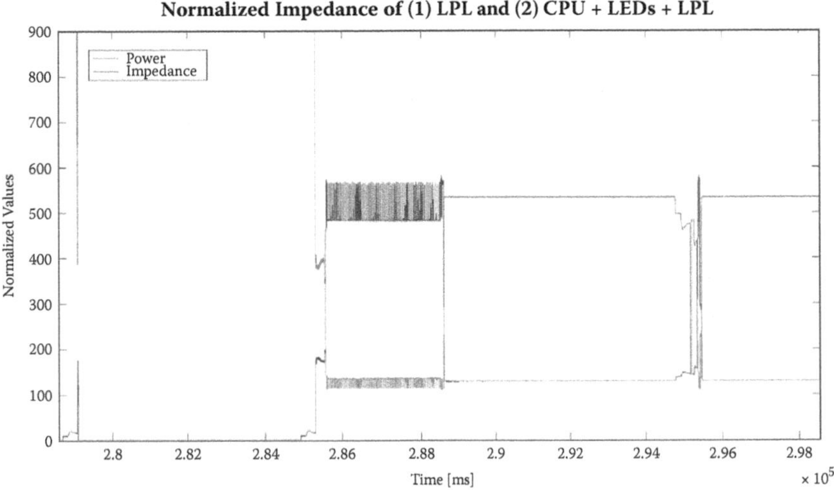

Figure 11.12 The normalized resistance values are shown at the beginning of the 25-day battery performance measurement and at the end. It can clearly be seen that the normalized calculated impedance values of the LEDs and the processor change with the supply voltage, while the according values of the CC1k stays constant.

The balancing current limits the maximum charging current of the system. For a charging process that is starting with empty DLCs, the average charging current should not exceed 200 mA. However, this value is the limit for a complete charging process while assuming a pretty high voltage difference of 0.2 V of the DLCs. This way, the maximum current that may come from the EHDs can be modeled.

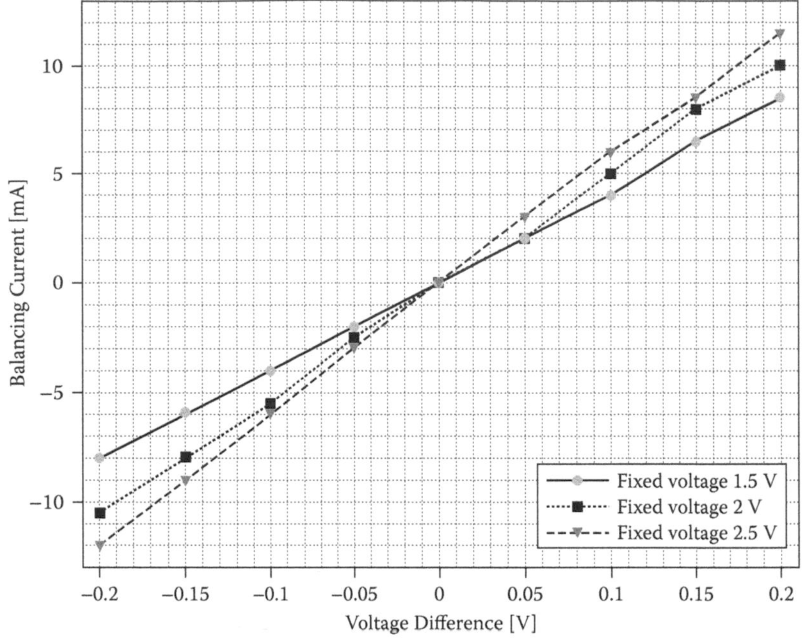

Figure 11.13 Measurement of the balancing current depending on the voltage difference of the DLCs.

Although DLC leakage and the balancing circuitry have carefully been evaluated, results cannot directly be mapped for completely describing RiverMote's overall leakage and BOS. Therefore, the system has been evaluated for its overall leakage and BOS after a charging process with solar cells attached. RiverMote 2 has been charged to 4.40 V by a solar cell and RiverMote 3 has been charged to 3.55 V. For assuming harsh conditions, the system leakage measurements are started immediately after the charging process with no stabilization of the DLCs. Figure 11.14 shows results of the overall EHS performance including DLC leakage, EHS leakage, and circuit efficiency. The EHS performs well with several days of BOS in both cases. At a shutdown voltage, the EHS completely shuts off the drain that comes from the EHS itself.

While Figure 11.14 shows the result of the two tests alone, Figure 11.15 depicts the complete BOS characteristic curve of the platform. It is still based on the measurements with harsh conditions with starting the blackout experiment immediately after the charging process. Results of the experiments have been plotted as a reference as well. It can be seen that the BOS can be up to three weeks for a fully charged system. In case the system might know about an approaching blackout period, the load could be reduced for stabilization of the DLCs and even longer BOS. The local information on how long RiverMote can sustain a blackout can be computed with locally evaluating the characteristic curve generating function. This relaxes the amount of

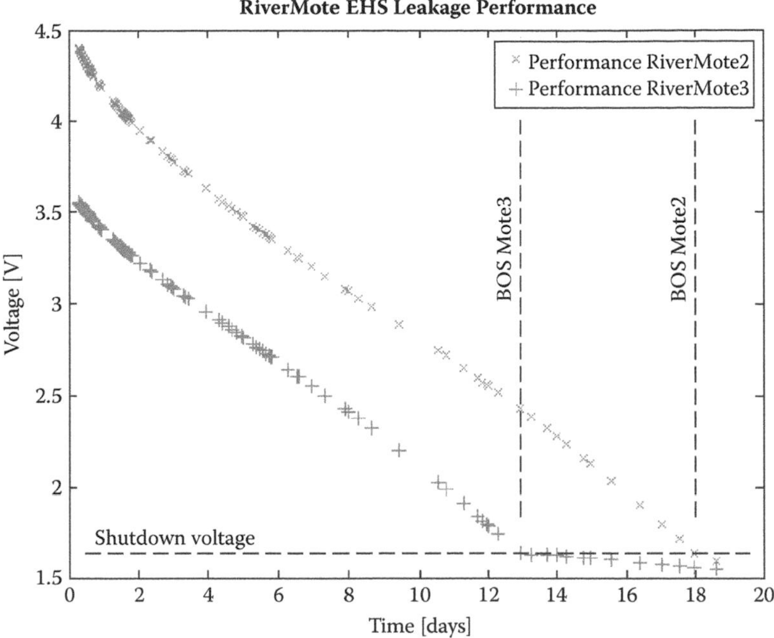

Figure 11.14 System voltage over time during a complete blackout.

energy that is needed for determining the BOS online and possibly sending an error message with a negligible amount of the overall energy budget.

11.5 Implementation

To be able to model partly precharged capacitors in the ECD, we have replaced the parallel capacitors in our model with capacitors, plus a voltage source for making the model applicable to using a Laplace transform.

Figure 11.16 outlines one implementation option that can be feasible for approximating the battery performance traces that have been profiled in experiments. For arriving at analytical expressions of the model in different domains, as has been indicated in Figure 11.9, we need to set up current equations at each of the nodes and voltage equations for each mesh. Equation (11.1) to equation (11.11) show how results can be obtained.

11.5.1 Analytical Derivation of Model Equations

Kirchhoff's current law for node 1:

$$I_{CC}(t) - I_{CAP}(t) - I_{CPL1}(t) - I_{CPU1}(t) = 0 \tag{11.1}$$

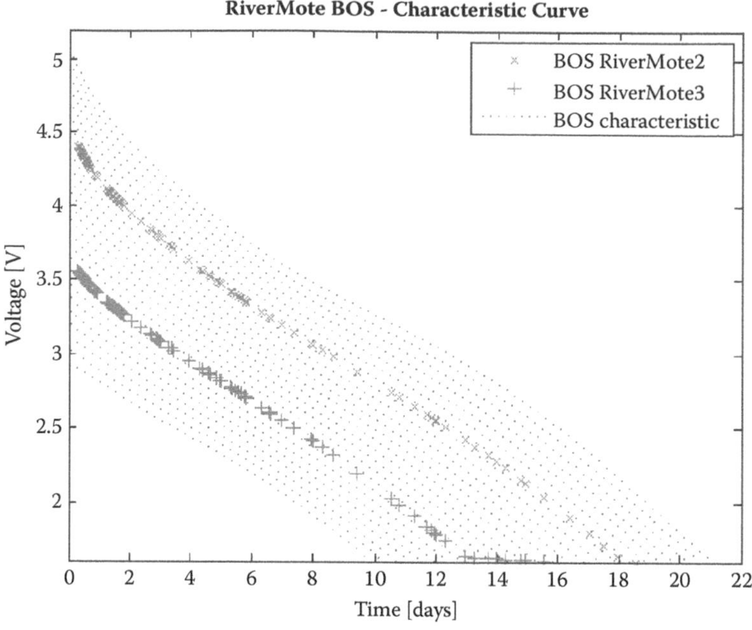

Figure 11.15 The RiverMote platform BOS characteristic.

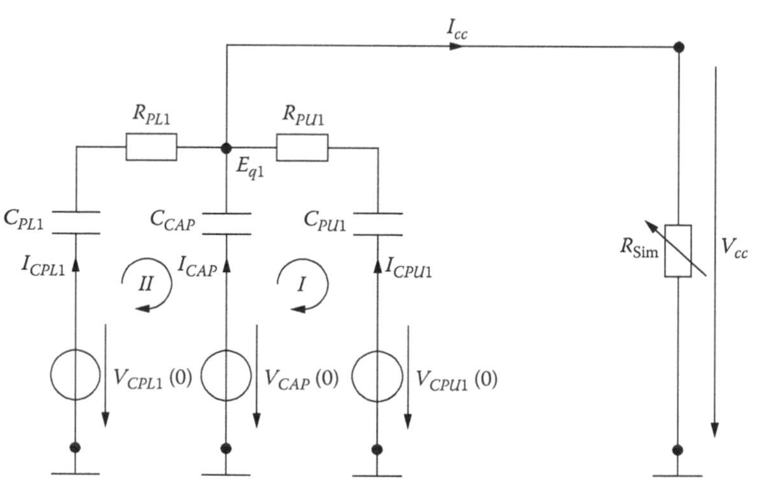

Figure 11.16 Circuit of the battery model.

Kirchhoff's voltage law for mesh I:

$$V_{CPU1}(0)-V_{CAP}(0)+V_{CAP}(t)-V_{RPU1}(t)-V_{CPU1}(t)=0 \tag{11.2}$$

Kirchhoff's voltage law for mesh II:

$$V_{CAP}(0)-V_{CPL1}(0)+V_{CPL1}(t)+V_{RPL1}(t)-V_{CAP}(t)=0 \tag{11.3}$$

These equations can be directly transformed with the help of Laplace transformation and the number of unknowns can be reduced by using the Ohmic law.

Kirchhoff's current law for node 1:

$$I_{CC}(s)-I_{CAP}(s)-I_{CPL1}(s)-I_{CPU1}(s)=0 \tag{11.4}$$

Kirchhoff's voltage law for mesh I:

$$\frac{V_{CPU1}(0)}{s}-\frac{V_{CAP}(0)}{s}+V_{CAP}(s)-I_{CPU1}(s)\cdot R_{PU1}-V_{CPU1}(s)=0 \tag{11.5}$$

Kirchhoff's voltage law for mesh II:

$$\frac{V_{CAP}(0)}{s}-\frac{V_{CPL1}(0)}{s}+V_{CPL1}(s)+I_{CPL1}(s)\cdot R_{PL1}-V_{CAP}(s)=0 \tag{11.6}$$

Now, one can use the Laplace transformed current-voltage relation for the capacitor to simplify the mesh equations.

Kirchhoff's voltage law for mesh I:

$$\frac{V_{CPU1}(0)}{s}-\frac{V_{CAP}(0)}{s}+\frac{I_{CAP}(s)}{C_{CAP}\cdot s}-I_{CPU1}(s)\left(R_{PU1}+\frac{1}{C_{PU1}\cdot s}\right)=0 \tag{11.7}$$

Kirchhoff's voltage law for mesh II:

$$\frac{V_{CAP}(0)}{s}-\frac{V_{CPL1}(0)}{s}+I_{CPL1}(s)\left(R_{PL1}+\frac{1}{C_{PL1}\cdot s}\right)-\frac{I_{CAP}(s)}{C_{CAP}\cdot s}=0 \tag{11.8}$$

With the help of equation (11.4), equation (11.7), equation (11.8), and the Ohmic law, it is possible to derive a solution for V_{CC} that only depends on R_{Sim}.

$$V_{CC}(s)=f(R_{Sim}(s),s) \tag{11.9}$$

The resulting solution for our Laplace model:

$$V_{CC}(s) = \left(R_{Sim}(s) \left(\frac{C_{PL1}V_{CAP}(0)}{C_{CAP}s(C_{PL1}R_{PL1}s+1)\left(\frac{C_{PL1}}{C_{CAP}(C_{PL1}R_{PL1}s+1)} + \frac{C_{PU1}}{C_{CAP}(C_{PU1}R_{PU1}s+1)} + 1\right)} \right.\right.$$

$$+ \frac{C_{PU1}V_{CAP}(0)}{C_{CAP}s(C_{PU1}R_{PU1}s+1)\left(\frac{C_{PL1}}{C_{CAP}(C_{PL1}R_{PL1}s+1)} + \frac{C_{PU1}}{C_{CAP}(C_{PU1}R_{PU1}s+1)} + 1\right)}$$

$$- \frac{C_{PL1}V_{CPL1}(0)}{C_{CAP}s(C_{PL1}R_{PL1}s+1)\left(\frac{C_{PL1}}{C_{CAP}(C_{PL1}R_{PL1}s+1)} + \frac{C_{PU1}}{C_{CAP}(C_{PU1}R_{PU1}s+1)} + 1\right)}$$

$$\left.\left. - \frac{C_{PU1}V_{CPU1}(0)}{C_{CAP}s(C_{PU1}R_{PU1}s+1)\left(\frac{C_{PL1}}{C_{CAP}(C_{PL1}R_{PL1}s+1)} + \frac{C_{PU1}}{C_{CAP}(C_{PU1}R_{PU1}s+1)} + 1\right)} - \frac{V_{CAP}(0)}{s} \right)\right)$$

$$/ \left(- \frac{1}{C_{CAP}s\left(\frac{C_{PL1}}{C_{CAP}(C_{PL1}R_{PL1}s+1)} + \frac{C_{PU1}}{C_{CAP}(C_{PU1}R_{PU1}s+1)}\right)} - R_{Sim}(s) \right) \tag{11.10}$$

With the help of the inverse Laplace transform, V_{CC} can be solved.

$$V_{CC}(t) = L^{-1}(V_{CC}(s)) \tag{11.11}$$

For completely solving the model, V_{CPU1} and V_{CPL1} have to be calculated the same way. For solving systems of differential equations in analytical form, one can use Mathematica, which also allows providing code output. Bash scripts for post-processing of this output leads to C-code that is automatically included by the environment that builds the license-free executable for running the model.

Although, it is a cumbersome task to set up all equations and solve them analytically with Mathematica and then implementing root-solving procedures in C-code, it finally pays off if an arbitrary number of instances can be created license-free to perform gradient descent on the model parameters for optimizing their fit to measured data traces.

11.6 Application of Battery-Aware Simulation to Load Balancing

First of all, training results and ECD settings of the battery model are shown that could best capture the measurement results' characteristics. Battery models that are trained for fitting battery performance with Mica2 loads at given DC periods and

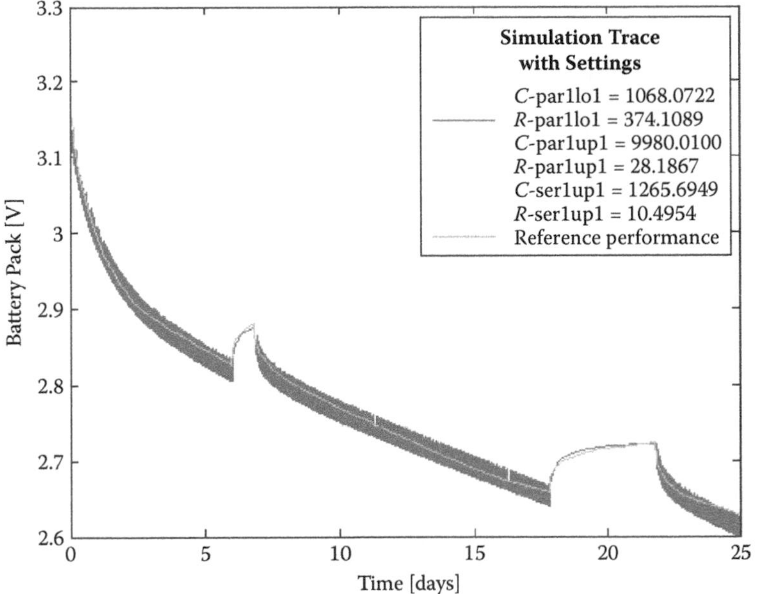

Figure 11.17 The model approximates the measurement trace that is taken as the battery pack's reference performance, which is excellent.

average discharge rates are then used for simulating programs that show the same basic DC and discharge characteristics.

11.6.1 Training a Battery Model with Gradient Descent

The best model error that could be achieved with a model mapped to all stages of the design flow is shown in Figure 11.17 including its model parameter settings. The gradient decent method tries optimizing the RMSE (root mean square error) of the quite peaky simulation curve that contains the application-level DC effects as well as compared to the averaged measurement curve (reference performance).

11.6.2 Evaluation of a Wireless Sensor Network Program

WSN goals usually include maximizing the lifetime by evenly draining the motes' energy reservoirs. A standard way of doing this is to evenly balance the load that comes from the burden to fulfill tasks for achieving a given end-user performance, among the motes in the whole network or in clusters. With battery effects in mind, the question has to be answered if load balancing schemes still work as expected with evenly draining energy reservoirs for maximizing the network lifetime.

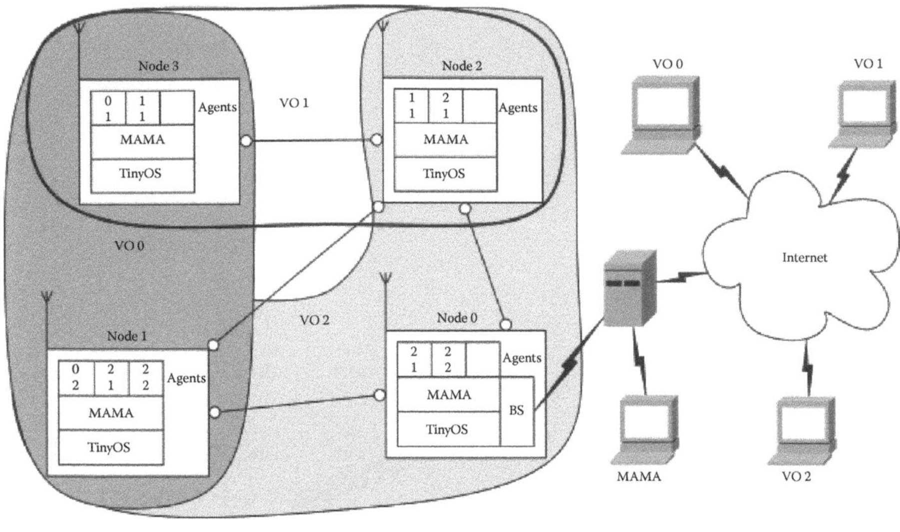

Figure 11.18 Overview of the concept of the MAMA middleware.

An example for the TinyOS implementation of a workload balancing WSN middleware is the agent-based, multiapplication middleware MAMA that is presented in Glatz, et al. [45]. Figure 11.18 and Figure 11.19 show how MAMA is dealing with several applications at the same time by using the concept of virtual organizations (VOs) and where the agents are implemented in the TinyOS protocol stack. It turns out that load balancing and state-of-the-art power profiling simulation alone are not sufficient for achieving the most efficient solution to the question of how the agents shall be grouped and how they shall migrate among motes. A sample evaluation of the ongoing project for energy storage-aware simulation by implementing Avrora extensions and interfaces to the modeling tools described above is performed with a simple load-balancing example where battery effects take place and impact power-aware optimizations.

<table>
<tr><td>Backend on the PC</td><td>Powerconsole with dutycycling</td></tr>
<tr><td rowspan="2">MAMA</td><td>Agents with effective workload balancing</td></tr>
<tr><td>Network - power-aware routing</td></tr>
<tr><td>TinyOS</td><td>MAC-Layer - B-MAC with LPL</td></tr>
</table>

Figure 11.19 Layers of power management in MAMA.

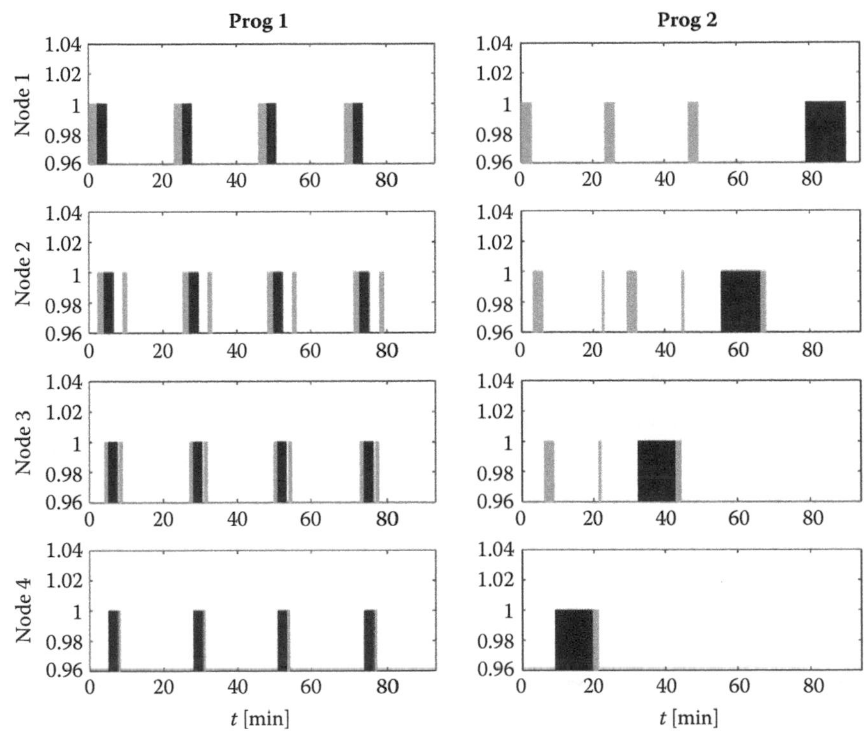

Figure 11.20 Plots show that the more "coarse-grain" resolution of the load-balancing scheme achieves better results on average.

The application scenario assumes a linear-topology (for ease of discussion) WSN that has load balancing applied. Possible application scenarios include applications where a small cluster of motes is responsible for monitoring a given area with a given spatial sampling rate per cluster and sending back its results to a sink. Figure 11.20 shows the burden at each node from radio activation (sending commands and sending back results) with gray bars and sensor and processor activation with black bars. Two different types of load balancing are applied to simplified nesC programs for sample evaluation with the modified Avrora including a battery model.

First (prog 1), each mote splits its work per 100 min. time slot into four smaller packets. Fifty percent energy drain from the radio and 50 percent energy drain from an energy-intensive sensor or algorithm are assumed, which is reasonable for using a GPS sensor as it comes with RiverMote or the Cyclops camera [46] that can be used with WSNs as well. The second type of load balancing (prog 2) has the same load applied on average. The difference in battery performance, that might occur, can come from different time spans of loads and times that are given for battery relaxation. We apply the same balancing among the motes, but this time the load is split into single packets per 100 min. time slot instead of four.

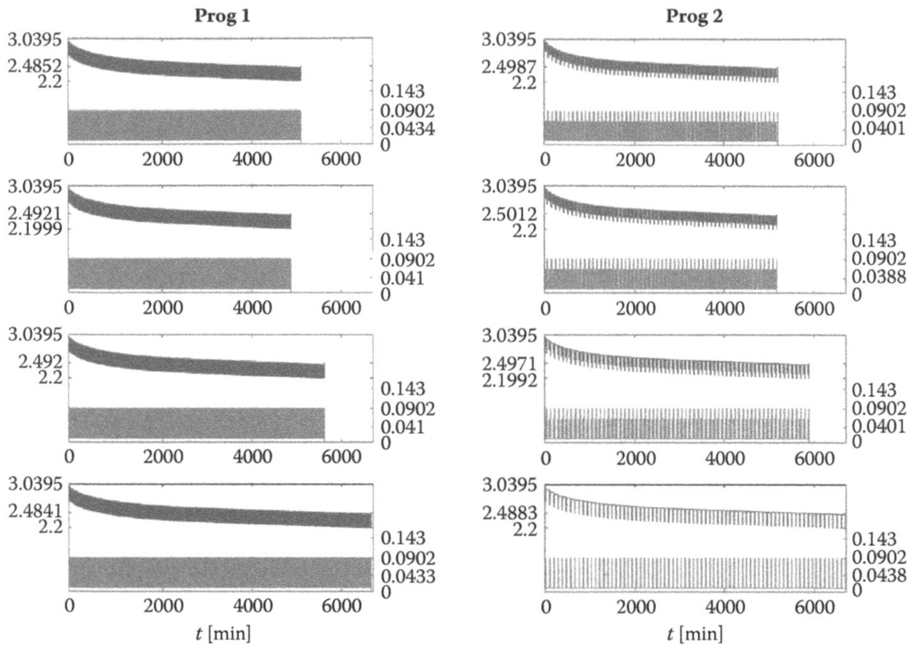

Figure 11.21 For a theoretical four-motes-in-a-line network, the burden at each mote is identified. The burden comes from balancing the activation periods for achieving the needed end-user performance, and it comes from multihop forwarding of results from motes that are farther away from the sink mote.

Results in Figure 11.21 show that the strategy with a single activation per 100 min. time slot achieves better results than splitting it up into a larger number of work load packets. Intuitively interpreting the results, one can say that the load is short enough such that it applies the same close to linear decrease every time the mote is active, but it pays off if more time is given for relaxation effects (Figure 11.21).

11.7 Conclusion and Future Work

This chapter has been written with the intention of shedding light on modeling aspects as well as energy budget simulation and power profiling hardware measurements of low-power and energy harvesting wireless embedded systems. It aims at surveying existing technology and modeling methodologies while providing novel insights, especially into issues related to battery and harvesting modeling.

Several case studies in the field of WSNs have been outlined. The scope of the chapter spans different aspects from different levels of abstraction for power profiling and networking with related aspects. As a main result, BOS modeling of a state-of-the-art EHS–WSN platform is shown and a common instruction-level, power profiling simulation environment is extended with a novel battery effect modeling and simulation environment. A special focus is put on the expressiveness, possible level of detail, and accuracy of LPL and networking optimization at the middleware level.

Future work will include further automation of the measurement setup that is being used for profiling measurements of energy storage characteristics. In addition, a state-of-the-art, high performance and low-power multicore DSP-FPGA board is planned to have its PSM described. The PSM shall then be transformed to an impedance model as has been shown in this chapter for WSNs. This shall allow implementing an adaptive battery modeling environment *in situ* on board a platform that will benefit from it.

REFERENCES

[1] D. Saha and A. Mukherjee, "Pervasive computing: A paradigm for the 21st century," *Computer*, vol. 36, no. 3, pp. 25–31, Mar. 2003.

[2] J. M. Kahn, R. H. Katz, and K. S. J. Pister, "Next century challenges: Mobile networking for "smart dust," in *Proceedings of the 5th annual ACM/IEEE International Conference on Mobile Computing and Networking*, ser. MobiCom '99. New York: ACM, 1999, pp. 271–278. Online at: http://doi.acm.org/10.1145/313451.313558

[3] I. F. Akyildiz, W. Su, Y. Sankarasubramaniam, and E. Cayirci, "Wireless sensor networks: A survey," *Computer Networks*, vol. 38, no. 4, pp. 393–422, 2002.

[4] J. Yick, B. Mukherjee, and D. Ghosal, "Wireless sensor network survey," *Comput. Netw.*, vol. 52, no. 12, pp. 2292–2330, 2008.

[5] S. Giordano, Mobile Ad Hoc Networks, in *Handbook of Wireless Networks and Mobile Computing*. New York: John Wiley & Sons, 2002, chap. 15.

[6] W. Du, D. Navarro, F. Mieyeville, and F. Gaffiot. Towards a taxonomy of simulation tools for wireless sensor networks. In *Proceedings of the 3rd International ICST Conference on Simulation Tools and Techniques (SIMUTools '10)*. ICST (Institute for Computer Sciences, Social-Informatics and Telecommunications Engineering), ICST, Brussels, Belgium, Article 52, 2010. Online at: http://dx.doi.org/10.4108/ICST.SIMUTOOLS2010.8659

[7] R. Rao, S. Vrudhula, and D. Rakhmatov, "Battery modeling for energy aware system design," *Computer*, vol. 36, no. 12, pp. 77–87, 2003.

[8] P. M. Glatz, L. B. Hörmann, C. Steger, and R. Weiss, "A system for accurate characterization of wireless sensor networks with power states and energy harvesting system efficiency," in *IEEE International Workshop on Sensor Networks and Systems for Pervasive Computing*, March 2010, pp. 468–473.

[9] L. B. Hörmann, P. M. Glatz, C. Steger, and R. Weiss, "Energy efficient supply of WSN nodes using component-aware dynamic voltage scaling," in *European Wireless 2011 (EW2011)*, pages 147–154, Vienna, Austria, 2011.

[10] J. Polastre, R. Szewczyk, and D. Culler, "Telos: Enabling ultra-low power wireless research," in *Fourth International Symposium on Information Processing in Sensor Networks (IPSN)*, 2005.

[11] P. Levis, N. Lee, M. Welsh, and D. Culler, "TOSSIM: Accurate and scalable simulation of entire tinyOS applications," in *SenSys '03: Proceedings of the 1st International Conference on Embedded Networked Sensor Systems*. New York: ACM, 2003, pp. 126–137.

[12] E. Perla, A. O. Cath′ain, R. S. Carbajo, M. Huggard, and C. McGoldrick, "PowerTOSSIM z: Realistic energy modelling for wireless sensor network environments," in *PM2HW2N '08: Proceedings of the 3rd ACM Workshop on Performance Monitoring and Measurement of Heterogeneous Wireless and Wired Networks*. New York: ACM, 2008, pp. 35–42.

[13] B. Titzer, D. Lee, and J. Palsberg, "Avrora: Scalable sensor network simulation with precise timing," in *Fourth International Symposium on Information Processing in Sensor Networks, 2005. IPSN* 2005, pp. 477–482.

[14] O. Landsiedel, K. Wehrle, and S. Gotz, "Accurate prediction of power consumption in sensor networks," in *EmNets '05: Proceedings of the 2nd IEEE Workshop on Embedded Networked Sensors*. Washington, D.C.: IEEE Computer Society, 2005, pp. 37–44.

[15] J. Hill and D. Culler, "A wireless embedded sensor architecture for system-level optimization," in Technical report, Berkeley, CA: Computer Science Department, University of California at Berkeley, 2001.

[16] J. Polley, D. Blazakis, J. McGee, D. Rusk, and J. Baras, "ATEMU: A fine-grained sensor network simulator," in *2004 First Annual IEEE Communications Society Conference on Sensor and Ad Hoc Communications and Networks, 2004 (IEEE SECON)*, 2004, pp. 145–152.

[17] S. Mahlknecht, J. Glaser, and T. Herndl, "PAWiS: Towards a power aware system architecture for a soc/sip wireless sensor and actor node implementation," in *Proceedings of 6th IFAC International Conference on Fieldbus Systems and Their Applications*, 2005, pp. 129–134.

[18] F. Osterlind, A. Dunkels, J. Eriksson, N. Finne, and T. Voigt, "Cross-level sensor network simulation with Cooja," in *Proceedings 2006 31st IEEE Conference on Local Computer Networks*, 2006, pp. 641–648.

[19] P. M. Glatz, P. Meyer, A. Janek, T. Trathnigg, C. Steger, and R. Weiss, "A measurement platform for energy harvesting and software characterization in WSNs," in *IFIP/IEEE Wireless Days*, Nov. 2008, pp. 1–5.

[20] P. M. Glatz, L. B. Hörmann, and R. Weiss, "Designing perpetual energy harvesting systems explained with RiverMote: A wireless sensor network platform for river monitoring," *Electronic Journal of Structural Engineering*, Special Issue: *Wireless Sensor Networks and Practical Applications*, pp. 55–65, 2010.

[21] L. Mateu and F. Moll, "System-level simulation of a self-powered sensor with piezoelectric energy harvesting," *International Conference on Sensor Technologies and Applications*, vol. 0, pp. 399–404, 2007.

[22] A. Seyedi and B. Sikdar, "Modeling and analysis of energy harvesting nodes in wireless sensor networks," in *46th Annual Allerton Conference on Communication, Control, and Computing*, 2008, pp. 67–71.

[23] G. Merrett, N. White, N. Harris, and B. Al-Hashimi, "Energy-aware simulation for wireless sensor networks," in *6th Annual IEEE Communications Society Conference on Sensor, Mesh, and Ad Hoc Communications and Networks, 2009. (SECON)*, 2009, pp. 1–8.

[24] P. De Mil, B. Jooris, L. Tytgat, R. Catteeuw, I. Moerman, P. Demeester, and A. Kamerman, "Design and implementation of a generic energy-harvesting framework applied to the evaluation of a large-scale electronic shelf-labeling wireless sensor network," *EURASIP J. Wireless Commun. Netw.*, vol. 2010, pp. 7:1–7:14, February 2010. Online at: http://dx.doi.org/10.1155/2010/343690

[25] P. M. Glatz, C. Steger, and R. Weiss, "Tospie2: Tiny operating system plug-in for energy estimation," in *IPSN '10: Proceedings of the 9th ACM/IEEE International Conference on Information Processing in Sensor Networks*. New York: ACM, 2010, pp. 410–411.

[26] Duracell, "Entire mno2 technical bulletin collection." Online at: http://www1.duracell.com/oem/Pdf/others/ATB-full.pdf

[27] V. Pop, H. Bergveld, P. Notten, and P. Regtien, "State-of-the-art of battery state-of-charge determination," *Measurement Science and Technology*, vol. 16, no. 12, pp. R93–R110, 2005. Online at: http://doc.utwente.nl/62192/

[28] B. Schweighofer, K. Raab, and G. Brasseur, "Modeling of high power automotive batteries by the use of an automated test system," *IEEE Transactions on Instrumentation and Measurement,* vol. 52, no. 4, pp. 1087–1091, 2003.

[29] S. Abu-Sharkh and D. Doerffel, "Rapid test and non-linear model characterisation of solid-state lithium-ion batteries," *Journal of Power Sources*, vol. 130, no. 1–2, pp. 266–274, 2004. Online at: http://www.sciencedirect.com/science/article/B6TH1-4BK2FX3-1/2/4938d1b3771bb71f4989b7010f4d160a

[30] Z. Salameh, M. Casacca, and W. Lynch, "A mathematical model for lead-acid batteries," *IEEE Transactions on Energy Conversion,* vol. 7, no. 1, pp. 93–98, Mar. 1992.

[31] D. Linden, *Handbook of Batteries*, 3rd ed. New York: McGraw-Hill, 2002.

[32] M. Pedram and Q. Wu, "Design considerations for battery-powered electronics," in *Proceedings of the 36th annual ACM/IEEE Design Automation Conference*, ser. DAC '99. New York: ACM, 1999, pp. 861–866. Online at http://doi.acm.org/10.1145/309847.310089

[33] P. Nuggehalli, V. Srinivasan, and R. Rao, "Energy efficient transmission scheduling for delay constrained wireless networks," *IEEE Transactions on Wireless Communications,* vol. 5, no. 3, pp. 531–539, March 2006.

[34] C. Ma and Y. Yang, "Battery-aware routing for streaming data transmissions in wireless sensor networks," *Mob. Netw. Appl.*, vol. 11, pp. 757–767, October 2006. Online at: http://dx.doi.org/10.1007/s11036-006-7800-2

[35] C.-F. Chiasserini and R. Rao, "Improving battery performance by using traffic shaping techniques," *IEEE Journal on Selected Areas in Communications*, vol. 19, no. 7, pp. 1385–1394, July 2001.

[36] YEG Components, "HC power series ultracapacitors datasheet," Maxwell Technologies, San Diego, CA.

[37] EPCOS, "Ultracap single cell 5 F/2.3V B49100A1503Q000 data sheet." Online at: http://www.epcos.com/inf/20/35/ds/B49100A1503Q000.pdf

[38] N. White and S. Beeby, "Energy Harvesting for Autonomous Systems," Boston: Artech House, June 2010.

[39] M. Belleville, E. Cantatore, H. Fanet, P. Fiorini, P. Nicole, M. Pelgrom, C. Piguet, R. Hahn, C. V. Hoof, R. Vullersand, and M. Tartagni, "Energy autonomous systems: Future trends in devices, technology, and systems," Paris: Cluster for Application and Technology Research in Europe on Nanoelectronics (CATRENE), 2009.

[40] B. Chu, *Selecting the Right Battery System for Cost-Sensitive Portable Applications While Maintaining Excellent Quality*, Chandler, AZ: Microchip Technology Inc.

[41] M. D. Stoller, S. Park, Y. Zhu, J. An, and R. S. Ruoff, "Graphene-Based Ultracapacitors," *Nano Letters*, vol. 8, no. 10, pp. 3498–3502, 2008.

[42] Z.-J. Fan, J. Yan, T. Wei, G.-Q. Ning, L.-J. Zhi, J.-C. Liu, D.-X. Cao, G.-L. Wang, and F. Wei, "Nanographene-Constructed Carbon Nanofibers Grown on Graphene Sheets by Chemical Vapor Deposition: High-Performance Anode Materials for Lithium Ion Batteries," *ACS Nano*, vol. 5, no. 4, pp. 2787–2794, 2011.

[43] P. M. Glatz, C. Steger, and R. Weiss, "Design, simulation and measurement of an accurate wireless sensor network localization system," in *The 5th ACM International Workshop on Performance Monitoring, Measurement and Evaluation of Heterogeneous Wireless and Wired Networks (PM2HW2N 2010)*, Bodrum, Turkey, October 2010.

[44] P. M. Glatz, L. B. Hörmann, C. Steger, and R. Weiss, "Designing sustainable wireless sensor networks with efficient energy harvesting systems," in *IEEE WCNC 2011–Service and Application (IEEE WCNC 2011)*, Cancun, Mexico, March 2011.

[45] P. M. Glatz, L. B. Hörmann, C. Steger, and R. Weiss, "MAMA: Multi-application middleware for efficient wireless sensor networks," in *2011 18th International Conference on Telecommunications (ICT 2011)*, pp. 1–8, Ayia Napa, Cyprus, May 2011.

[46] M. Rahimi, R. Baer, O. I. Iroezi, J. C. Garcia, J. Warrior, D. Estrin, and M. Srivastava, "Cyclops: In situ image sensing and interpretation in wireless sensor networks," in *Proceedings of the 3rd International Conference on Embedded Networked Sensor Systems*, SenSys '05. New York: ACM, 2005, pp. 192–204. Online at: http://doi.acm.org/10.1145/1098918.1098939

Chapter 12

Energy Consumption Profile for Energy Harvested WSNs

T. V. Prabhakar, R. Venkatesha Prasad,
H. S. Jamadagni, and Ignas Niemegeers

Contents

12.1 Introduction

Miniaturization and improvements in the communication and computation capabilities of devices is bringing about steep growth in wireless communication devices. The exploding number of devices comes with the immediate task of supplying them with energy. Thus, energy harvesting has become an important aspect. Energy that is harvested from sources such as solar panels, vibration, and thermogenerators, provides varying instantaneous power. We focus on the performance of energy-harvesting embedded sensors and study the problems associated with applications that assume constant power harvesting. We begin the chapter by taking a simple environment-sensing application and show its baseline performance under the trivial assumption of constant power availability. We then propose simple mechanisms specific to harvested power-enabled wireless sensor nodes to constantly adjust their operational schedule based on the incoming energy profile. In the initial part, we begin with system-related node operations and further extend this to a network of wireless sensor nodes. Since energy availability can be different across the nodes in the network, even if similar harvesting is used, network setup and collaboration is a nontrivial task. At the same time, in the event of excess harvested energy, the collaboration between nodes is possible, which is exciting, but also challenging. Operations such as sensing, computation, storage, and communication are required to achieve the common goal for any sensor network.

The chapter introduces the example of a "smart application" assisted by a decision engine that transforms itself into an "energy-matched" application. The chapter also explains the architecture of such a decision engine. We provide a few performance results of such applications. The results are based on measurements using Crossbow's IRIS motes [1] as well as custom motes running on solar energy. To accurately measure the energy consumption, we have done away with batteries; instead, we used low-leakage supercapacitors to store harvested energy. The last part of the chapter has a section with results of a "distributed smart application" where network-related information is fed as an additional input to provide networking-related decisions.

12.2 Energy Harvesting

Energy harvesting devices capture small amounts of energy over a long time from sources, such as ambient light, wind, vibration, linear motion, temperature differential, radio frequency (RF) energy, etc. Such harvested energy, when converted to an electric charge, can be stored in storage devices, such as batteries and supercapacitors. Although the modern definition of energy harvesting means conversion from one form of energy to electricity, this is not necessarily true in a broader context. For example, the well-known mechanical watches and wall clocks of the past have generated mechanical energy and stored it in a coiled spring. There was no requirement for conversion into electricity. Moreover, while there are many harvesting sources, not all could be regarded as "energy for free" sources. For instance, energy scavenging from sources, such as waste heat from industrial plants, vibrations from machinery in large industrial manufacturing plants, and temperature differentials in automobiles and aircrafts, are all examples of secondary sources of energy harvesting. One may regard them as sources where energy is made to work twice [2]. Other examples, some quite fanciful, are those gadgets that use human energy conversion to electricity. Energy generated from human action, such as running, walking, cycling, lifting, and pressing is commonly cited. We might have seen many people, especially little kids, wearing shoes with inserts where a light glows for every step walked. The bicycle dynamo, batteryless winding radio, and the hand-cranked flashlight are other examples. Sometimes environmental energy conversion into electricity can happen in two steps. As an example, a wind belt created with a speed of 1 to 2 meters/second can generate a vibration. This vibration can be converted into electricity that will be sufficient to power a wireless sensor node. In our present context, we will examine conversion from one form of energy into electricity. We particularly concentrate on energy requirements to drive embedded communication devices used for environmental monitoring and control. We regard electricity as the purest form of energy and its efficient harvesting from the environment is necessary to provide relief from the ever-growing number of embedded devices used by humans. In a recent survey, it is predicted that by the year 2017, the world's 7 billion population will be strapped with 7 trillion embedded devices [3]. Thus, it is important to find other ways to power a large portion of those devices. Furthermore, in the context of this chapter, we limit our discussions on power generated to be several hundreds of milliwatts (mW), sufficient to drive low-power electronics used for embedded applications. Such embedded applications could be in healthcare, industrial monitoring and control, security, agriculture, structural monitoring, automotive, and infotainment.

12.2.1 Motivations for Energy Harvesting

We have mentioned above about the trillion devices that are likely to be used by 2017 [3]. Let us assume for the time being that this estimate is true. We then will have to address a few issues. Firstly, what are the associated technologies to support such a large number? For instance, each device should have a unique address in a manner that it would be accessible from anywhere at anytime. While the address problem seems to be well geared toward large-scale IPv6 (Internet protocol, v. 6) deployments, the battery technology to support perennial or everlasting operation is unlikely to support all the devices. The well-known and commercially available lithium–manganese dioxide coin cell-2450 (use and discard) used in sensor devices costs a little over $2 apiece and supports a nominal capacity of about 600 mA/h. They have a shelf life of about five years at room temperature. This figure does not have credence since sensor nodes might be deployed in outdoor settings and, thus, subject to large temperature swings. If we assume that 7 billion batteries are fitted to the devices, the effect of their replacement and disposal can be enormous—cost wise as well as environmentally. If one now replaces the power source with rechargeable lithium batteries, complicated electronics supporting overvoltage, short circuit protection, and a cell temperature monitored design are required to prevent battery explosions. Moreover, we need power to charge the battery and this manual activity itself would be cumbersome. Such batteries also suffer from limited charge–discharge cycles not exceeding around 500 to a maximum of about 1,000 cycles. Again, cost of replacement of these batteries is an issue. Furthermore, a recent survey indicates that by the year 2020, the European Union is committed to reducing carbon emissions by 20 percent. It is expected that this figure is possible due to use of information communication technologies (ICTs). Also, over 17 percent of power generated worldwide is required to power ICTs [4]. In summary, even if we assume a lesser increase in the growth of the number of devices per person (about 100), the explosion of devices by the year 2017 and a need to equip each one with a battery is a serious challenge. Given this, the main focus of this chapter is how to guarantee lifetime support from a sensor device.

12.2.2 Energy Harvesting: A Possible Solution

To tackle the limitation of battery power and its replacement, and to guarantee a perennial operation of the sensor node, we envisage energy harvesting to ensure a self-powered node. A state-of-the-art survey in micro power generators for powering wireless sensor networks (WSNs) is covered in Borca-Tasciuc, et al. [5]. The initial part of the survey takes us through several mechanical to electrical energy converters, such as electromagnetic, piezoelectric, and electrostatic converters. The second half proposes a dielectrophoretic (DEP) actuation-based electrostatic harvester. Such capacitance-based recent harvesters, which switch their dielectric constant between air and liquid, are promising to offer at least four magnitudes increase in energy conversion. Thus, harvesting efficiency needs to be improved vastly to see its application complete. It is now reasonable to assume that if environmental energy (solar and wind) is perennial, then our sensor node also is perennially powered. Energy from vibrations has recently been promising. Researchers at Cornell [6] have created the "Piezo-tree" technology where low wind speed is sufficient to convert wind energy into vibration energy. Bio-energy generated due to pH imbalance in several parts of a tree and the surrounding soil is being explored by researchers from Massachusetts Institute of Technology [7]. The scavenged energy is sufficient to power a network of temperature and humidity sensors that are expected to aid in fire management. The motivation for this effort is replacement of batteries, which is expensive and impractical. The project is titled Early Wildfire Alert Network (EWAN). The bio-energy harvester also is commercially available from commercial vendors now, such as Voltree. Figure 12.1 depicts a made up diagram of harvested energy by a device that varies with time. For example, the diurnal cycle of a solar energy incident on a

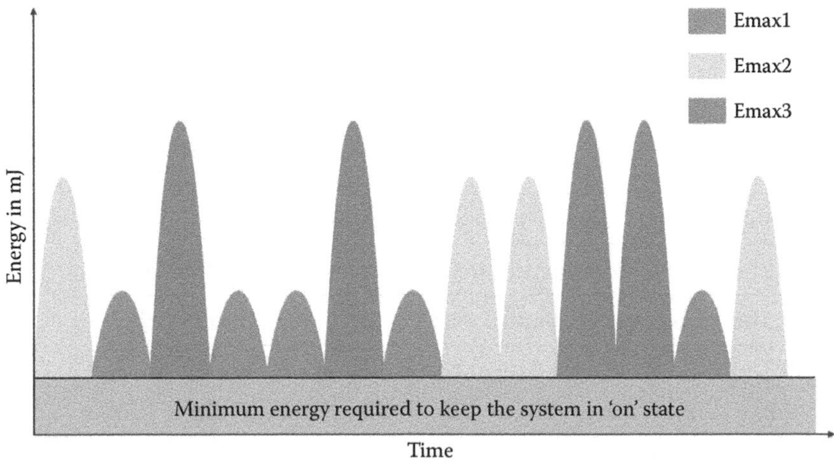

Figure 12.1 Depiction of harvested energy varying with time. A minimum energy is required to keep the WSN node functional.

photovoltaic panel is maximal at noon during the summer months. Similarly, wind energy harvested at a given time depends on the wind speed. One may attribute similar variations with other sources, such as thermo energy generators where varying temperature differential creates a voltage that varies with time.

12.3 Energy Harvesting: Beyond the Solar Harvester—Is It a Viable Option?

Let us now look at the possibility of energy harvesting technology being a viable option to drive wireless sensor networks (WSNs). At this point, we do not consider a solar energy harvester for two reasons. Firstly, solar energy harvesting is a definite possibility if the panel size is dimensioned appropriately. Secondly, researchers and vendors are concentrating on many indoor applications where solar energy is not a possible solution. Moreover, solar photovoltaic panels are expensive to manufacture.

To look beyond solar energy harvesting, let us see the energy harvesting capacities of other methods. Table 12.1 [8] shows the power densities from different energy harvesters. It shows that the thermoelectric generator provides more power density compared to the solar source. Vibration harvesters also are promising with power sufficient to drive embedded communication devices when an electric charge from the vibration harvester is collected in a supercapacitor for a sufficiently long time (15–20 min.).

Thus, let us now turn our attention to secondary energy harvesting sources and their suitability to drive embedded wireless communication devices. Energy harvesting sources are available from several vendors. Micropelt [9] offers a Seebeck

Table 12.1 Power Densities of Different Energy Harvesters

3 V flexible solar cell 1,000 lux	7 mW/kg
3 V flexible solar cell 10,000 lux	280 mW/kg
Vibration generator (60 Hz) a = 0.24 m/sec^2	2.78 mW/kg
Vibration generator (60 Hz) a = 0.98 m/sec^2	37 mW/kg
Thermoelectric generator delta T = 10 K	8 W/kg
Thermoelectric generator delta T = 40 K	131 W/kg

Source: Becker, et al. Power Management for Thermal Energy Harvesting in Aircrafts, from the *Proceedings of IEEE Sensors*, 2008.

effect-based thermo energy generator (TEG) harvester. Seebeck effect is a phenomenon that generates a voltage when a temperature differential is applied across two dissimilar metals, such as tellurium and bismuth. The MPG-0751 is a popular thin film technology-based TEG. Figure 12.2 shows the setup of a TEG subjected to a temperature differential. One side of the semipackaged TEG is connected to a heater reaching 50°C. The other side of the TEG has an ice-filled container to develop the thermal gradient. The output power is conditioned with a low startup DC-DC converter and other maximum power point control functionality to harvest the maximum amount of energy. A DC-DC converter is a highly efficient (greater than 90 percent) electronic circuit that has the ability to provide the required load voltage from very low harvested voltages. For example, the DC-DC converter (Texas Instruments' TPS 63031) can be configured to provide the required output voltage from an input voltage of 1.8 V. The energy is used

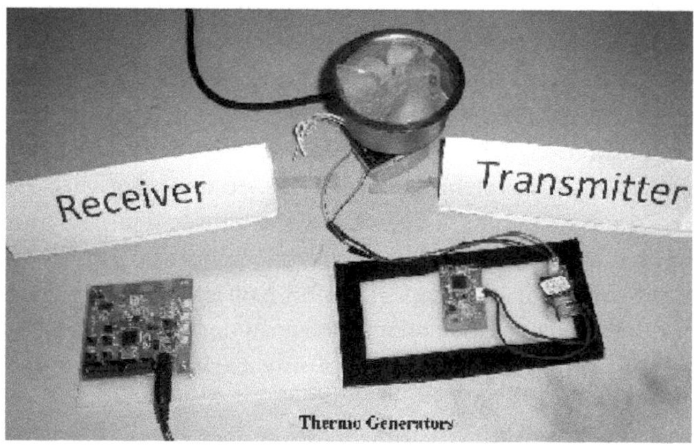

Figure 12.2 Thermo energy generation from Micropelt's MPG–0751.

to charge a 15 mF supercapacitor. Once the output voltage across the capacitor crosses a critical value, the transmitter sends a data packet. The data packet has a typical length of about 128 bytes. The system shown in Figure 12.2 is built using a 16-bit RISC (reduced instruction set computing) microcontroller MSP430 and a radio transceiver CC2520 from Texas Instruments. The radio conforms to IEEE 802.15.4 standard that defines the lower layers (MAC and PHY layers) for low rate wireless personal area networks (LR-WPAN)s.

V21BL vibration harvesters from Mide Volture [10] are attractive because they have two piezo fibers packaged for serial or parallel connection. The parallel connection offers a peak-to-peak of 20 V. Figure 12.3 shows the fiber mounted on the box that houses a motor and cam for generating vibrations. To ensure that the system resonates effectively with the source, a suitable tipping mass is applied. The piezo fiber generates peak power at resonance.

Enoceon [11] offers ECO-100 linear motion harvester for wireless switch applications. Figure 12.4 shows the 2.3 msec. pulse generated from a single operation of

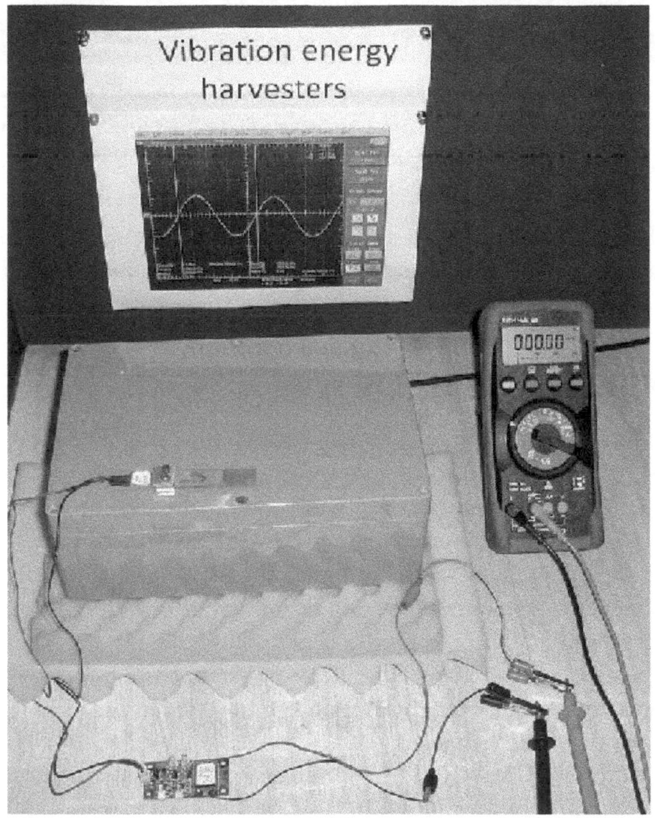

Figure 12.3 Vibration energy generation from Mide Volture's V21BL–piezo fiber.

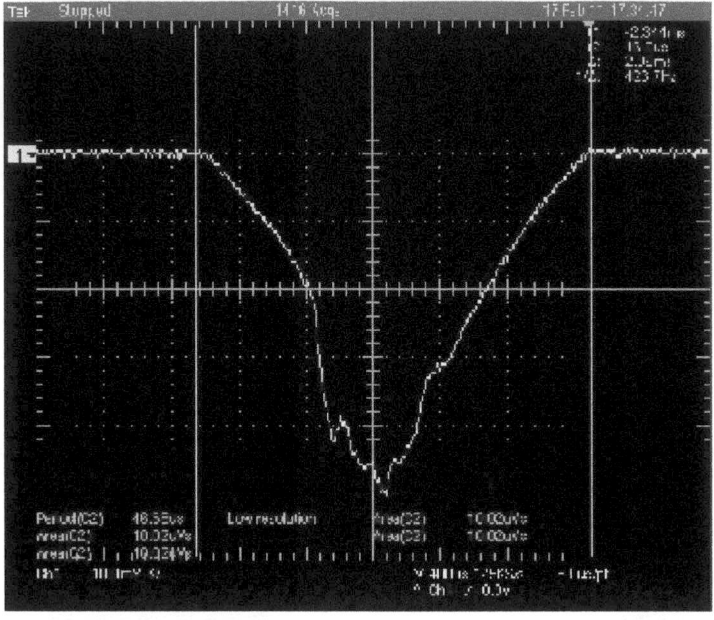

Figure 12.4 Current waveform from Enoceon's ECO-100 linear motion harvester (negative sign only indicates direction).

a mechanical switch. The current peak reaches 50 mA, although momentarily. The energy generated is found to be sufficient to transmit three data packets of size 50 bytes.

12.4 Storing Harvested Energy

Having now increased the basket of energy harvesting sources beyond the solar harvester, the next problem is to explore the energy storage options. As we stated earlier, one of the biggest drawbacks of battery storage is the number of charge–discharge cycles that are supported. The battery beyond these cycles becomes a source of energy leakage instead of exhibiting the ideal storage characteristics. In recent times, vendors have started offering thin film batteries for storing harvested energies. However, the nominal capacity is not comparable to that of normal batteries. The Enerchip CBC-050 rechargeable solid state battery from Cymbet [12] offers 50 microampere-hour capacities with over five thousand charge–discharge cycles if the depth of discharge is about 10 percent. The thin film battery from Excellatron [13] offers 1 mA/h and 10 mA/h batteries. STMicroelectronics's [14] EnFilm is another thin film battery.

Recent advances in supercapacitors have provided another alternative to store energy. Perhaps one of the biggest and best known drawbacks with capacitors is the problem of their extremely low energy density compared to the batteries. Nowadays, with the use of electric double layer capacitors (EDLC) in these ultracapacitors, the energy density is reasonable enough for energy harvested WSNs to consider them as a viable option. A 10 F capacitor can be easily mounted on a printed circuit board (PCB). However, supercapacitors suffer from high leakage and, thus, result in high discharge. It is observed that the capacitors discharge significantly in the first couple of hours after charging. However, one of the major advantages of supercapacitors is the efficiency of storage, which is exceptional, and a capacitor ideally has an infinite number of charge–discharge cycles. These advantages outweigh the discussed disadvantages. If we now consider an embedded communication system with associated harvesting electronics and supercapacitor storage, we question whether this is a viable system. If not, what are the means and ways to make the system viable? We demonstrate the problem at a higher level and then discuss the options available to make the sensor nodes sustainable under varying energy profiles.

12.4.1 Energy Harvesting System

Figure 12.5 shows the block diagram of typical harvesting electronics. Because harvesting sources provide variable power output, it becomes necessary to operate the system at the maximum power availability point. However, it is easy to

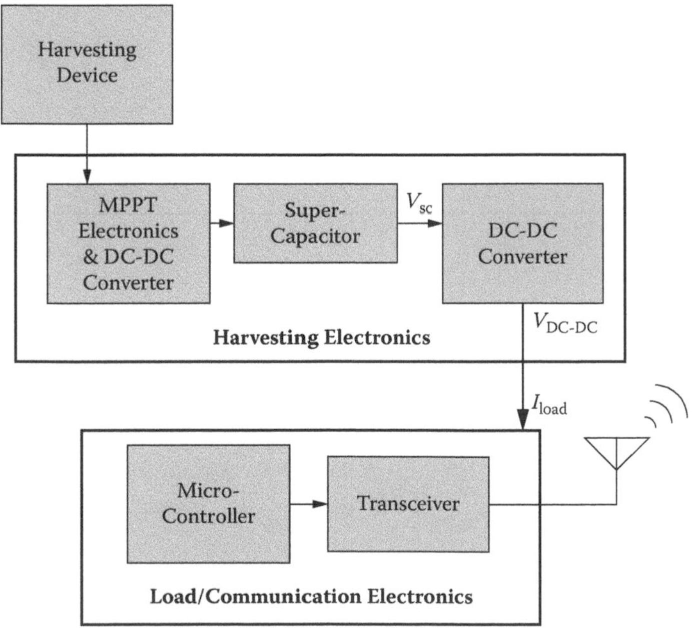

Figure 12.5 Block diagram of an energy harvested embedded communication system.

see that, in order to perform maximum power point tracking (MPPT), the harvesting electronics require power. Therefore, the harvesting electronics initially perform power conditioning and store a small amount of energy to generate their own power. Once this power is available, the input power stored is conditioned successfully and stored in supercapacitors. The small input circuit matches the impedance of the energy source, rectifies the voltage whenever necessary, and delivers the power to the supercapacitor. This output voltage is shown as V_{sc}. The output DC-DC converter is a voltage regulator that feeds power to the load and is shown as $V_{dc\text{-}dc}$. MPPT algorithms are required to ensure that the source impedance variations due to the varying input power and load impedance are matched for transferring the maximum power from the source to the storage device. Due to efficiency issues in harvesting electronics with DC-DC converters, a critical input power is required to go over the *tipping point* for perennial functioning of the load. Figure 12.6 shows the voltages across the DC-DC converter and the supercapacitor (as shown in Figure 12.5) as well as the current drawn (I_{load} , load current) by a communication system in a low input power scenario. The initial start-up current requirement is approximately 23 mA, as seen in Figure 12.6. Each time the output DC-DC converter switches on, the load attempts to draw I_{load} indicated by the voltage waveform. The voltage builds up across the storage device; the super capacitor is also shown. As soon as the voltage across the supercapacitor builds up to the minimum input voltage for the DC-DC converter (1.8 V for TPS63031), the output switches on (3.3 V), and almost immediately loads the input harvester. This behavior repeats itself as long as the input power is low, leading to a large

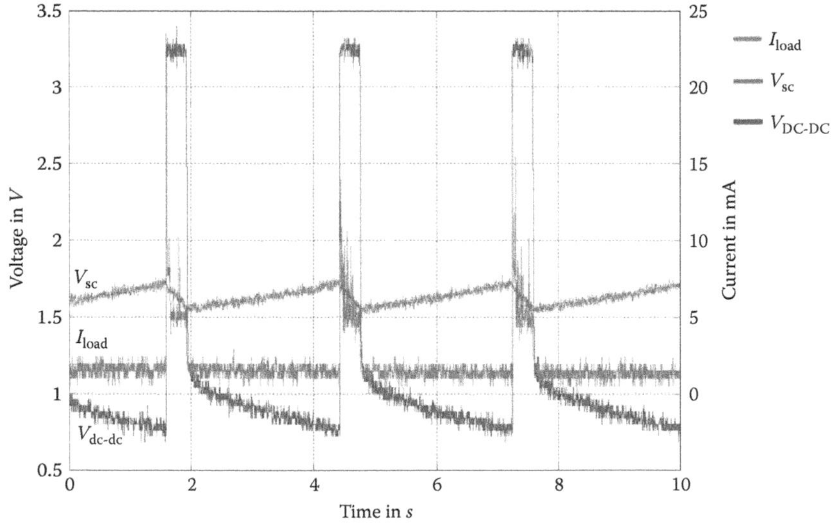

Figure 12.6 Current and voltage waveforms when the embedded communication attempts to power on.

wastage of energy that could otherwise be accumulated for later use. This is a typical problem that requires adequate attention from system designers. Now, an important task is to search for simple solutions. Assuming for a moment that the system does power up, we now try to find how to ensure sustainable operation of the sensor node.

Many chip manufacturers and vendors recently have addressed this specific problem. For instance, Linear Technologies' [15] LTC 3108 is a DC-DC converter ideally suited for energy harvesting applications, particularly when harvester output is in the range of about 20 mV. The IC additionally provides a "power good" signal. The power good is a logic high signal that can be connected to any of the central processing units (CPUs) General Purpose Input/Output (GPIO) or Interrupt pins. The power good signal goes to logic high whenever the output voltage buildup of about 92 percent of the target value is achieved, This signal can then be used to bring the CPU out of the deep sleep modes at the exact time when there is energy for a specific activity. The power good signal indicates that the output voltage from the harvester electronics is within regulation.

$$E_{tot} = P_{tot}\Delta t = C_{tot}V_{dd}^2 f\,\Delta t + V_{dd}I_{leak}\Delta t \tag{12.1}$$

Equation (12.1) shows that the total energy consumed by the system has two components. The first part is the dynamic power consumption where the operating frequency, supply voltage, and device capacitance of the specific technology contribute. The second part is the static leakage power dissipated in every electronic component. This means that one may now have to go beyond the low power modes by monitoring and altering the supply voltage and altering the system operating frequency [16]. These are two important handles to adapt the system to the time varying, available power. Almost all low power microcontrollers including the MSP430 [17] used in our experiments work over a range of operating frequencies. The radio (CC2520) used in our communication experiments is IEEE 802.15.4 compliant [18]. As one can observe, higher operating voltage and frequency increases the power dissipation. The next question that arises is when one should alter these system parameters. From equation (12.1), since the product of f*t is a constant for a given task, one way of reducing power dissipation is by varying the voltage. However, if the processor is awake for a longer time, perhaps altering the frequency is the right approach. Several IC manufacturers provide power and clock gating technique options to ensure that dynamic and static power to processor subsystems is minimized. An example of clock gating is the serial peripheral interface (SPI) clock used between a microcontroller and its peripherals, such as a radio transceiver. The clock circuit is started only when the data are available. Other examples include a camera clock that may not be required when there is no image processing. Similarly, a USB clock is not required when the system is not communicating. Thus, powering embedded communication systems with energy harvesting brings

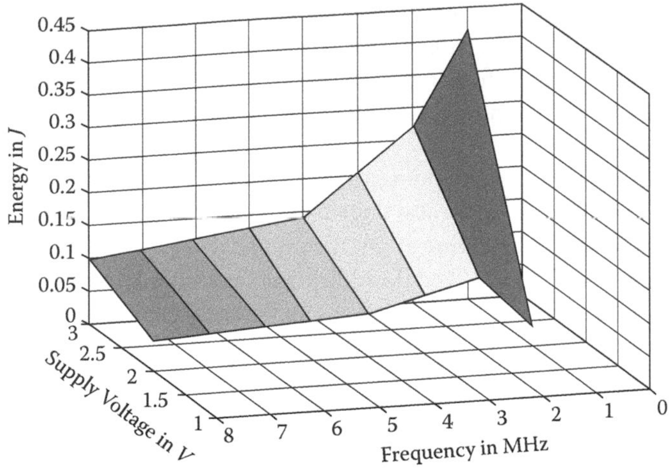

Figure 12.7 Energy variation with frequency and voltage.

to focus dynamic tuning of system parameters well beyond the system's simple linearly polarized modes (LPMs).

12.4.2 Experimental Measurement

A laboratory measurement of energy consumed for various frequencies and operating voltages was recorded and is shown in Figure 12.7. The application running on the microcontroller was a quick-sort algorithm that sorted 10 numbers repeatedly about 7,000 times. The energy was calculated using a current probe amplifier. Equation (12.1) together with Figure 12.7 shows that, when the operating frequency is fixed, the total energy consumption of the system increases with the increase in operating voltage. Furthermore, at higher voltages, as frequency decreases, energy increase is near exponential compared to lower operating voltages. We now show one possibility of "on-the-fly" change in the system's operating voltage. Today's harvesting electronics are capable of generating multiple outputs based on input commands from a microcontroller. The LTC 3588 can accept input from a microcontroller to set its output voltage. Figure 12.8 shows the block diagram of the circuit where the microcontroller's GPIO pins D0 and D1 can signal the required output voltage from the harvester electronics. Four output voltages, 1.8 V, 2.5 V, 3.3 V, and 3.6 V, can be chosen by the microcontroller based on the D0 and D1 logic level combinations. The lithographic test chip (LTC) is capable of providing a continuous current of 100 mA.

Earlier, we mentioned about supercapacitors as energy storage buffers. While the capacitor value determines the quantum of energy storage, embedded communication systems running on supercapacitors have to wait longer for higher

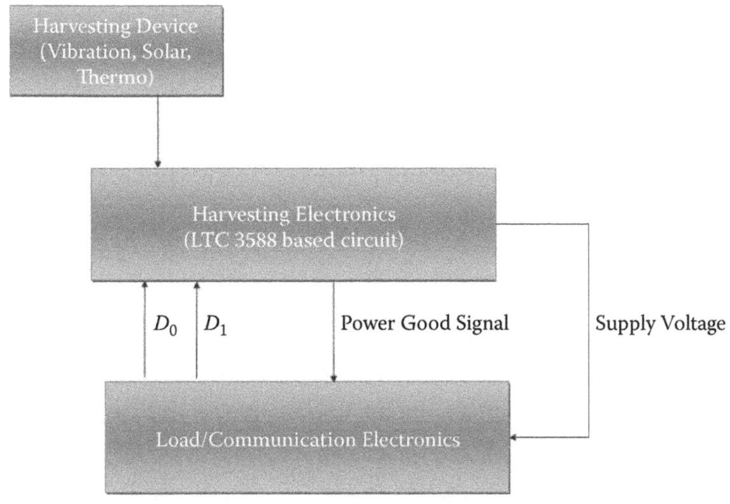

Figure 12.8 A schematic block diagram of an energy harvested communication system with commands to generate a specific output voltage.

value capacitors to build up the required output voltage. To enable the capacitor to build faster, one option is to run the system at a lower frequency. To illustrate the impact of the frequency scaling on sustainable operation of the sensor node, we measured the time required for the supercapacitor to reach the operating voltage of 3 V from the tipping voltage of 1.8 V for various operating frequencies. Figure 12.9 shows the time taken to reach the operating voltage of 3V across the supercapacitor versus frequency. The node is running a simple analog-to-digital conversion (ADC) sensing application with a 2 sec sampling interval. The figure clearly shows the benefits of dynamically adjusting the frequency to the input power from the harvesting source.

12.5 Energy Budgets: System and Network Operations

So far, we have looked at energy harvesting technologies and the viability of such energies on commercially available, embedded communication systems. Let us now see how energy expenditure can be minimized so that the harvested energy can always remain at a sufficiently high threshold and, thus, ensure perennial operation.

To complete a packet transmission, a node has to expend some amount of energy. This energy will have to ensure that packet transmission is successful without any midair collisions with data from other sources. Thus, our goal is to ensure that packet retransmissions are almost always avoided. This requires that we transmit the packet after making sure that the channel is free. We conducted an experiment to study the energy requirements for transmitting packets under conditions such as

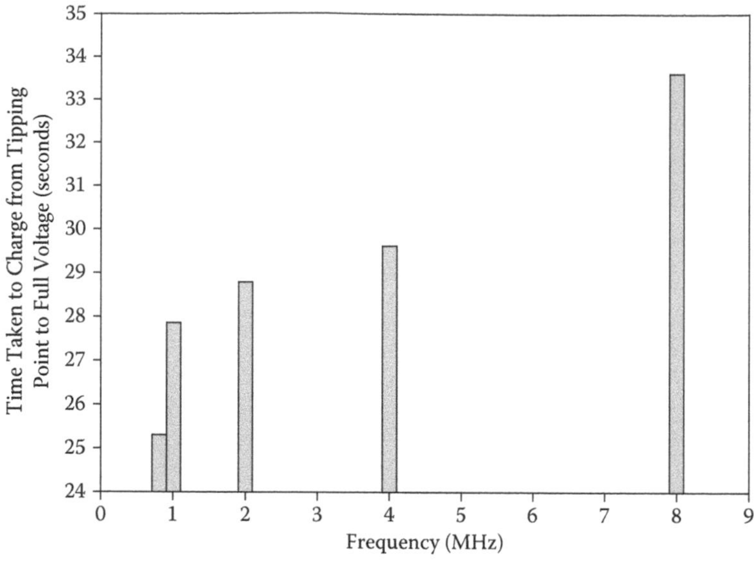

Figure 12.9 Effect of frequency scaling on harvested power.

with and without channel sensing, and with and without enabling acknowledgments from the receiver. Table 12.2 shows the summary of measurements. It turns out that by enabling acknowledgments, the transmitter consumes a significant amount of energy waiting for an acknowledgment. This is quite understandable, in the sense that the transmission is completed soon (in less than 4 msec) and then the system switches to the reception mode and waits for an acknowledgement. Hence, waiting in reception state is energy intensive and should be used carefully by applications. For example, there might be some information that does not require a high degree of reliability. Such packets can be transmitted without expecting an acknowledgment.

Table 12.2 shows the energy spent for -18 dBm transmitted power. In this measurement, we considered many factors to be included when estimating the energy

Table 12.2 Energy Requirement for Channel Sensing and Acknowledgements

Operation	*Energy in µJ*
A: Transmit 30 bytes without channel sensing (-18 dBm)	65.0
B: Energy required for channel sensing (without backoff)	3.2
C: Transmit 30 bytes without enabling ACKs (-18 dBm)	65.0
D: Transmit 30 bytes with ACKs (-18 dBm)	110.5
Extra energy for acknowledgements: (D–C)	45.5

spent. These are channel sensing and overhead for reliability in terms of acknowledgment (ACK), backoff, and transmission. Table 12.2 shows a snapshot of energy measurements conducted for channel sensing and packet transmission with and without ACKs.

Assuming that a sensor monitoring two environment parameters, such as temperature and pressure, requires about 30 bytes for communication with a header and payload, to transmit this data, the table shows that channel sensing requires about 3.2 μJ of energy "each time" a carrier-sensing operation is carried out. On an average, at least four channel-sensing strobes have to be issued to ensure one packet transmission. This value can be substantially high if we consider a multinode scenario with backoff mechanisms. Similarly, packet ACK is at an additional cost in energy. When the transmitter requests an ACK from its destination, it has to change its state from transmit state to receive state and wait for an ACK. Over and above the 65 μJ incurred for packet transmission, Table 12.2 shows there is an overhead of 45.5 μJ when ACKs are requested. The energy harvesting node could benefit by disabling channel sensing and ACKs for small packets.

12.5.1 Energy Harvested Applications: Challenges

Networking with harvested energies is fundamentally different compared to battery-driven systems because one has to look at the maximum rate at which energy can be used and not the limit on energy available from the source [19]. Applications designed over embedded communication devices have to ensure they are aware of the energy available in the system. Thus, applications have to be enabled to be "smart" to ensure that data gathering and application functioning are perennial. In our example, let us assume a base application, as shown in Figure 12.9. The application has to sense two environment parameters, namely light and sound. How does the node manage all the activities, such as sensing, computing, and storing, and data transmission? To answer this, we propose a decision engine running on a sensor node to complete this task successfully.

The goal for the decision engine (DE) is to maximize the number of operations. We have done away with batteries and, instead, we have used low-leakage supercapacitors to harvest the energy. The capacitor (which acts as an energy buffer) is divided into two halves. The lower half of the energy is used for routine activities, such as neighbor discovery, route establishment, channel sensing, and other housekeeping activities. The upper half of the energy is used by the application and DE. The first step is to find whether the energy is above the lower threshold is checked. Time is divided into fixed slots and energy harvested in a slot is used by the engine to provide its recommendations. The DE utilizes two pieces of data to provide its recommendations. First, a time series, energy prediction model helps to identify how energy could be harvested. The second input is the energy cost database for operations required for energy budgeting. This database was built using real measurements. Figure 12.10 shows the architectural design of the engine.

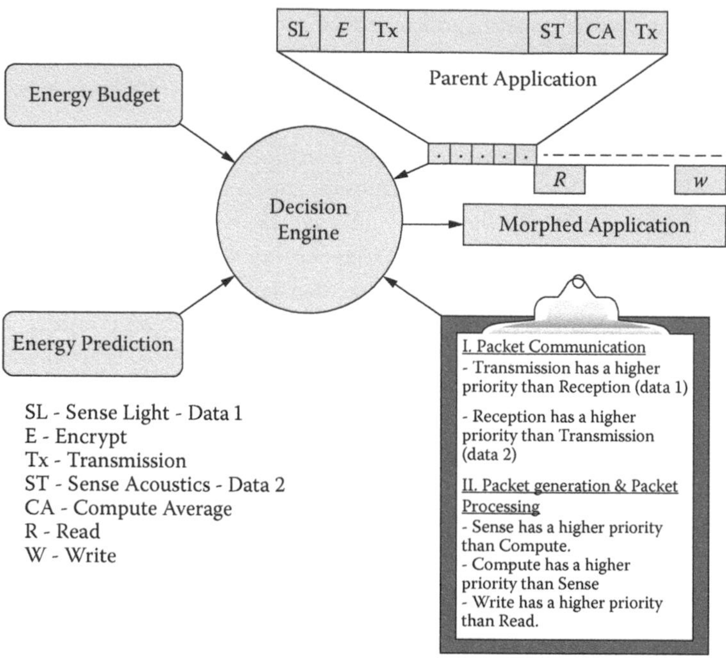

Figure 12.10 Architectural overview of the decision engine.

All operations associated with the base application are shown. The base application is comprised of two environment parameters that sense data, such as light and acoustic. Since the read from flash "R" and write "W" to flash are performed on a longer time scale, the figure displays them separately. While the light data are required to be encrypted, acoustic data can go as plain text. A forwarding node for multiple sensors has the additional role of aggregating acoustic data. Each block in the base application indicates a timer trigger to complete an activity. The energy-aware DE calculates the total energy requirement in each time slot and, assisted by a "heuristic rule book," decides prioritization of the operations. The system finally recommends the best possible set of operations matching the energy harvested in the current time slot.

The other tool used by the DE is the energy budget calculator, which looks up a table of values of energy consumed per operation by Atmel microcontroller-based IRIS motes. The experimental setup includes a current probe amplifier connected to an oscilloscope to measure the current consumed by a node for an operation. The experimental results are documented in Table 12.3. It can be seen that the communication operation, together with writing to flash, requires significantly higher energy compared to other operations. The sum of energy consumed by the next operation is calculated every time the timer expires for a scheduled operation. Scheduled operation is performed only if the predicted energy inflow is higher than

Table 12.3 Energy Consumed by IRIS Motes

Operation	*Energy in J*
Average of 50 samples	7.056 μJ
Finding peak among 50 samples	7.392 μJ
Sensing once from ADC	16.128 μJ
Writing 1 byte to Flash	0.136 mJ
Reading 1 byte from Flash	28.224 μJ
Transmitting @ 0 dBm 28 bytes once	0.784 mJ
Receiving 28 bytes once	0.672 mJ

the energy required. If the predicted inflow is lower than the needed amount, the scheduled operation is ignored and other operations that can be accommodated are performed. To decide between multiple matches, the heuristic rule book is used to decide the priority. At the next scheduled timer trigger, the node checks for previously ignored operations and goes back to perform them if harvested energy is sufficient in this time slot.

Figure 12.11 illustrates the operation of the exponentially weighted moving average (EWMA) time series forecasting. The EWMA time series was proposed by Raghunathan, et al. [20]. It can be seen from the figure that the predicted value seems to be adapting to the changes in the actual measurement. A weight

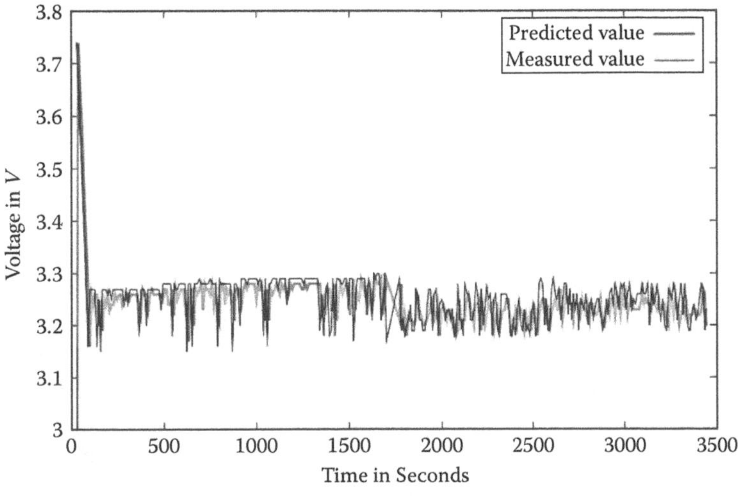

Figure 12.11 Measured and predicted solar panel terminal voltage.

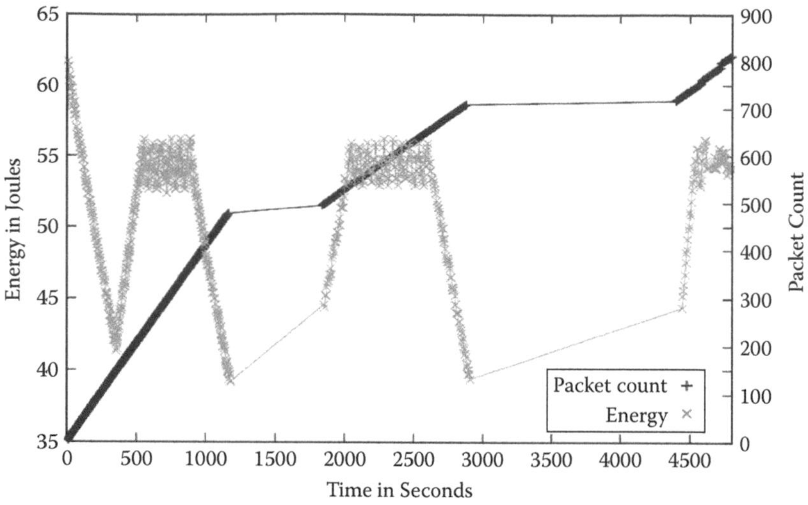

Figure 12.12 Variation in packet transmission while harvested energy depletes and replenishes.

α of 0.5 also ensured smoothing the curve as a correction for measurement noise. Figure 12.12 shows a plot of energy contained in the supercapacitor and illustrates the application response to the harvested energy. The lower limit prevents energy depletion from the supercapacitor.

We conducted all the experiments outdoors under bright sunlight. During the course of the experiment, to emulate low light conditions, we blocked the light to the solar panels. This blocking is apparent at the times 0 to 350 sec, 935 to 1,450 sec, and 2,600 to 3,770 sec. During periods of bright light, the node performed the operations matching closely with the base application.

During low light conditions, it modified its operation states and also stopped all operations when the supercapacitor energy level crossed the threshold for "no light" condition. When energy is available again, the node allowed for a certain amount of energy to accumulate before resuming energy-matched operations. Each point in Figure 12.12 represents a packet transmission. The oscillatory behavior during energy harvesting was found to be attributed to the heliomote. Figure 12.12 also shows the transmitted packet count (right "Y" axis) at the node. One can observe three distinct slopes, indicating the modification of the transmission interval from 2 sec to 4 sec and 5 sec. The three slopes may be attributed to fluctuation in the availability of energy.

Figure 12.13 illustrated the behavior of an application. We conducted specific experiments to study the response of the application when energy was excess, when energy was constantly replenished, and otherwise. We performed a comparison of the operation of the sensor node in two cases: (1) when the node is flooded

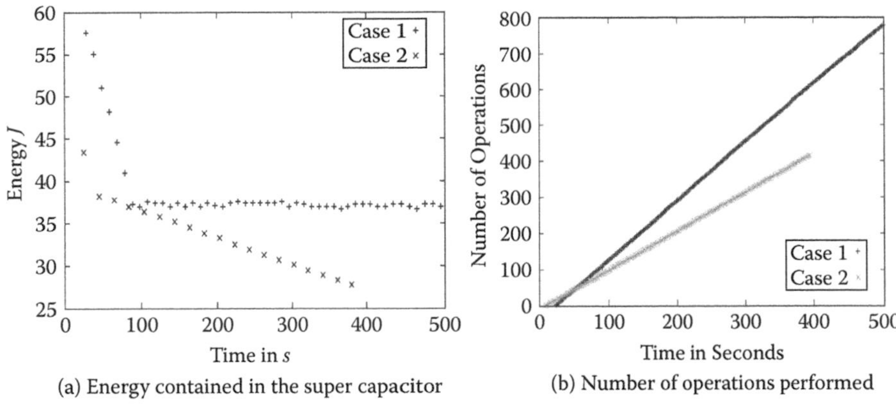

(a) Energy contained in the super capacitor (b) Number of operations performed

Figure 12.13 Comparison of application's behavior at high energy inflow versus low inflow.

with harvested energy, and (2) when there is no energy to harvest, but the supercapacitor has a high amount of stored energy. While Figure 12.13a shows the energy contained in the supercapacitor in both cases throughout the experiment, Figure 12.13b illustrates the cumulative count of the number of operations that the node performed in each case. It can be seen that in case (2), the application modified itself and performed a lesser number of operations than in case (1). The supercapacitor energy level remained constant in case (1), which clearly shows that the node performed energy-matched operations. For example, we looked at the behavior of the application between instances of time t = 200 to 300 sec. The number of operations performed by the node in case (1) was 165, whereas, for the same interval of time in case (2), the node performed 108 operations. Of the 165 operations in case (1), sense and compute operations alone accounted for 54 and the remaining 111 operations were transmit operations. In case (2), however, the number of transmissions dropped to 57 and the number of sense and compute operations was 51. The application found an opportunity in case (1) to transmit as many packets as possible compared to case (2) in which it continued to operate, depleting slowly from the energy storage buffer. Calculation of energy consumption against the available energy, found manually, verified the behavior of the application from its energy consumption.

12.5.2 Storage and Retrieval of System State

We observe in Table 12.3 that writing data to a nonvolatile flash memory is an energy-intensive operation. At the same time, the unpredictability of the energy harvesting source might suddenly shut down the system and, thus, bring about data loss. RF energy harvesting at extremely low power levels, such as –20 dBm, and discrete energy sources, such as linear switches, are examples where the energy harvested can be regarded as a discrete source. In addition to data loss, there might be certain parameters set dynamically during the course of the sensor's active state. For example, a sensor node was perhaps requested to change its sampling interval to say from "x" to "y" minutes. Thus, rollback to a set value also is lost when the node is subjected to a sudden shutdown. This problem of storage and retrieval of system state is a challenge in a network of energy harvesting nodes. Route discovery information, such as aggregator's address, are important data and "energy hard" to rediscover. Mikhaylov and Tervonen [21] describe this problem in detail and analyze the conditions under which data and system state can be stored and restored successfully when a node shutdown occurs. Thus, investigation into fast and reliable data storage and retrieval is required, particularly because recent nonvolatile memory technologies, such as Ferro Random Access Memory (FRAM), are promising in terms of speed and energy requirements.

12.5.3 Toward a Distributed Smart Application: Challenges

In the near future, reality will be where embedded sensor network communication is handled by energy harvesting wireless nodes. This also brings the smartness required by applications, as discussed previously. However, imagine a situation where nodes in the network are harvesting from different harvesting sources. The question is how nodes should schedule their packets. Moser, et al. [22] show that classical scheduling algorithms are not directly suitable for energy harvesting sensor nodes.

Figure 12.14 shows a typical mesh topology established by energy harvested sensor nodes, and illustrates that a collocated, solar harvesting sensor node cannot assume energy availability from a neighboring thermo energy harvesting node. Given this scenario, we need to ensure that an application running on a system functions and data packets reach a sink node reliably. Table 12.4 shows the four possible cases that establish a node's application performance with respect to available energy and the link quality. While case 1 assures a good performance, case 4 forces a node into hibernation, and recovery from this mode is possible only when the harvested energy is sufficient to complete basic activities, such as sensing the sensor field, sensing the channel, and so on.

Sufficient attention should be given toward the process of the system, which impacts the application performance. Figure 12.15 shows three process sets, namely: Urgent, Periodic, and Random-Time. An example of urgent set includes sense or read, followed by compute and transmit. The process becomes an urgent process due to emergency events.

For example, events such as fire detection, intruder detection, failure of a subsystem, etc. have to be done in real time. Figure 12.15 shows that when an urgent process is started, priority toward its completion is highest and, therefore, energy

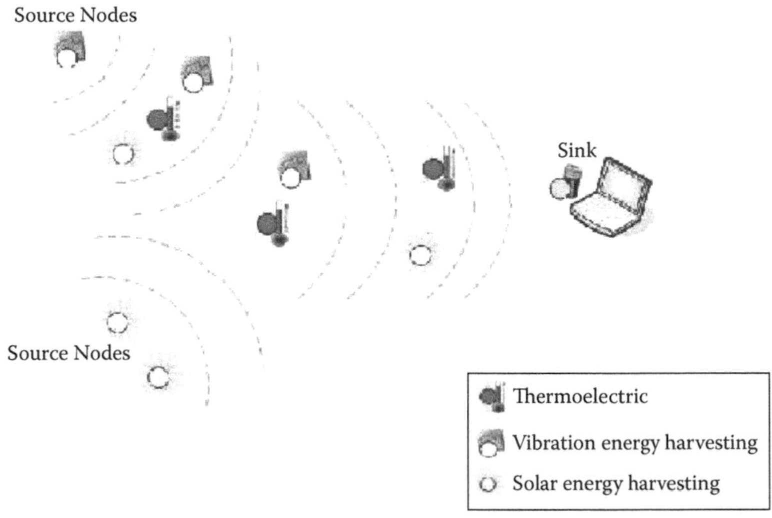

Figure 12.14 A network of energy harvesting sensor nodes powered with solar, thermo, and vibration harvesters.

Table 12.4 Impact of Available Energy and Link Quality on the Performance of Applications

Possible Cases	*Node's Energy*	*Parent's Energy*	*Link to Parent*
Case 1	Good	Good	Good
Case 2	Good	Good	Bad
Case 3	Good	Bad	Good
Case 4	Bad	Cannot be determined	Cannot be determined

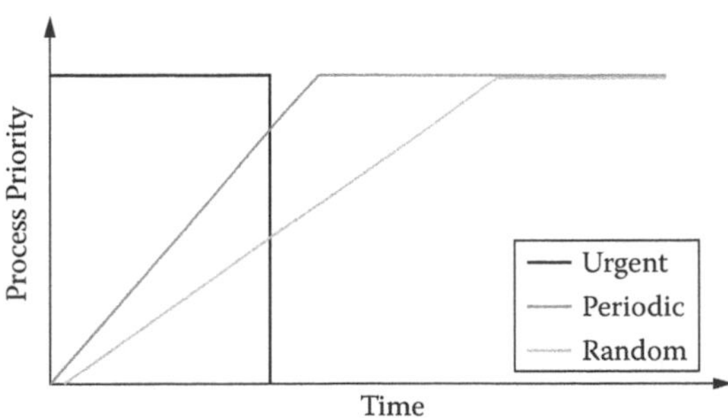

Figure 12.15 The priority setting of three types of processes.

usage preemption toward this is required. The choice of routes to relay such packets should ensure that packets do not get dropped or put into a queue in a waiting state at intermediate nodes. Therefore, choice of a relay and performing cooperative communication is, perhaps, a way to ensure that an energy harvested network of nodes run a distributed smart application successfully.

12.6 Summary

We provided various ways and avenues for energy harvesting for ICT devices. We also discussed the amount of energy each type of harvesting could provide. It was shown that the varying energy levels could throw tough challenges for using them in the nodes, especially WSNs. The networking with energy harvesting devices is nontrivial. It calls for several algorithms that are of low cost in terms of overheads and also in the *energy sense.* However, these issues could be circumvented by using necessary intervention when the input energy is time varying. We explained a system with energy harvesting and how to use the varying sources of energy in WSNs. We provided a basic method for the energy harvested WSNs with measurement results. The ideas, though, are basic if performed well in terms of available energy and the required packet transmission. The ideas presented here are useful for not only the researchers, but also for realistic implementation of energy harvested devices and network. The input power adjustment software strategies coupled with the hardware options to work with lower clock frequency makes it sufficiently clear that energy harvesting technology is here to stay.

REFERENCES

[1] Crossbow IRIS motes. Online at: http://www.xbow.com/Products/productdetails.aspx?sid=264

[2] David Lindley, The Energy Should Always Work Twice. *Nature*, vol. 458, March 12, 2009.

[3] Nigel Jefferies, "Global vision for a wireless world," *18th Meeting of WWRF*, Helsinki, Finland, June 2007.

[4] Online at: http://www.eubusiness.com/topics/internet/ict-low-carbon/

[5] D.-A. Borca-Tasciuc, M. M. Hella, and A. Kempitiya, "Micro-power generators for ambient intelligence applications," *4th International Workshop on Soft Computing Applications*, Arad, Romania, June 15–17, 2010.

[6] Online at: http://www.news.cornell.edu/stories/May10/VibroWind.html

[7] Online at: http://web.mit.edu/newsoffice/2008/trees-0923.html

[8] T. Becker, M. Kluge, J. Schalk, T. Otterpohl, and U. Hilleringmann, "Power management for thermal energy harvesting in aircrafts," *Proceedings of IEEE Sensors*, 2008.

[9] Online at: www.micropelt.com/

[10] Online at: www.mide.com/products/volture/v21bl.php

[11] EnOcean, The EnOcean GmBH, Germany. Online at: http://www.enocean.com/en/home/
[12] Online at: www.cymbet.com/
[13] Online at: www.excellatron.com
[14] Online at: http://www.st.com/stonline/products/literature/ds/17370/efl700a39.pdf
[15] Online at: www.linear.com/
[16] Amit Sinha, Energy Aware Software, master's thesis, Massachusetts Institute of Technology, Cambridge, MA, December 1999.
[17] Texas Instruments, MSP430. Online at: http://focus.ti.com/docs/prod/folders/print/msp430f1612.html
[18] IEEE 802.15.4-2006 Standard. Online at: http://standards.ieee.org/getieee802/download/802.15.4-2006.pdf
[19] A. Kansal, J. Hsu, S. Zahedi, and M. B. Srivastava, "Power Management in Energy Harvesting Sensor Networks," *ACM Transactions on Embedded Computing Systems*, vol. 6, no. 4, September 2007.
[20] Vijay Raghunathan, Aman Kansal, Jason Hsu, Jonathan Friedman, and Mani Srivastava, "Design considerations for solar energy harvesting wireless embedded systems," *Proceedings of 4th International Symposium on Information Processing in Sensor Networks (TPSN)*, April 15, 2005, Los Angeles, California.
[21] Konstantin Mikhaylov and Jouni Tervonen. "Energy efficient data restoring after power-downs for wireless sensor networks nodes with energy scavenging," *Proceeding of the 4th IFIP International Conference: New Technologies, Mobility and Security (NTMS)*, Paris, France, February 2011.
[22] Clemens Moser, David Brunelli, Lothar Thiele, and Luca Benini, "Real time scheduling for energy harvesting sensor nodes," *ACM Real Time Systems*, vol. 37, no. 3, December 2007.

Chapter 13

Radio Frequency Energy Harvesting and Management for Wireless Sensor Networks

Adamu M. Zungeru, Li-Minn Ang,
S. R. S. Prabaharan, and Kah Phooi Seng

Contents

13.1 Introduction

Radio frequency (RF) energy harvesting holds a promising future for generating a small amount of electrical power to drive partial circuits in wireless communicating electronics devices. Reducing power consumption has become a major challenge in wireless sensor networks. As a vital factor affecting system cost and lifetime, energy consumption in wireless sensor networks (WSNs) is an emerging and active research area. This chapter presents a practical approach for RF energy harvesting, and management of the harvested and available energy for WSNs using the improved energy efficient ant-based routing algorithm (IEEABR) as our proposed algorithm. The chapter looks at measurement of the RF power density, calculation of the received power, storage of the harvested power, and management of the power in WSNs. The routing uses IEEABR technique for energy management. Practical and real-time implementations of the RF energy using Powercast™ harvesters and simulations using the energy model of our Libelium Waspmote to verify the approach were performed.

The chapter is organized in the following format: The first part of the chapter, provided in section 13.2, covers a general perspective and the objective of the chapter. Section 13.3 reviews energy harvesting systems and power consumption in WSNs. Section 13.4 gives a detailed explanation of our RF energy harvesting method using the Powercast harvesters. Section 13.5 looks into the management of the harvested energy in our WSNs. Section 13.6 presents experimental setup and results, while also looking at the simulation results and its environment. Finally, section 13.7 concludes the chapter with an open research problem and future work to be done, and a comparative summary of our results with the EEABR algorithm, and ad hoc on-demand distance vector (AODV), which form strong energy management protocols.

13.2 RF Energy Harvesting

Finite electrical battery life is encouraging companies and researchers to come up with new ideas and technologies to drive wireless mobile devices for an enhanced period of time. Batteries add to size and their disposal adds to environmental pollution. For mobile and miniature electronic devices, a promising solution is available in capturing and storing the energy from external ambient sources, a technology known as energy harvesting. Other names for this type of technology are power harvesting, energy scavenging, and free energy, which are derived from renewable energy [1]. In recent years, the use of wireless devices is growing in many applications like mobile phones and sensor networks [2]. This increase in wireless applications has generated an increasing use of batteries. Many research teams are working on extending the battery life by reducing the consumption of the devices. Other teams have chosen to recycle ambient energy like in microelectromechanical systems (MEMS) [3]. The charging of mobile devices is convenient because the user can do it easily, as for mobile phones. But for other applications, like wireless sensor nodes that are located in difficult to access environments, the charging of the batteries remains a major problem. This problem increases when the number of devices is large and they are distributed in a wide area or located in inaccessible places. The research on RF energy harvesting provides reasonable techniques of overcoming these problems.

The rectification of microwave signals to DC power has been proposed and researched in the context of high-power beaming. It has been proposed for helicopter powering [4], solar power satellites [5], and the SHARP system [6]. The DC power depends on the available RF power, the choice of antenna, and frequency band. An energy harvesting technique using electromagnetic energy, specifically radio frequency, is the focus of this chapter. Communication devices generally have omnidirectional antennas that propagate RF energy in most directions, which maximizes connectivity for mobile applications. The energy transmitted from the wireless sources is much higher, up to 30 W for 10 GHz frequency [7], but only a small amount can be scavenged in the real environment. The rest is dissipated as heat or absorbed by other materials. The RF power harvesting technique also is used in radio frequency identification (RFID) tags and implantable electronics devices. Most commonly used wireless sensor nodes consume few μW in sleep mode and hundreds of μW in active mode. A great factor contributing to energy harvesting research and development is ultralow-power components.

The management of power available for sensor nodes has been dealt with to an extent using ant-based routing [8]–[16], which utilizes the behavior of real ants searching for food through pheromone deposition while dealing with problems that need to find paths to goals. The simulating behavior of an ant colony leads to optimization of network parameters for the WSN routing process to provide a maximum network lifetime.

The main goal of this chapter is to propose practical harvesting of RF energy using Powercast harvesters while managing the harvested and available energy of the sensor networks using our proposed algorithm (IEEABR), which helps in the optimization of the available power. The objective is to efficiently power sensor networks with or without batteries to maintain network lifetime at a maximum without performance degradation.

13.3 Review of Energy Harvesting Systems and Power Consumption in WSNs

For proper operation of sensor networks, a reliable energy harvesting technique is needed. Over the years, much work has been done on the research from both academic and industrial researchers on large-scale energy from various renewable energy sources. Less attention has been paid to small-scale energy harvesting techniques, although, quite a number of work has been carried out on energy scavenging for WSNs. The efficient far-field energy harvesting [17] uses a passively powered RF-DC conversion circuit operating at 906 MHz to achieve power of up to 5.5 µW. In related work [18]–[21], all consider the little available RF energy while utilizing it to power the sensor networks. Bouchouicha, et al. [2] studied ambient RF energy harvesting in which two systems, the broad band without matching and narrow band, were used to recover the RF energy. Among but not all of the available energy harvesting system for wireless sensors are: solar power, electromagnetic energy, thermal energy, wind energy, salinity gradients, kinetic energy, biomedical, piezoelectric, pyroelectric, thermoelectric, electrostatic, blood sugar, and tree metabolic energy. These could be further classified into three [22]: thermal, radiant, and mechanical energy. Based on these, Table 13.1 and Table 13.2 show the comparison of the different and common energy scavenging techniques.

Beside the harvested energy for the sensor network, the consumption of the harvested power for the different mode of the network has to be looked upon before choosing a power harvesting source. A review of some power consumption in some selected sensor nodes can be found in Gilbert and Balouchi [23]. For some commercial sensor network nodes, the consumption differs, as shown in Table 13.3; power consumption of the nodes differs among manufacturers.

13.3.1 Ambient RF Sources and Available Power

A possible source of energy comes from ubiquitous radio transmitters. Radio waves, a part of the electromagnetic spectrum, consist of magnetic and electrical components. Radio waves carry information by varying a combination of the amplitude, frequency, and phase of the wave within a frequency band. On contact with a conductor, such as an antenna, the electromagnetic (EM) radiation induces electrical current on the conductor's surface, known as skin effect. The communication devices use the antenna for transmission and/or reception of data by utilizing the different frequencies spectrum from 10 to 30 Kz. The maximum theoretical power available for RF energy harvesting is 7.0 µW and 1.0 µW for 2.4 GHz and 900 MHz frequency, respectively, for free space distance of 40 m. The path loss of signals will be different in an environment other than free space [29], although, for our work using the Powercast harvester, the power available for P2110, which operates at 915 MHz, is 3.5 mW before conversion and 1.93 mW after conversion at a distance of 0.6 m, and 1 µW at a distance of 11 m [30].

Table 13.1 Comparison of Energy Harvesting Sources for WSNs

Energy Source	*Classification*	*Performance (Power Density)*	*Weakness*	*Strength*
Solar Power	Radiant energy	100 mW/cm^3	Requires exposure to light, and low efficiency if device is in building	Can use without limit
RF Waves	Radiant energy	0.02 μW/cm^2 at 5 Km from AM radio	Low efficiency inside a building	Can use without limit
RF Energy	Radiant energy	40 μW/cm^2 at 10 m	Low efficiency if out of line of sight	Can use without limit
Body Heat	Thermal energy	60 μW/cm^2 at 5°C	Available only when temperature difference is high	Easy to build using thermocouple
External Heat	Thermal energy	135 μW/cm^2 at 10°C	Available only when temperature difference is high	Easy to build using thermocouple

13.4 RF Energy Harvesting and the Use of the Powercast Harvester

RF power harvesting is a process whereby RF energy emitted by sources that generate high electromagnetic fields, such as TV signals, wireless radio networks, and cell phone towers, but through power generating circuits linked to a receiving antenna, are captured and converted into usable DC voltage. Most commonly used is an application for RFID tags in which the sensing device wirelessly sends a radio frequency to a harvesting device that supplies just enough power to send back identification information specific to the item of interest. The circuit systems that receive the detected RF from the antenna are made on a fraction of a micrometer scale, but can convert the propagated electromagnetic waves to low voltage DC power at distances up to 100 m. Depending on concentration levels that can differ through the day, the power conversion circuit may be attached to

Table 13.2 Comparison of Energy Harvesting Sources for WSNs (Continued)

Body Motion	Mechanical energy	800 μW/cm^3	Dependent on motion	High power density not limited on interior and exterior
Blood Flow	Mechanical energy	0.93 W at 100 mmHg	Energy conversion efficiency is low	High power density not limited on interior and exterior
Air Flow	Mechanical energy	177 μW/cm^3	Efficiency is low inside a building	High power density
Vibrations	Mechanical energy	4 μW/cm^3	Has to exist at surroundings	High power density not limited on interior and exterior
Piezoelectric	Mechanical energy	50 μJ/N	Has to exist at surroundings	High power density not limited on interior and exterior

a capacitor that can disperse a constant required voltage for the sensor and circuit when there is not a sufficient supply of incoming energy. Most circuits use a floating gate transistor as the diode, which converts the signal into generated power, but is linked to the drain of the transistor, and a second floating gate transistor linked to a second capacitor can enable a higher output voltage once the capacitors reach full potential [31].

The effectiveness of energy harvesting depends largely on the amount and predictable availability of energy sources, whether from radio waves, thermal differentials, solar or light sources, or even vibration sources. There are three categories for ambient energy availability: intentional, anticipated, and unknown, as shown in Figure 13.1.

Our research relies basically on the intentional use of the Powercast harvester.

13.4.1 Intentional Energy Harvesting

The designs rely on an active component in the system, such as an RF transmitter that can explicitly provide the desired type of energy into the environment when the device needs it. Powercast supports this approach with an energy source of 3 W, 915 MHz RF transmitters; the P1110 and P2110 also are used along with it as

Table 13.3 Comparison of Power Consumption of Some Selected Sensor Network Nodes

Operating Conditions	*Manufacturers*			
	Crossbow MICAz [24]	*Waspmote [25-26]*	*Intel IMote2 [27]*	*Jennic JN5139 [28]*
Radio Standard	IEEE 802.15.4/ Zigbee	IEEE 802.15.4/Zigbee	IEEE 802.15.4	IEEE 802.15.4/Zigbee
Typical Range	100 m (outdoor), 30 m (indoor)	500 m	30 m	1 km
Data Rate	250 kbps	250 kbps	250 kbps	250 kbps
Sleep Mode (deep sleep)	15 μA	62 μA	390 μA	2.8 μA
Processor Consumption	8 mA active mode	9 mA	31–53 mA	2.7+0.325 mA/ MHz
Transmission	17.4 mA (+0 dBm)	50.26 mA	44 mA	34 mA (+3 dBm)
Reception	19.7 mA	49.56 mA	44 mA	34 mA
Supply Voltage (min)	2.7 V	3.3 V	3.2 V	2.7 V
Average Power	2.8 mW	1 mW	12 mW	3 mW

receivers. The intentional energy approach is appropriate for other types of energy as well, such as placing an energy harvester on a piece of industrial equipment that vibrates when it is operating. Using an intentional energy source allows designers to engineer a consistent energy solution. A quick look at the basic operation of the transmitter and receiving circuit is discussed below.

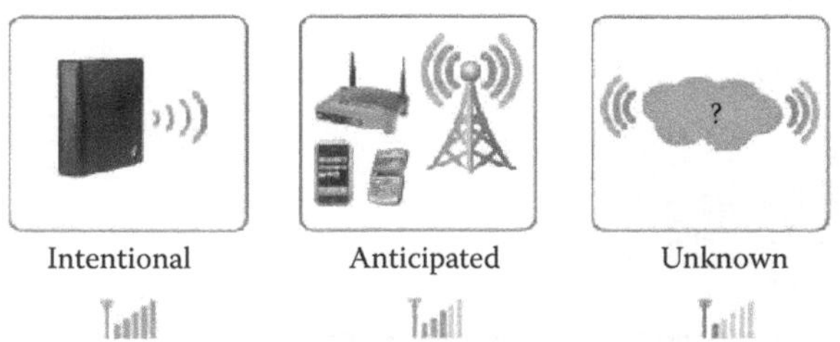

Figure 13.1 Pictorial view of intentional, anticipated, and unknown energy sources.

13.4.2 The Powercast TX91501 Powercaster Transmitter

The Powercast TX91501 is an RF power transmitter specifically designed to provide both power and data to end devices containing the Powercast P2110 or P1110 power harvester receivers [30]. The transmitter is housed in a durable plastic case with mounting holes. It is powered by a regulated 5 V DC voltage, mostly from a power source of 240 V AC, rectified and regulated to its accommodated voltage of 5 V DC from its built-in internal circuitry. The transmitter has a factory set, fixed power output, and no user adjustable settings. Also, a beautiful control feature is the status LED that provides a feedback on a functional state. It provides a maximum of 3 W EIRP (equivalent or effective isotropic radiated power). A side view, real view, and its transmission state are shown in Figure 13.2.

The Powercast transmitter transmits power in the form of direct sequence spread spectrum (DSSS) and data in the amplitude shift keying (ASK) modulation and at a center frequency of 915 MHz. The power output is 3 W EIRP and vertically polarizes for optimal transmission. For data communication, it has an 8-bit factory set, TX91501 identification (ID) number broadcast with random intervals up to 10 ms using ASK modulation. Its operating temperature is within the range of -20°C to 50°C at the power input from mains of 5 VDC/1A.

13.4.3 Powercast Power Harvester Receiver

The Powercast receivers can harvest directly a directed or ambient RF energy and convert it to DC power for remotely recharging batteries or battery-free devices. The two modules available for our research are P1110 and P2110 and both have similarities and differences in their area of applications, as shown in Table 13.4.

The Powercast P2110 power harvester receiver is an RF energy harvesting device that converts RF energy to DC voltage. It has a wide RF operating range, and provides

Figure 13.2 (a) A side view and (b) a real view of a TX91501 Powercaster transmitter in its transmission state.

Table 13.4 Comparison of the Two RF Energy Powercast Eeceivers

Receivers	*Differences*	*Similarities*
P2110	1. Design for battery charging and direct power applications 2. Provide intermittent/pulsed power output 3. Configurable, regulated output voltage up to 5.25 V 4. Power management and control I/O for system optimization	1. Harvesting range from 850–950 MHz 2. Works with standard 50-ohm antennas
P1110	1. Design for battery charging and direct power applications 2. Configurable over voltage protection up to 4.2 V 3. Connect directly to rechargeable batteries including alkaline, lithium ion, and Ni-MH.	1. Harvesting range from 850–950 MHz 2. Works with standard 50-ohm antennas

RF energy harvesting, and power management for battery-free micropower devices. It converts the RF energy to DC and stores it in a capacitor as well as boosting the voltage to the set output voltage level and enables the voltage output.

13.4.4 Measurement of RF Power Received and Gains

Utilized in our experiment were power meters that provide the most accurate measurement of RF power of any of the types of RF measurement equipment, and the simplified Friis equation that provides a reasonable estimate of the amount of power that is received and available for use.

13.4.4.1 Friis Transmission Equation

The Friis transmission formula is used solely for studying RF communication links [32]. The formula can be used in situations where the distance between two antennas is known, and a suitable antenna needs to be found. Using the Friis transmission equation, one can solve for the antenna gains needed at either the transmitter or receiver in order to meet certain design specifications.

$$\frac{P_r}{P_t} = G_t G_r \left(\frac{\lambda}{4\pi R} \right)^2 \tag{13.1}$$

where P_r is the received power in watts (W), P_t is the transmitted power, G_t is the transmitting antenna's gain, G_r is the receiving antenna's gain, λ is the wavelength of the transmitted and received signal in meters, and R is the distance between the antennas in meters. The gain of the antennas, usually measured in decibels, can be converted to a power ratio using;

$$G = 10^{\frac{G_{DB}}{10}} \tag{13.2}$$

$$\lambda = \frac{c}{f} \tag{13.3}$$

where C is the speed of light in meters per second, and f is the frequency in Hz. Hence, C is equal to 3×10^8 m/sec.

A simplified version of the Friis equation [33] is provided by the Powercast company for quick and easy calculation on a spreadsheet, where a reasonable estimate of the amount of power generated, received, and available for use are calculated.

13.4.4.2 Power Density

RF propagation is defined as the travel of electromagnetic waves through or along a medium. For RF propagation between approximately 100 MHz and 10 GHz, radio waves travel very much as they do in free space and travel in a direct line of sight and a slight difference in the dielectric constants of air and space [34]. For air, the dielectric is one and 1.000536 at sea level. In antennas theory, an isotropic radiator is a theoretical, lossless, omnidirectional (spherical) antenna [34] [35]. That is, it radiates uniformly in all directions. The power of a transmitter that is radiated from an isotropic antenna will have a uniform power density (power per unit area) in all directions. Power density at any distance from an isotropic antenna is the ratio of the transmitted power by the surface area of a sphere ($4\pi R^2$) at that distance. The surface area of the sphere increases by the square of the radius, therefore, the power density, P_D, (watts/square meter) decreases by the square of the radius.

$$P_D = \frac{P_t}{4\pi R^2} \tag{13.4}$$

where P_t is the peak or average power, P_D the power density, and R the distance between the transmitter and the receiving antenna. Radars use directional antennas to channel most of the radiated power in a particular direction. The gain (G_t) of an antenna is the ratio of power radiated in the desired direction as compared to the power radiated from an isotropic antenna, or:

$$G_t = \frac{\text{Maximum radiation intensity of actual antenna}}{\text{Radiation intensity of isotropic antenna with same power input}} \tag{13.5}$$

The power density at a distant point from the radar with an antenna gain of G_t is the power density from an isotropic antenna multiplied by the radar antenna gain. Power density from the radar:

$$P_D = \frac{P_t G_t}{4\pi R^2} \tag{13.6}$$

13.4.5 Energy Storage

The most common energy storage device used in a sensor node is a battery, either nonrechargeable or rechargeable. A nonrechargeable battery (e.g., alkaline) is suitable for a microsensor with very low power consumption (e.g., 50 μW). Alternatively, a rechargeable battery (e.g., lithium ion) is used widely in sensor nodes with energy harvesting technology [36]. A battery is used not only for storage of energy generated by the harvesting device, but also to regulate the supply of energy to a sensor node. A wireless sensor node is powered by exhaustible batteries [37]. Several factors affect the quality of these batteries, but the main factor is cost. In a large-scale deployment, the cost of hundreds and thousands of batteries is a serious deployment constraint. Batteries are specified by a rated current capacity, *C*, expressed in ampere-hours. This quantity describes the rate at which a battery discharges without significantly affecting the prescribed supply voltage (or potential difference). Practically, as the discharge rate increases, the rated capacity decreases. Most portable batteries are rated at 1 *C*. This means a 1,000 mAh battery provides 1,000 mA for 1 hour, if it is discharged at a rate of 1 *C*. Ideally, the same battery can discharge at a rate of 0.5 *C*, providing 500 mA for 2 hours, and at 2 *C*, 2,000 mA for 30 minutes and so on. 1 *C* is often referred to as a 1-hour discharge. Likewise, a 0.5 *C* would be a 2-hour and a 0.1 *C* a 10-hour discharge. In reality, batteries perform at less than the prescribed rate. Often, the Peukert equation is applied to quantifying the capacity offset (i.e., how long a battery lasts in reality):

$$T = \frac{c}{I^n} \tag{13.7}$$

where C is the theoretical capacity of the battery expressed in ampere-hours, I is the current drawn in ampere (A), T is the time of discharge in seconds, and n is the Peukert number, a constant that directly relates to the internal resistance of the battery.

The value of the Peukert number indicates how well a battery performs under continuous heavy current. A value close to 1 indicates that the battery performs well; the higher the number, the more capacity is lost when the battery is discharged at high current. The Peukert number of a battery is determined empirically. For example, for lead acid batteries, the number is typically between 1.3 and 1.4. Drawing current at a rate greater than the discharge rate results in a current consumption rate higher than the rate of diffusion of the active elements in the electrolyte. If this process continues for a long time, the electrodes run out of active material even though the electrolyte has not yet exhausted its active materials. This situation can be overcome by intermittently drawing current from the battery and also proper power management techniques.

13.5 Energy Management in WSNs

Despite the fact that energy scavenging mechanisms can be adopted to recharge batteries, e.g., through Powercast harvesters [30], solar panels [2], or piezoelectric or acoustic transducers [21], energy is a limited resource and must be used judiciously. Hence, efficient energy management strategies must be devised at the sensor nodes to prolong the network lifetime as much as possible. Many routing, power management, and data dissemination protocols have been specially designed for WSNs [38]. The EAGRP (energy-aware geographic routing protocol) [39], an enhanced AODV (ad hoc on-demand distance vector) [40], and an EEABRA for WSNs [41] all have developed different protocols in order to manage the available energy in WSNs. In a related work, Alippi, et al. [42] use energy-hungry sensors in trying to manage the available energy in WSNs. Reducing power consumption has become a major challenge in wireless sensor networks. As a vital factor affecting system cost and lifetime, energy consumption in WSNs is an emerging and active research area. The energy consumption of WSNs is of crucial concern due to the limited availability of energy. Whereas energy is a scarce resource in every wireless device, the problem in WSNs is more severe for the following reasons [37]:

1. Compared to the complexity of the task they carry out, sensing, processing, self-managing, and communication, the nodes have been very small in size to accommodate high-capacity power supplies.
2. While the research community is investigating the contribution of renewable energy and self recharging mechanisms, the size of nodes is still a constraining factor.
3. Ideally, a WSN consists of a large number of nodes. This makes manually changing, replacing, or recharging batteries almost impossible.
4. The failure of a few nodes may cause the entire network to fragment prematurely.

The problem of power consumption can be approached from two angles. One is to develop energy efficient communication protocols (self-organization, medium access, and routing protocols) that take the peculiarities of WSNs into account. The other is to identify activities in the networks that are both wasteful and unnecessary and mitigate their impact. Wasteful and unnecessary activities can be described as local (limited to a node) or global (having a scope network-wide). In either case, these activities can be further considered as accidental side effects or results of nonoptimal software and hardware implementations (configurations). For example, observations based on field deployment reveal that some nodes exhausted their batteries prematurely because of unexpected overhearing of traffic that caused the communication subsystem to become operational for a longer time than originally intended [43].

Similarly, some nodes exhausted their batteries prematurely because they aimlessly attempted to establish links with a network that had become no longer accessible to them. Most inefficient activities are, however, results of nonoptimal configurations in hardware and software components. For example, a considerable amount of energy is wasted by an idle processing or a communication subsystem. A radio that aimlessly senses the media or overhears while neighboring nodes communicate with each other consumes a significant amount of power. A dynamic power management (DPM) control strategy is aimed at adapting the power/performance of a system to its workload. The DPM, having a local or global scope, or both, aims at minimizing power consumption of individual nodes by providing each subsystem with the amount of power that is sufficient to carry out a task at hand [37]. Hence, it does not consider the residual energy of neighboring nodes. IEEABR as the proposed algorithm considers the available power of nodes and the energy consumption of each path as the reliance of routing selection. It improves memory usage and utilizes the self organization, self-adaptability, and dynamic optimization capability of the ant colony system to find the optimal path and multiple candidate paths from source nodes to sink node. The protocol avoiding using up the energy of nodes on the optimal path and prolongs the network lifetime while preserving network connectivity. This is necessary because, for any WSN protocol design, the important issue is the energy efficiency of the underlying algorithm due to the fact that the network under investigation has strict power requirements.

It has been proposed by Kalpakis, Dasgupta, and Namjoshi [44] that forward ants be sent directly to the sink node; the routing tables only need to save the neighbor nodes that are in the direction of the sink node. This considerably reduces the size of the routing tables and, in essence, the memory needed by the nodes. Since one of the main concerns in WSN is to maximize the lifetime of the network, which means saving as much energy as possible, it would be preferable that the routing algorithm could perform as much processing as possible in the network nodes, rather than transmitting all data through the ants to the sink node to be processed there. In fact, in huge sensor networks where the number of nodes can easily reach more than thousands of units, the memory of the ants would be so large that it would be unfeasible to send the ants through the network. To implement these ideas, the memory M_k of each ant is reduced to just two records, the last two visited nodes [41]. Since the path followed by the ants is no longer in their memories, a memory must be created at each node that keeps records of each ant that was received and sent. Each memory record saves the previous node, the forward node, the ant identification, and a timeout value. Whenever a forward ant is received, the node looks into its memory and searches the ant identification for a possible loop.

If no record is found, the node saves the required information, restarts a timer, and forwards the ant to the next node. If a record containing the ant identification is found, the ant is eliminated. When a node receives a backward ant, it searches its memory to find the next node to where the ant must be sent. In this section, we modify the EEABR to improve the energy consumption in the nodes of WSNs and also to improve the performance and efficiency of the networks. The main focus of this chapter is on IEEABR power management strategies in WSNs.

The algorithm of our proposed power management techniques is as follows.

1. Initialize the routing tables with a uniform probability distribution:

$$P_{ld} = \frac{1}{N_k} \tag{13.8}$$

   where P_{Id} is the probability of jumping from node l to node d (destination), and N_k is the number of nodes.
2. At regular intervals, from every network node, a forward ant k is launched with the aim of finding a path to the destination. The identifier of every visited node is saved onto a memory M_k and carried by the ant.
   Let k be any network node; its routing table will have N entries, one for each possible destination.
   Let d be one entry of k routing table (a possible destination).
   Let N_k be the set of neighboring nodes of node k.
   Let P_{kl} be the probability with which an ant or data packet in k, jumps to a node l, $l \in N_k$, when the destination is d ($d \neq k$). Then, for each of the N entries in the node k routing table, it will be n_k values of P_{ld} subject to the condition:

$$\sum_{l \in N_k} P_{ld} = 1; \quad d = 1, \ldots, N \tag{13.9}$$

3. At every visited node, a forward ant assigns a greater probability to a destination node d for which falls to be the destination among the neighbor nodes, d $\in N_k$. Hence, initial probability in the routing table of k is then:

$$P_{dd} = \frac{9N_k - 5}{4N_k^2} \tag{13.10}$$

   Also, for the rest of the neighboring nodes among the neighbors for which m $\in$ N_k will then be:

$$P_{dm} = \begin{cases} \dfrac{4N_k - 5}{4N_k^2}, & \text{if } N_k > 1 \\ 0, & \text{if } N_k = 1 \end{cases} \tag{13.11}$$

Of course, equation (13.10) and equation (13.11) satisfy equation (13.9). But, if none of the neighbors is a destination, equation (13.8) applies to all the neighboring nodes. Or

4. The forward ant selects the next hop node using the same probabilistic rule proposed in the ACO (ant colony optimization) metaheuristic:

$$P_k(r,s) = \begin{cases} \dfrac{[\tau(r,s)]^{\alpha}.[E(s)]^{\beta}}{\sum\limits_{u \notin M_k} [\tau(r,u)]^{\alpha}.[E(s)]^{\beta}}, & s \notin M_k \\ 0, else \end{cases} \tag{13.12}$$

where $p_k(r,s)$ is the probability with which ant k chooses to move from node r to node s, τ is the routing table at each node that stores the amount of pheromone trail on connection (r,s), E is the visibility function given by $\frac{1}{(C-e_s)}$ (c is the initial energy level of the nodes and e_s is the actual energy level of node s), and α and β are parameters that control the relative importance of trail versus visibility. The selection probability is a trade-off between visibility (which says that nodes with more energy should be chosen with high probability) and actual trail intensity (which says that if on connection (r,s) there has been a lot of traffic, then it is highly desirable to use that connection).

5. When a forward ant reaches the destination node, it is transformed into a backward ant whose mission is now to update the pheromone trail of the path it used to reach the destination and that is stored in its memory.
6. Before backward ant k starts its return journey, the destination node computes the amount of pheromone trail that the ant will drop during its journey:

$$\Delta\tau = \frac{1}{C - \left[\frac{EMin_k - Fd_k}{EAvg_k - Fd_k}\right]} \tag{13.13}$$

And, the equation used to update the routing tables at each node is

$$\tau(r,s) = (1-\rho) * \tau(r,s) + \left[\frac{\Delta\tau}{\phi.Bd_k}\right] \tag{13.14}$$

where ϕ is a coefficient and Bd_k is the distance travelled (the number of visited nodes) by the backward ant k until node r, in which the two parameters will force the ant to lose part of the pheromone strength during its way to the source node. The idea behind the behavior is to build a better pheromone distribution (nodes near the sink node will have greater pheromone levels) and will force remote nodes to find better paths. Such behavior is important when the sink node is able to move, because pheromone adaptation will be much quicker [41].

7. When the backward ant reaches the node where it was created, its mission is finished and the ant is eliminated.

By performing this algorithm in several iterations, each node will be able to know which are the best neighbors (in terms of the optimal function represented by equation (13.14)) to send a packet toward a specific destination. The flow chart describing the action of movement of a forward ant for our proposed algorithm is shown in Figure 13.3. The backward ant takes the opposite direction of the flow chart.

13.5.1 Algorithm Operations

After the initialization of the routing table and setting up a forward ant for hopping from node to node in search of the sink, at every point in time, a node becomes a source holding, in its stack or memory, information about an event around itself (neighbors). The information gathered in its memory is transferred or disseminated toward the sink node with the help of neighbor nodes behaving as repeaters. Associated raw data generated at each source (nodes) are divided into M pieces, known as data parts. An integer value M also represents the number of ant agents involved in each routing task. This raw data provided by the source node about an event contains information, such as source node identification, event identification, time, and the data about the event. The data size is chosen based on the sensor nodes deployed and the size of the buffer. After the splitting of the raw data, each part is associated with routing parameters to build a data packet ready to transfer. These parameters are code identification, describing the code following as data, error, or acknowledge: C_{ID}. Next is the node identification to which the packet is transferred: N_{ID}. The packet number also represents the ant agent k; S_N is the sequence number, N_k, which contains the number of visited nodes so far and the kth data part shown in Figure 13.4. In this figure, the group of the first four fields is named the data header. When delivery of all data packages is accomplished, the base combines them into raw data.

When a node participating in a routing receives a data packet whose agent number is given, it makes decisions about the next destination for that packet of data. The decision on the next node or destination for which the packet of data should be transferred will depend on equation (13.10), otherwise equation (13.12) with the highest probability. The pheromone level of the neighbors that is the first determining factor follows the energy levels of the neighbor nodes, which are most important on the decision rule. For any of the neighbors chosen, the N_{ID} field of the node is updated and the packet is then broadcast. The remaining neighbors among the chosen node also hear the broadcast; they check the N_{ID} field and understand that the message is not made for them. They then quickly discard the packet immediately

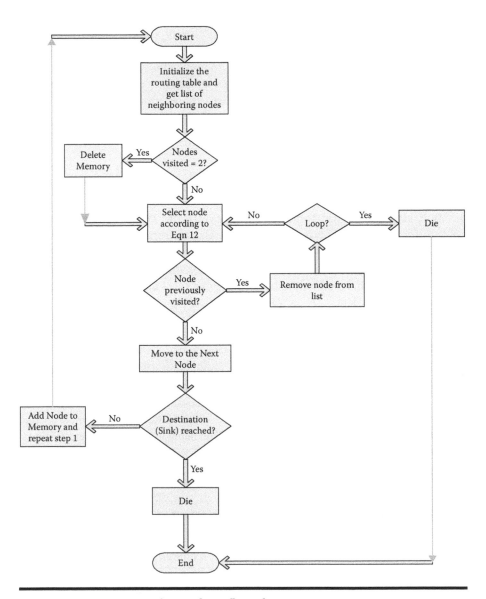

Figure 13.3 An IEEABR forward ant flow chart.

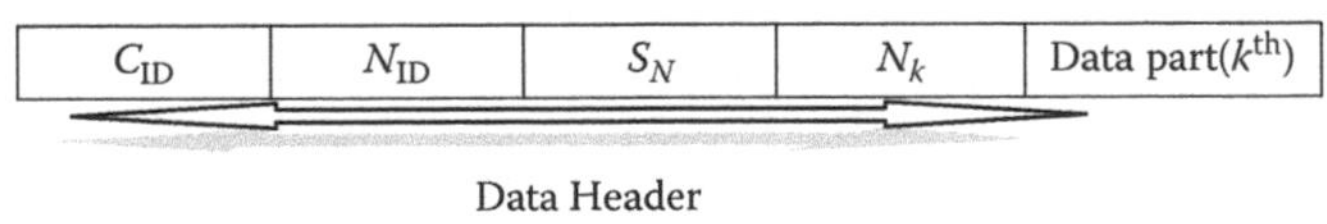

Figure 13.4 Data packet content.

after only listening to the N_{ID} field of the packet. The N_k is updated with an increment by one after ensuring that S_N is not in the list of the tabus (routing table) of the chosen node. The next node is determined to update the N_{ID} field by performing the same operation as performed earlier by the first node and the sequence continues until the packet gets to the sink node. The reversed operation is done for the backward ant as for the acknowledgment, which gets to the source: now the last bus stop for the backward ant, and it dies off after reaching the source.

13.6 Experiment and Simulation Results

Different experiments are conducted to measure the circuit's parameter and the influence of the RF power source. Simulation results based on the performance of the circuit with differences in distance of the harvester from the power sources, the energy usage, and energy management using our proposed IEEABR, are all analyzed below while showing the harvesting setup and the simulation environment.

13.6.1 Experimental Results

Using the Powercast calculator and setting components P2110 at 1.2 V to 915 MHz, battery capacity at 1150 mAh for P2110, and P1110 at 4.0 V to 915 MHz for the same battery capacity, while varying the distance between the transmitters, the readings will be as shown in Table 13.5 to Table 13.8, with differences in the receiver and antenna used in the experiment. The behavior of the packets received with time is shown in Figure 13.5a,b , while the packets received with distance for different harvesters and antennas are compared in Figure 13.6a,b.

Table 13.5 Amount of Power Harvested by P2110 Harvester Using a Dipole Antenna

Distance (ft)	*P (μW)*	*I (μA)*	*Recharge Time (hrs)*
2	3687	3073	22.08
5	523	436	155.04
10	135	112	602.64
12	85	71	952.32
15	37	31	2169.12
18	11	9	7360.56
20	1	1	68339.28

Table 13.6 Amount of Power Harvested by P2110 Harvester Using a Patch Antenna

Distance (ft)	*P (μW)*	*I (μA)*	*Recharge Time (hrs)*
5	1925	1604	42.24
10	386	322	210.50
15	189	158	429.40
18	131	109	618.5
20	102	85	797.50
25	50	41	1639.00
30	19	16	4353.00
35	5	4	15517.00
36	1	1	70019.00

Table 13.7 Amount of Power Harvested by P1110 Harvester Using a Dipole Antenna

Distance (ft)	*P (μW)*	*I (μA)*	*Recharge Time (hrs)*
2	3688	922	62.40
4	1085	271	211.92
6	259	65	888.72
7	86	22	2659.92

13.6.2 Simulation Results

We use event-driven network simulator-2 (NS-2) [45] based on the network topology to be able to evaluate the implementation of the proposed energy management protocol. This software provides a high simulation environment for wireless communication with detailed propagation, MAC (media access control), and radio layers. AntSense (an NS-2 module for Ant Colony Optimization) [46] was used for the EEABR. The simulation parameters are shown in Table 13.9. We assume that all nodes have no mobility since the nodes are fixed in the application of most wireless sensor networks. Simulations were run for 60 minutes (3,600 seconds) each time the simulation started, and the remaining energy of all nodes were taken and recorded at the end of each simulation. The average energy was calculated while also noting the minimum energy of the nodes. This helps in recording tracks

Table 13.8 Amount of Power Harvested by P1110 Harvester Using a Patch Antenna

Distance (ft)	*P (μW)*	*I (μA)*	*Recharge Time (hrs)*
2	16115	4029	14.16
4	3070	768	74.88
6	1551	388	148.30
8	810	203	283.90
10	366	92	627.60
12	93	23	2475.00
13	26	7	8750.00

of the performance of the management protocols in term of the network's energy consumption.

As energy is the key parameter to be considered when designing protocol for power management to enhance the maximum lifetime of sensor networks, we use:

1. The minimum energy that gives the lowest energy amount of all nodes at the end of simulations
2. The average energy that represents the average of energy of all nodes at the end of simulation
3. The simulation was done on a static WSN where sensor nodes were randomly deployed with the objective of monitoring a static environment

Nodes were responsible for monitoring and sending the relevant sensor data to the sink node where nodes near the phenomenon will easily depreciate in energy because

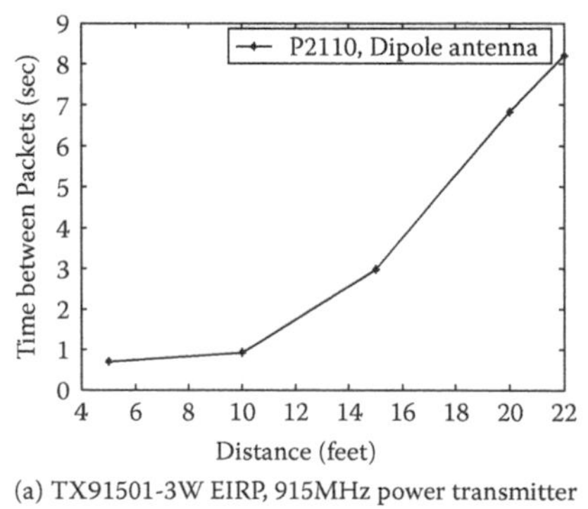

(a) TX91501-3W EIRP, 915MHz power transmitter

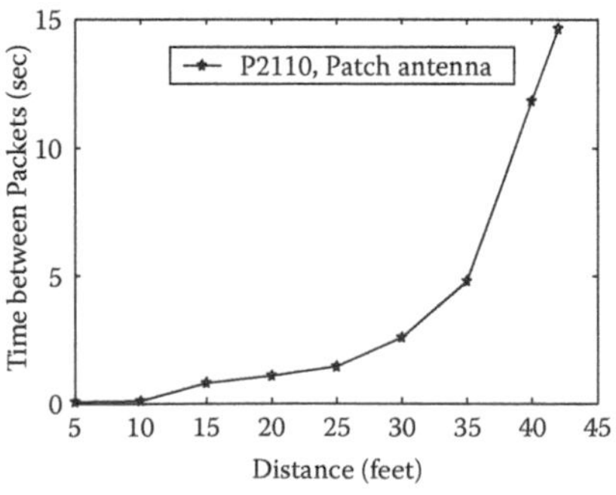

(b) TX91501-3W EIRP, 915MHz power transmitter

Figure 13.5 Variation in time between Packets received and distance of harvesting.

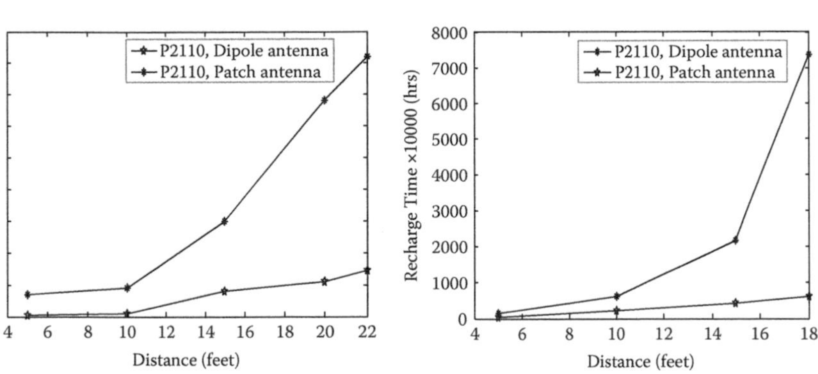

(a) Comparison of packet received with distance using different antenna

(b) Comparison of battery recharge time, using dipole and patch antennas with P2110

Figure 13.6 Comparison of power harvesting using dipole and patch antennas with P2110.

Table 13.9 Simulation Parameters

Parameters	*Values*
Routing Protocols	AODV, EEABR, IEEABR
MAC Layer	IEEE 802.15.4
Frequency	2.4 GHz
Packet Size	1 Mb
Area of Deployment	200 × 200 m^2 (10 nodes), 300 × 300 m^2 (20 nodes), 400 × 400 m^2 (30 nodes), 500 × 500 m^2 (40 nodes), 600 × 600 m^2 (50–100 nodes),
Data Traffic	Constant Bit Rate (CBR)
Simulation Time	3,600 sec.
Battery power	1,150 mAH, 3.7 V
Propagation Model	Two-ray ground reflection
Data Rate	250 Kbps
Current Draw in Sleep Mode	62 μA
Current Draw in Transmitting Mode	50.26 mA
Current Draw in Receiving Mode	49.56 mA
Current Draw in Idle Mode (Processor)	9 mA

they will be forced to periodically transmit data. Simulations were run for 60 minutes (3,600 seconds) each time the simulation started, and the remaining energy of all nodes was taken and recorded at the end of each simulation. The average energy was calculated while also noting the minimum energy of the nodes. Figure 13.8 presents the results of the simulation for the studied parameters: the average energy, and minimum energy of AODV, EEABR, and IEEABR. As can be seen from the results presented in Figure 13.8, the IEEABR protocol had better results in both the average energy of the nodes and the minimum energy of nodes experienced at the end of simulation. The AODV as compared to the EEABR performs worst in all cases. In term of average energy levels of the network, IEEABR as compared with EEABR average energy values varies between 2 and 8 percent, while AODV is in the range of 15 to 22 percent also to the minimum energy of the nodes. Figure 13.7 shows a screenshot of a NAM window of the simulation environment for 10 nodes randomly deployed, while the results of the simulations are as shown in Figure 13.8.

13.6.3 Real-Time Implementation of RF Powercast Energy Harvester

The real-time implementation in harvesting RF energy from Powercast harvesters and the management of the power that is harvested, using the IEEABRA, is

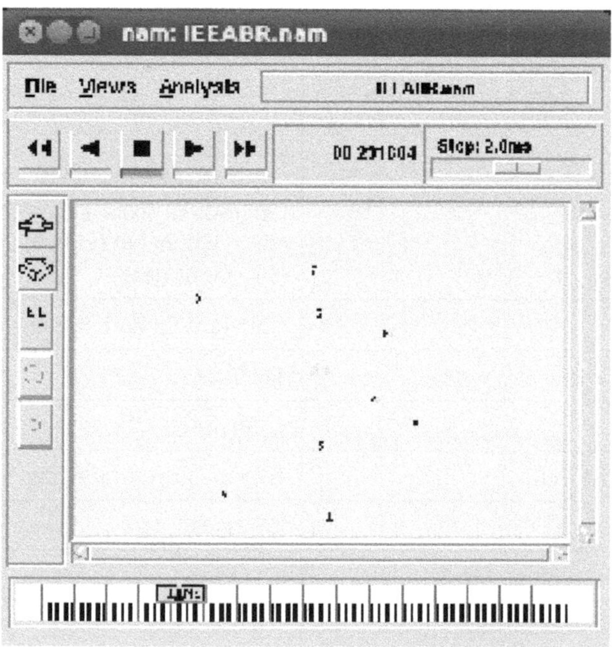

Figure 13.7 Graphical representation of the simulation environment in NS-2.34 with 10 nodes.

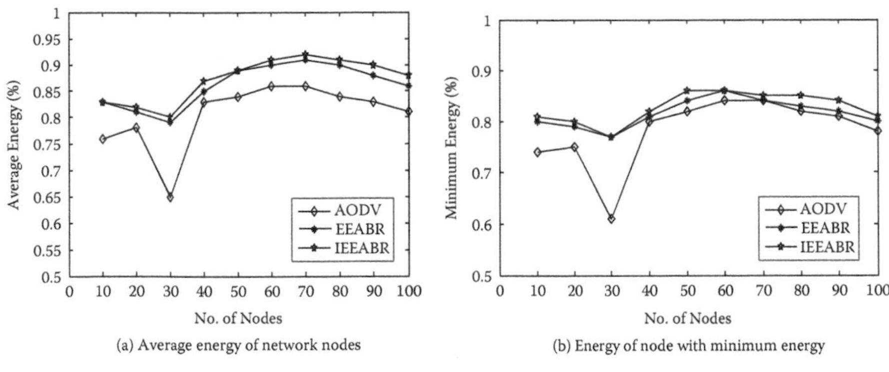

Figure 13.8 Performance analysis of AODV, EEABR, and IEEABR energy management protocols.

presented in Figure 13.9. The experiment was set up and available power measured while also varying the distance between the harvester and transmitter. The time between each packet delivery, i.e., the harvesting period, were noticed and recorded. The power consumption of the Waspmote under consideration can be found in Waspmote: *Technical Guide–Libelium* [25] and Zigbee Protocol [26].

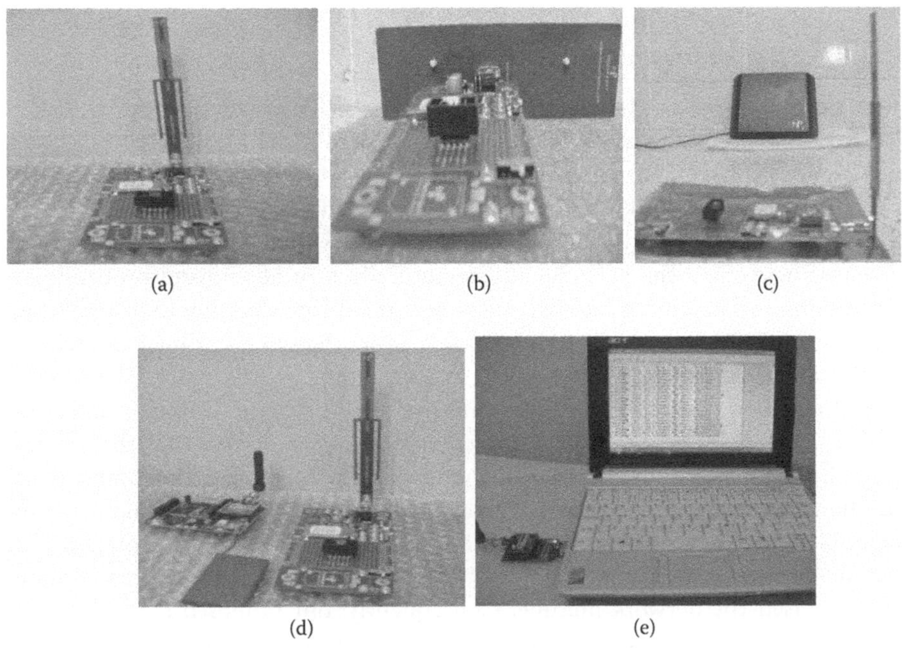

Figure 13.9 Hardware setup of the real-time implementation.

The battery powering of the Waspmote is 1,150 mAh at 3.7 V, which can sufficiently power each of the nodes separately under constant transmission or reception for 19.39 h. For our management protocol that was applied, the maximum energy consumed was found to be 23 percent of the supply energy amounting to total current drawn to be 264.5 mA for 1 h. It then means that the battery can sufficiently power the node with the minimum energy for 4.35 h without recharging. For the recharging of the battery at 15 ft, as shown in Table 13.6, it takes 429.4 h to fully recharge when empty, and 91.9 h to replenish the drawn current of 264.5 mA. However, with constant harvesting, it then means that the total energy of the battery remains without reduction, which can then sustain the network for the required number of years needed for sensing. A quick look at the receiver in its receiving and conversion state with dipole and patch antenna connected for the application of harvesting from the Powercast transmitter and the harvesting mode of the receiver 3 ft (0.914 m) away from the transmitter, respectively, are shown in Figure 13.9a-c. Shown are the P2110 Powercast harvester receiver with (a) dipole (omnidirectional) antenna, (b) patch (directional) antenna, (c) TX95101 Powercaster transmitter in its harvesting mode, (d) with Waspmote, and (e) Gateway connected to the sink. The results of the measurements of the harvested RF energy are presented in Table 13.5 to Table 13.8 and Figure 13.5a,b.

13.7 Conclusions and Future Work

In this chapter, research based on the application of RF energy harvesting, using Powercast harvesters to support the limited available energy of wireless sensor networks, and its management using Ant Colony Optimization metaheuristic was adopted. In this work, we proposed an Improved Energy Efficient Ant Based routing Algorithm energy management technique, which improves the lifetime of sensor networks. The IEEABR utilizes initialization of uniform probabilities distribution in the routing table while giving special consideration to neighboring nodes that fail to be the destination (sinks) in other than to save time in searching for the sink, leading to reduced energy consumption by the nodes. The experimental results showed that the algorithm leads to very good results in different WSNs. Also looking at the harvested energy, the time of charging the battery powering the sensor nodes is drastically reduced, while requiring time intervals of 91.9 hours to recharge the battery. The protocol considers the residual energy of nodes in the network after each simulation period. Based on NS-2 simulation, the IEEABR approach has effectively balanced the WSN node power consumption and increased the network lifetime. Consequently, our proposed algorithm can efficiently extend the network lifetime without performance degradation. This algorithm focused mainly on energy management and the lifetime of wireless sensor networks.

As for future work, we intend to build a linking circuit so as to directly charge the Waspmote battery from the Powercast harvesters, harvest the useless energy from the Waspmote, study a dual approach in the selection of sink and the self destruction of the backward ants should there exist a link failure, and find an alternate means of retrieving the information carried by the backward ant to avoid loss of information. We also intend to design a maximum power point tracker (MPPT), so as to dual power the Waspmote, and model both sources for perpetual operation of the sensor networks.

REFERENCES

[1] F. E. Little, J. O. McSpadden, K. Chang, and N. Kaya, "Toward Space Solar Power: Wireless Energy Transmission Experiments Past, Present and Future." *AIP Conference Proceedings of Space Technology and Applications International Forum*, vol. 420, pp. 1225–1233, 1998.

[2] D. Bouchouicha, F. Dupont, M. Latrach, and L. Ventura, "Ambient RF Energy Harvesting," *International Conference on Renewable Energies and Power Quality* (ICREPQ'10), Granada (Spain), March, 2010.

[3] S. P. Beeby, M. J. Tudor, and N. M. White, "Energy Harvesting Vibration Sources for Microsystems Applications," *Measurements Science and Technology*, vol. 17, pp. 175–195, 2006.

[4] R. M. Dickinson, "Evaluation of a microwave high-power reception-conversion array for wireless power transmission," Tech. Memo 33-741. Jet Propulsion Laboratory, California Institute of Technology, Pasadena, CA, Sept. 1975.

[5] H. Hayami, M. Nakamura, and K. Yoshioka, "The Life Cycle CO_2 Emission Performance of the DOE/NASA Solar Power Satellite System: A Comparison of Alternative Power Generation Systems in Japan," *IEEE Transactions on Systems, Man, and Cybernetics Part C: Applications and Reviews*, vol. 35, no. 3, August 2005.

[6] T. W. R. East, "Self-Steering, Self-Focusing Phased Array for SHARP," *Antennas and Propagation Society International Symposium*, 1991, AP-S, Digest 24–28, pp. 1732–1735, vol. 3, June 1991.

[7] RF HAMDESIGN Microwave equipments and parts. Online at: http://www.rfham-design.com/products/parabolicdishkit/1682909a390cc1b03/index.html

[8] Y. Zhang, L. D. Kuhn, and M. P. J. Fromherz, "Improvements on Ant Routing for Sensor Networks," Ant Colony, in *Optimization and Swarm Intelligence*, Lecture Notes Computer Science, 2004, 3172, pp. 289–313.

[9] P. X. Liu, "Data Gathering Communication in Wireless Sensor Networks Using Ant Colony Optimization," *2004 IEEE International Conference on Robotics and Biomimetics*, 2004, pp. 822–827.

[10] Y.-Feng Wen, Y.-Quan Chen, and M. Pan, "Adaptive ant-based routing in wireless sensor networks using Energy*Delay metrics," *Journal of Zhejiang University SCIENCE A*, vol. 9, Mar. 2008, pp. 531–538.

[11] R. GhasemAghaei, M. A. Rahman, W. Gueaieb, and A. El Saddik, "Ant Colony-Based Reinforcement Learning Algorithm for Routing in Wireless Sensor Networks," *2007 IEEE Instrumentation & Measurement Technology Conference (IMTC)*, May 2007, pp. 1–6.

[12] X. Wang, L. Qiaoliang, X. Naixue, and P. Yi, "Ant Colony Optimization-Based Location-Aware Routing for Wireless Sensor Networks," in *Proceedings of the Third International Conference on Wireless Algorithms, Systems, and Applications* (WASA'08), Springer-Verlag, Berlin, Heidelberg, vol. 5258, 2008, pp. 109–120.

[13] M. Paone, L. Paladina, M. Scarpa, and A. Puliafito, "A multi-sink swarm-based routing protocol for Wireless Sensor Networks," in *IEEE Symposium on Computers and Communications,* July 2009, pp. 28–33.

[14] G. De-Min, Q. Huan-Yan, Y. Xiao-Yong, and W. Xiao-Nan, "Based on ant colony multicast trees of wireless sensor network routing research," *Journal of iet-wsn.org*, vol. 2, 2008, pp. 1–7. Online at http://www.iet-wsn.org/webeditor/UploadFile/20106261139547047 2.pdf.

[15] S. Xia and S. Wu, "Ant Colony-Based Energy-Aware Multipath Routing Algorithm for Wireless Sensor Networks," in *2009 Second International Symposium on Knowledge Acquisition and Modeling*, Nov. 2009, pp. 198–201.

[16] G. Wang, Y. Wang, and X. Tao, "An Ant Colony Clustering Routing Algorithm for Wireless Sensor Networks," in *2009 Third International Conference on Genetic and Evolutionary Computing,* Oct. 2009, pp. 670–673.

[17] T. Le, K. Mayaram, and T. Fiez, "Efficient Far-Field Radio Frequency Energy Harvesting for Passively Powered Sensor Networks," *IEE Journal of Solid-State Circuits*, vol. 43, no. 5, pp. 1287–1302, May 2008.

[18] H. Jabbar, Y. S. Song, and T. D. Jeong, "RF Energy Harvesting System and Circuits for Charging of Mobile Devices," *IEEE Transaction on Consumer Electronics*, Jan. 2010.

[19] C. Lu, V. Raghunathan, and K. Roy, "Micro-Scale Harvesting: A System Design Perspective," in *Proceedings of the 2010 Asia and South Pacific Design Automation Conference* (ASPDAC'10), IEEE Press: Piscataway, NJ, 2010.

[20] L. Tang and C. Guy, "Radio Frequency Energy Harvesting in Wireless Sensor Networks," in *Proceedings of the 2009 International Conference on Wireless Communications and Mobile Computing* (ICWCMC'09): Connecting the World Wirelessly, ACM: New York, 2009.

[21] J. A. Hagerty, T. Zhao, R. Zane, and Z. Popovic, "Efficient Broadband RF Energy Harvesting for Wireless Sensors," Colorado Power Electronics Center (COPEC), Boulder, CO, 2003.

[22] W. Seah and Y. K. Tan, "Review of Energy Harvesting Technologies for Sustainable Wireless Sensor Network," in *Sustainable Wireless Sensor Networks*, Y. K. Tan (ed.), pp. 15–43. Tech Publishing, Rijeka, Croatia, Dec. 2010.

[23] J. M. Gilbert and F. Balouchi, "Comparison of Energy Harvesting Systems for Wireless Sensor Networks," *International Journal of Automation and Computing*, vol. 5, no. 4, pp. 334–347, Oct. 2008.

[24] MICAz Datasheet. Online at: http://courses.ece.ubc.ca/494/files/MICAz_Datasheet.pdf

[25] Waspmote: *Technical Guide-Libelium*. Online at: http://www.libelium.com/documentation/waspmote/waspmote-technical_guide_eng.pdf

[26] Zigbee Protocol. Online at: http://wiki.kdubiq.org/kdubiqFinalSymposium/uploads/Main/04_libelium_smfkdubiq.pdf

[27] Imote2 Documents—WSN. Online at: http://www.cse.wustl.edu/wsn/images/e/e3/Imote2_Datasheet.pdf

[28] JN5139—Jennic Wireless Microcontrollers. Online at: http://www.jennic.com/products/wireless_microcontrollers/jn5139

[29] T. T. Le, "Efficient Power Conversion Interface Circuits for Energy Harvesting Applications," PhD disser., Oregon State University, Corvallis, 2008.
[30] Powercast Documentation. Online at: http://Powercastco.com/
[31] B. Dixon, "Radio Frequency Energy Harvesting." Online at: http://rfenergyharvesting.com/
[32] J. A. Shaw, "Radiometry and the Friis Transmission Equation." Online at: http://www.coe.montana.edu/ee/rwolff/EE548/sring05%20papers/Friis_Radiometric_2005Feb9.pdf
[33] Wireless Power Calculator. Online at: http://Powercastco.com/wireless-power-calculator.xls
[34] Power Density. Online at: http://www.phys.hawaii.edu/~anita/new/papers/military-Handbook/pwr-dens.pdf
[35] R. Struzak, "Basic Antenna Theory." Online at: http://wirelessu.org/uploads/units/2008/08/12/39/5Anten_theor_basics.pdf
[36] N. Dusit, H. Ekram, M. R. Mohammad, and K. B. Vijay, "Wireless Sensor Networks with Energy Harvesting Technologies: A Game-Theoretic Approach to Optimal Energy Management," *IEEE Wireless Communications*, vol. 14, no. 4, pp. 90–96. September 2007.
[37] D. Waltenegus and P. Christian, *Fundamental of Wireless Sensor Networks: Theory and Practice*, New York: Wiley Series on Wireless Communication and Mobile Computing, pp. 207–213, 2010.
[38] A. Kansal, J. Hsu, S. Zahedi, and M. B. Srivastava, "Power management in energy harvesting sensor networks," in *Proceedings of ACM Transactions on Embedded Computing Systems*, vol. 6, September 2007, pp. 32–66.
[39] A. G. A. Elrahim, H. A. Elsayed, S. El Rahly, and M. M. Ibrahim, "An Energy Aware WSN Geographic Routing Protocol," *Universal Journal of Computer Science and Engineering Technology*, vol. 1, no. 2, Nov. 2010, pp. 105–111.
[40] W. Li, M. Chen, and M-M. Li, "An Enhanced AODV Route Protocol Applying in the Wireless Sensor Networks," *Fuzzy Information and Engineering*, vol. 2, AISC 62, pp. 1591–1600, 2009.
[41] T. C. Camilo, Carreto, J. S. Silva, and F. Boavida, "An Energy-Efficient Ant Based Routing Algorithm for Wireless Sensor Networks," In *Proceedings of 5th International Workshop on Ant Colony Optimization and Swarm Intelligence*, Brussels, Belgium, pp. 49–59, 2006.
[42] C. Alippi, G. Anastasi, M. D. Francesco, and M. Roveri, "Energy management in Wireless Sensor Networks with Energy-Hungry Sensors," *IEEE Instrumentation and Measurements Magazine*, vol. 12, April 2009, pp. 16–23.
[43] X. Jiang, J. Taneja, J. Ortiz, A. Tavakoli, P. Dutta, J. Jeong, D. Culler, P. Levis, and S. Shenker, "Architecture for Energy Management in Wireless Sensor Networks," *ACM SIGBED Review*, Special issue on the workshop on wireless sensor network architecture, vol. 4, no. 3, pp. 31–36, July 2007.
[44] K. Kalpakis, K. Dasgupta, and P. Namjoshi, "Maximum Lifetime Data Gathering and Aggregation in Wireless Sensor Networks," *in Proceedings of IEEE International Conference on Networking*, vol. 42, no. 6, August 2003.
[45] NS2 installation on Linux. Online at: http://paulson.in/? p=29
[46] NS-2 Module for Ant Colony Optimization (AntSense). Online at: http://eden.dei.uc.pt/~tandre/antsense/

Index

K

L

M

T

U

V

Z

Milton Keynes UK
Ingram Content Group UK Ltd.
UKHW031126141024
449569UK00006B/420